S0-BJU-234

Navigating Strategic Decisions

The Power of Sound Analysis and Forecasting

ENDORSEMENTS FOR *NAVIGATING STRATEGIC DECISIONS*

An indispensable guide for strategic project managers and a "must-read" book for strategic planners aspiring to advance their careers and become trusted advisors to decision makers, written by an expert in strategic decision forecasting as a comprehensive practical guide that lays the foundations toward competitive advantage. It will revolutionize your company's strategic thinking by showing how to bring together methods, processes, and techniques toward sound analysis, evaluation, and forecasting for superior strategic decisions.

George S. Vozikis, PhD
Institute for Family Business
and
Department of Management
Craig School of Business
California State University, Fresno, California

Already a published authority on managing successful acquisition and joint venture projects, John Triantis has now focused on marking out a realistic, yet crucial, role for strategic forecasters participating in strategic projects. Forecasters following his precepts will prove themselves indispensable to the project team, while at the same time contributing to reduced time-to-decisions, better/sounder decisions, and increased chances of realizing forecasted project values. This is a "must-read" book for forecasters aspiring to make a difference in the world of strategic project decisions.

John K. Hendricks
Former AT&T General Attorney
International M&A and Joint Ventures, Strategic Alliances, Basking Ridge, New Jersey

Finally—a book that demonstrates the connection between the art and science of strategic forecasting and successful strategic projects is here! John Triantis uncovers the truths about strategic decisions to everyone in this field and is in the forefront of helping organizations understand the importance of and implement strategic forecasting. All forecasters and decision participants will benefit from John's 40+ years of hands-on experience and practical insights.

Hu Song, MD, PhD
WW Business Insights
Ortho Clinical Diagnostics, Johnson & Johnson, Raritan, New Jersey

Dr. John Triantis is the consummate translator of the vernacular strategic analysis. *Navigating Strategic Decisions: The Power of Sound Analysis and Forecasting* is written for the professional who deals with both analysis and forecasting for strategic projects. His long experience in the real world allows him to draw from the things that have worked and enables him to make sound recommendations for industry analysts and forecasters. The reader will benefit immensely from the practical insights. The information presented is understandable and transparent. The strength of the book is that it is based on workable and practical methodologies and not theoretical concepts. An excellent book!

Dean H. Stamatis, PhD, CQE, CMfgE, MSSBB
Contemporary Consultants, Southgate, Michigan
Anhui University of Finance and Economics in Bengbu, Anhui, People's Republic of China

Managers too often predict demand emotionally, despite the importance of the decisions being made as a result of good fact-based analysis. Subsequently, forecasts are often extrapolations of historical volume trends, rather than well-thought-out projections. This has been the accepted process for strategic forecasting and planning for the past 30 years. This is the first book that provides a proven framework for business executives to translate fact-based decisions into viable strategic forecasts.

Charles Chase, MA, PhD (Pursuing)
Supply Chain Global Practice
SAS Institute, Inc., Cary, North Carolina

Understanding how to transform data into useful information is critical in a world of ever increasing noise. This is an essential reference book for organizations considering entering new markets, evaluating product portfolio mix or developing new products. It serves as a comprehensive resource of the critical methods and techniques needed to enable strategic decisions.

Demetra Simos Paguio, MBA
Former Sales and Marketing Operations Manager
Ryerson Inc., Chicago, Illinois

Navigating Strategic Decisions

The Power of Sound Analysis and Forecasting

John E. Triantis

CRC Press is an imprint of the
Taylor & Francis Group, an **informa** business
A PRODUCTIVITY PRESS BOOK

CRC Press
Taylor & Francis Group
6000 Broken Sound Parkway NW, Suite 300
Boca Raton, FL 33487-2742

Printed on acid-free paper
Version Date: 20130102

International Standard Book Number: 978-1-4665-8598-0 (Hardback)

Library of Congress Cataloging-in-Publication Data

Triantis, John E., 1944-
Navigating strategic decisions : the power of sound analysis and forecasting / John E. Triantis.
pages cm
"A Productivity Press book."
Includes bibliographical references and index.
ISBN 978-1-4665-8598-0
1. Strategic planning. 2. Decision making. I. Title.

HD30.28.T748 2013
658.4'012--dc23 2012046265

Visit the Taylor & Francis Web site at
http://www.taylorandfrancis.com

and the CRC Press Web site at
http://www.crcpress.com

To the memory of my father

Contents

PART II A New Paradigm for Strategic Decision Forecasting

PART III Strategic Decision Forecasting Applications

List of Figures

List of Tables

Preface

EXPERIENCES THAT MOTIVATED THE BOOK

The nature of strategic decisions, large capital investments, and long-range planning necessitates forecasts 5, 10, even 20 years out in order to calculate the value of these projects. However, the focus of the forecasting profession has been on short-term sales forecasts, usually six months to two years out. These forecasts are based on statistical forecasting models that are often mechanized to accommodate a large number of forecasts and updates needed for supply chain management operations. As a result, most of the training of forecasters is in those areas, and the majority of them do not operate outside their statistical training comfort zone.

Most senior managers know that long-term statistical forecasts are not to be trusted; and in the absence of other alternatives, they rely on their own intuition. Other senior managers realize that forecasting for strategic decisions requires a lot of sophistication well beyond models and practical industry experience; and for that reason, they tend to outsource forecasting support for strategic projects. Outsourcing strategic forecasts is necessary in some instances. However, not only can valuable proprietary information be leaked to competitors and expected results rarely materialize, but also one-way valuable knowledge and experiences are transferred to outside experts.

We recognized a long time ago the need for a systematic approach to generate sound long-term forecasts and provide support for strategic decisions and projects. Our literature research showed that there was nothing to guide forecasting managers in creating strategic decision forecasting organizations, processes, methods, analyses, and long-term forecasts to drive decision making. While learning from participation in strategic projects and experimenting with different approaches, we also realized that to do justice to the discipline of providing forecasting support in strategic decisions and projects requires a lot of industry and practical knowledge and confluence of unique experiences. Thus, writing the book was delayed until sufficient knowledge and experiences were accumulated; and in the next section, we share, in brief, 15 actual experiences that provided the motivation for this book.

Assembling material for this book is the product of actual project experiences over forty years, researching the outcome of strategic projects, and recording lessons learned in different projects. Also, extensive interactions with strategic project team members and industry analysts added to our understanding of issues and challenges in different types of projects. Our experiences were augmented through collaboration with experts and seasoned consultants in the areas that impact strategic forecasting and by discussions with senior managers, clients, forecasters, and managers of strategic forecasting groups. Lastly, project assignments in benchmarking best-in-class processes and practices and reviewing accounts of competitor responses to strategic projects provided additional ideas and incentive to undertake this task.

A. MOTIVATING EXPERIENCES

1. *The client was the expert.* The author's first encounter with the need for long-term forecasting came when asked to generate a 20-year statistical forecast of freight rates for a foreign shipping company to be used in valuing an acquisition target company. We knew that anything beyond three years out would not have much credence because so many things can change to render the forecast useless. We also explained to the client that the necessary foreign trade and ship capacity variables projected out 20 years to generate a forecast of freight rates were not available. Instead of producing a causal model forecast, we recommended using expert opinions, analogs of previous 20-year

involving several underdeveloped countries. What we got was a "consensus" forecast created by the suppliers to the project: a forecast that had no country economic conditions evaluation or inputs, service pricing assumptions, ability to pay studies, or any real basis for their forecast. The suppliers' motives underlying the forecast became clear when it was demonstrated that their forecast did not reflect the reality of developing country environments and market demand trends nor did it include any country, commercial, or political risk adjustments to the overly optimistic forecast.

Despite efforts to help the project sponsor representatives understand the issues and create reasonable long-term forecasts to assess the project economics, they thought our organization was not being cooperative. The validity of our position, however, was confirmed in meetings with the World Bank, which refused to participate in the financing of the project as did all the commercial banks that were invited to participate in the project. Why? Because of the poor quality of the forecasts underlying the projected financials was the joint response.

9. *Forecast request to drive the strategic plan.* A request came to a forecasting group in a healthcare-related company which the author was advising during the 2000–2001 debate in Washington to create a drug reimbursement program. The request came from a client group responsible for developing strategic plans with the following instructions to the forecasting group: (a) provide a 20-year forecast for Medicare reimbursements for a certain class of drugs and (b) the forecast is needed in two weeks for input to the Spring Strategic Plan.

The client response to push back from the forecasting group on the timeline was "Just give us annual numbers for the next 20 years; we'll look at them, change them to include actions we will be initiating, and develop the strategy to deal with this issue." A number of scenarios were considered by the forecasting group and a middle-of-the-road, compromise scenario was picked and presented to the client with supporting facts and evidence. A summary of the client response: "Don't bother us with details; our strategy will drive the financial outcome in this mess." It did. And they are still living with the consequences of that mess.

10. *Divergence of partner objectives.* Why is our long-term demand forecast so much lower than our prospective partner's? This question was addressed to the author and the client, a seasoned forecaster with outstanding credentials and broad industry experience in the healthcare industry. A new product forecast had been created for a drug licensing project using what we considered a sound methodology and a balanced approach. The forecasting group had done extensive product research, evaluation, and input from industry experts inside and outside the company. Also, they had considered and simulated several alternative scenarios with guidance from the author; performed sanity checks; and assessed the implications of the forecasts using analogs, industry standards, and benchmarks, among other things.

The client was a product manager who initially was not convinced that the company forecaster supporting him was as good a strategic decision forecaster as the one of the potential partner, a small biotech firm. The prospective partner's forecast was substantially higher than that of the client firm, a large and well-regarded industry leader. And, in the course of discussions and negotiations over the following three months, it became apparent that the prospective partner wanted the project no matter what. Their key objective was to get brand recognition through a strategic alliance with a large, well-respected company; and to achieve that, they generated more optimistic forecasts than those of the large company's forecasting group.

11. *Focus on second-order factors.* The author was the Business Development organization's representative and served as the project team leader in a foreign JV project. Our responsibilities included coordination of partner contributions and screening of all assessments for the business case to be submitted to each partner company's Board of Directors for approval. At one of the partners' insistence, development of the 15-year demand forecast was assigned to seasoned Finance managers

from two of the major partner companies. Discussions on demand growth rates dragged on and on because of difference of opinion on a 4.5% versus a 6.2% growth rate to apply to calculate revenues 15 years out.

The Finance managers missed the deadline for submitting the forecast and upon inquiring about the source of the difference in perspective, we were told by one that they had different experiences than the other Finance manager and that they were representing companies with different objectives and perspective. When we asked if they had verified the basic assumption of guaranteed exclusivity of operations in the host country over the forecast horizon, they answered in the negative. At that point, our question to them was "If you don't really know if the JV has an exclusive license and operation rights in the forecast period, does it matter if you assume a 4.5% or a 6.2% revenue growth rate?" With no hesitation, they both agreed to use an average growth rate of 5%; no problem. Geniuses, indeed.

12. *Long-term forecasting group responsibility.* The author met an old acquaintance who had recently been promoted to a senior management position in the energy industry at a conference where he was presenting a position paper. He related to us that he wanted to get some help to create a long-term, strategic project forecasting function to be implemented in-house. He approached the forecasting group manager and asked him to think about that idea and come back to him with his own plan outlining the strategic forecasting function, the organization, its scope, roles and responsibilities, headcount and costs, and the analyses, evaluations, modeling, and forecasting support his group would provide.

The forecasting group manager came back in four days with the following response:

> We have considered your request and concluded that we are unable to do forecasting for strategic projects because that kind of forecasting is not taught in schools and we lack the experience. The consulting company which has been doing long-term forecasting has done a good job and it is not that expensive. However, we believe that the best use of the forecasting group is to take on the responsibility for updating and maintaining a data base of assumptions used to generate long term project forecasts and monitor their performance.

We ended up helping the forecasting group manager create a strategic forecasting team, put in place processes, and assisted in the development of the first strategic project forecast. It is truly amazing how people pass on opportunities to make a difference!

13. *The financials validate the forecast.* This is an experience the author encountered as a consultant brought in to help a company with strategic forecasting, strategic planning, and product portfolio management (PPM) issues. One of the business units of this company had major issues and growth challenges over a number of years. In one of the first meetings with the head of that business unit, he set the tone of the relationship by declaring: "I have my strategy in place. Why do I need strategic forecasts to develop strategies? What will they tell me that I don't already know?"

We explained to the head of this business unit how good forecasts for strategic planning are done, how they help companies develop sound strategies, and how they help to make decisions and execute them successfully. His response was "Look, Finance has validated the forecasts and determined that we have a winning strategy in place." After doing some joint analysis and evaluations with the CFO staff, we pointed out to this senior manager the holes in the logic, modeling, and risk analysis which was performed earlier. When he saw the deficiencies in the evidence that the long-term forecasts were based on and the implications of the scenario used, he began questioning the whole approach he had initially espoused. What he at first considered a validated forecast became a suspect forecast, and we started the process of developing sound long-term forecasts from ground zero. Were there ups and downs on the way? There were, every day, all the way to the end of the process.

What is a sound strategic forecast? *This question comes up all the time and for our purposes, a sound strategic, long-term forecast is one that is characterized by the following key properties:*

- *It follows a structured and well-balanced approach in identifying forecasting needs and selecting appropriate methods, processes, and techniques to meet those needs.*
- *It identifies what needs to be known and knowable and is based on a comprehensive, model of company operations following preestablished processes.*
- *It validates and uses all the information and learning from company, industry, and external expert sources.*
- *It creates common assumptions driving project planning, which are subjected to sanity checks and consistency throughout the project.*
- *It incorporates external environment assessments, industry and market analyses, and company and competitor strength, weakness, opportunity, and threat (SWOT) evaluations.*
- *It has identified project uncertainty and risks and evaluated the forecast implications for company resources.*
- *It uses scenario planning to describe the events, timing, and transition from the current to the future state and has project team buy-in and client and senior management support.*
- *It has performance measures attached to it, is monitored closely, and includes a competitor reaction and future-state realization (FSR) plan.*

14. *Outsourcing of strategic project forecasts.* More than a decade ago, a well-respected consumer products forecasting organization hired the author to assess and help with demand modeling and forecasting for all types of projects and forecast horizons. We first tackled tactical forecasts, had some successes, and the progress achieved was recognized by the senior management team. We then proceeded to address long-term market potential modeling and new product development (NPD) forecasting problems and issues.

At that point, we realized that the organization from the president of the company down to individual product managers and long-range planners did not consider the company's forecasting group qualified to sit at the table with them. They considered them technocrats who lacked business management skills and who did not have relationships with project clients to be asked to participate in strategic projects. Instead of using the internal forecasting group, the clients would invite an outside consultant to generate long-term forecasts for some existing products and forecasts for all projects of strategic nature. When we pointed out the inefficient use of human capital and corporate resources and several intellectual property and confidentiality issues, the reaction was "We don't trust them (*their own forecasting group*), we want a third party view."

15. *Conference themes and client inputs.* Over the past 30 plus years the author has participated in a number of industry analyst meetings. He has led forecasting conferences, industry workshops, and NPD, PPM, strategic planning, and project financing seminars across various industries both in the United States and abroad. A lot of good work, methods, and approaches are shared concerning demand analysis and tactical forecasting. In fact, progress made in the short-term, quantitative forecasting area has been truly impressive. However, when it comes to analysis and forecasting for strategic projects, that is, forecasting support for projects with long horizons and large impact, little is presented or shared. Everybody though is talking about the lack of a systematic approach, methods, and models, and techniques for strategic forecasting. Complaints are voiced openly about severe problems in this area, the lack of experience and qualified people, unreliable processes, lack of management support, lack of credibility, and so on.

Recently, we reviewed results of earlier client surveys, opinions of conference and workshop participants, and discussions with strategic project team members conducted over the past 15 years. The major recurring themes and key points made are the following:

1. There is a wide recognition of the need for a strategic decision, long-term forecasting discipline, but nobody knows how to do it or does anything about it.
2. The concepts and value of long-term forecasting for strategic projects are not well articulated nor are they understood by forecasters and clients alike.
3. The perception is that strategic forecasting is an expensive proposition with uncertain returns. It is an unknown quantity.
4. Strategic forecasting should be closely linked with strategic planning and other planning functions to be effective, but it requires new processes and skills not currently possessed by most forecasters.
5. Ideally, strategic forecasting should be done in-house to maintain the firm's intellectual property and competitive standing, but forecasting groups are not prepared to handle this challenge.
6. Forecasting groups do not have credibility and management support in forecasting for strategic decisions due to lack of sophistication, limited interpersonal skills, and undeveloped technical competencies and qualifications.
7. Strategic forecasting methodology and tools are either lacking or proprietary and not much of practical value is written about it or presented in professional forecasting forums.
8. Forecasters are good at tactical forecasting, but not trained or experienced in the area of supporting strategic project decisions. Somebody needs to create a curriculum to teach it, develop strategic forecasters, and advance the discipline.
9. Strategic forecasting is a discipline of complex networks, processes, methods, and experiences that come together to create long-term forecasts. It requires a confluence of industry knowledge, broad management skills, and specialized functional expertise that is rare.
10. Strategic forecasting groups belong in the Strategic Planning or the Portfolio Management organizations, but people do not know how to use them to create value.

B. OTHER REASONS AND MOTIVES

The author's involvement in business research, decision analysis, and demand evaluation and forecasting dates back to 1971 when we took on the first forecasting assignment. Since then, we have participated in many strategic decision projects in different capacities across several industries. We saw the absence of training, skills, methods, models, and tools to handle strategic decision forecasts as impediments to forecasters making valuable contributions. But, we did not undertake the challenge of writing a book on this topic until we felt qualified with sufficient practical experiences dealing with analysis, evaluations, developing long-term forecast solutions, establishing and training SDF teams, and supporting strategic decisions effectively with actionable recommendations.

Highlights of directly applicable experiences and knowledge obtained include the following positions, jobs, and projects both in the United States and abroad that enabled the author to pull together all the experiences required to do strategic decision forecasting successfully:

1. *Macroeconomic and international expertise.* This includes country research, economic modeling, and investment analysis experience with focus on adding value to international strategic projects:

a. Country macroeconomic analysis, evaluation, and forecasting of business trends by major industry groups
b. International macro, trade, and monthly foreign exchange forecasts for strategic investment decisions

c. Model development and production of quarterly macro and international economic forecasts used in corporate planning functions and published in professional journals
d. International project evaluations and financing of large infrastructure projects requiring understanding of government agency operations and the equipment supplier, project operator, and the commercial and investment banking industries
e. Evaluation of privatization opportunities in developing countries in a lead consultant capacity

2. *Environmental assessment experiences.* These experiences came about through active participation or lead roles in strategic planning, portfolio management, and M&A and JV projects, the major ones being the following:

a. In-depth business environment, industry, and market analysis for strategic and financial planning
b. Assessment of megatrends and subtrends impacting a company's future operations
c. Political, economic, social, technological, legal, educational, and demographic environment evaluations for product portfolio adjustments
d. Analysis and evaluations for new market or business entry, business development, and materials sourcing from projects abroad
e. Competitor and internal SWOT analyses to determine levels of competencies necessary to execute strategic projects successfully

3. *Forecasting experiences.* These are extensions expected of the author's educational background and encompass the following work or project experiences:

a. Business research, competitive analysis, demand analysis, and forecasting in several industries
b. Lifecycle management project support for product extensions and enhancements
c. Managing forecasting for M&A, JV, product licensing projects, and PPM applications
d. Creation of forecasting methodology for highly specialized applications, such as methodology to evaluate Rx to over-the-counter product switching projects
e. Project managing the outsourcing function of lifecycle management and strategic decision forecasts and market research studies
f. Establishing, training, managing, and developing tactical and strategic forecasting and business research organizations
g. Advising clients on strategic forecasting issues, helping to create organizations to support strategic decisions, and leading the development of strategic decision forecasts
h. Leading workshops on forecasting for strategic decisions, forecasting due diligence, Six Sigma forecasting, and forecast realization planning

4. *Strategic project experiences.* The author acquired and developed expertise supporting strategic projects through the following types of engagements:

a. Creating or evaluating strategic, corporate, and new entity business plans and performance targets in different industries
b. Structuring and project management of strategic projects along with guiding the strategic forecasting function
c. Analytical and implementation support for NPD and portfolio management projects
d. Creation, training, and development of project financing and portfolio management organizations
e. Lead roles in M&A, JV, product licensing, and project financing undertakings
f. Creation of trade-offs and negotiations support for M&A, JV, and product licensing projects

g. Establishing, developing, and coaching strategic decision support groups and introducing Six Sigma and total quality management (TQM) principles to strategic forecasting projects
h. Advising clients on benchmarking best practices to use in creating and developing strategic forecasting processes and organizations
i. Developing large infrastructure project strategy, project structuring, and project financing for a multisponsor, multiuser project
j. Developing management dashboards, EWSs, risk management plans, and development of competitive response plans
k. Evaluations and lead consultant roles in evaluating and implementing privatizations, restructuring, and turnaround projects

We have been a recipient of forecasts in projects across a number of industries in several capacities, such as using forecasts for portfolio management, M&A, JV, product licensing, and financing projects. We have evaluated forecasts used in strategic planning, business development, long-range planning and have used business unit forecasts to create consensus forecasts for corporate planning. Other related experiences include helping forecasters and clients evaluate project uncertainty and forecast risks, create alternative scenarios, and manage decision risks. Also, we have been providing guidance to forecasting groups on strategic decision forecasting methodology, tools, and models and applying sanity checks to existing forecasts to rank project valuations, identify project risks, and select the higher value creation projects.

Knowledge obtained from assisting groups to create forecasts for strategic projects was transferred to others by demonstrating how to adapt known approaches and existing models to simulate alternative futures. We have also guided clients in assessing the feasibility of large projects, analyzing forecasting models to determine controllable levers, and creating the conditions needed for forecasts to materialize. Also, our independent verification and validation of long-range forecasts used in investment evaluations has produced numerous experiences in pricing, optimizing advertising spending, and assessing support requirements and reactions to market conditions to ensure successful product launches and forecast realization.

Fully realizing that no one has a monopoly on truth when it comes to forecasting for strategic decision making, we have humbly undertaken the project of writing this book with little guidance from the literature. However, we have had ample encouragement, support, and knowledge transfer from colleagues and clients in different disciplines and industry groups. Our expectation is that the book will help establish and broaden the acceptance of the discipline of analysis and forecasting for strategic decisions. It is our belief that this will increase forecaster effectiveness and provide the basis to improve the quality of the decision-making process through better long-term forecasts. And, in some cases, strategic forecasting forms a basis that can enable a company to create a competitive advantage through consistently making strategic decisions based on sound forecasts. This is worth the effort required to produce this book.

Acknowledgments

The concept of strategic decision forecasting as a discipline evolved from my experiences in market analysis, forecasting, strategic planning, financial management, business development, project financing, competitive analysis, and consulting projects as a business economist. The quest for sound processes, analysis, and forecasting for strategic decision making has been an integral part of my professional life, and it is difficult to be certain from whom or from which projects I have learned what.

My skills have been augmented a great deal by knowledge transfer from management consultants and the expertise of colleagues I worked with. I have also profited from presentations made in industry conferences and professional workshops I organized or attended, from valuable information and feedback provided by clients, and from cases shared by fellow business consultants.

In preparing this book, I have benefited by the constructive criticism and comments of Charlie Chase, John Hendricks, Hu Song, Dean Stamatis, and George Vozikis. I am thankful for their many useful comments and suggestions on an earlier draft of the manuscript. The ideas offered by Mark Covas and Dick Ross are also acknowledged. However, all errors and omissions are my own responsibility.

The encouragement and support of my wife during the preparation of the manuscript has been invaluable and enabled me to focus on completing this task. I also thank my friend and former mentor Dr. Elizabeth C. Bogan for her guidance and help through the years. Lastly, I acknowledge the help and support of the late James Sotirhos; without it, none of this would have been possible.

Author

John E. Triantis is a retired consulting economist, business strategist, and strategic decision forecaster with a track record of introducing effective processes and practices to strategic project decision making. He is an experienced business development project leader and a trusted advisor in large investment, international infrastructure financing, and organizational restructuring projects. His passion is helping clients minimize decision uncertainty and project risk by integrating analysis and forecasting with strategic planning, maximizing project value using unique methods and techniques, and creating world-class organizations, applying best practices, processes, and performance measures.

His broad experience encompasses customized analysis for strategic decisions, investment evaluations, project structuring, performance evaluations, and productivity enhancements. He develops practical methods, assessment tools, and early warning systems (EWSs) and uses well-balanced approaches to minimize time to decision, address uncertainty and risk management, and increase the chances of project success. And, he provides knowledge and hands-on coaching to clients on strategic forecasting, planning, pricing, and plan realization, opportunity assessments, business cases, and business planning to increase value creation.

John is currently managing director of Long Range Planning Associates, attending to strategic planning and forecasting, organization restructure and capability development, and business analysis and planning needs of decision makers. Prior to this, he served as a director of business analysis and forecasting in Gillette's oral care division and established forecasting and strategic business planning functions, processes, and models supporting Office of the Chairman needs. His international experience spans 30 plus years, further broadened while managing Forerunner Consultants, a startup company that served market assessment and forecasting, strategic planning, and business development needs of clients.

John's diverse background includes experiences as a senior advisor on international economics, strategic investment analysis, product licensing, product portfolio management, government working capital fund assessment and restructuring, and developing strategic competency groups. His 20-year AT&T experience includes industry competitive analysis, project financing, M&As and strategic alliances, privatizations, financial and strategic planning and pricing, macroeconomic analysis, country assessments, and market forecasting. He has served as an advisory board member at the Institute of Business Forecasting, editor of international economic affairs at the *Journal of Business Forecasting*, and advisor to the institute's professional forecasting certification program. He was also a long-standing editorial board member at the *Review of Business*.

To address needs and fill gaps in client understanding, he authored numerous position papers and published the book *Creating Successful Acquisition and Joint Venture Projects* and articles in professional domestic and international journals and books. He is currently publishing the book *Navigating Strategic Decisions: The Power of Sound Analysis and Forecasting* and working on a manuscript titled *Myths and Realities of Project Financing*. John holds a PhD in macroeconomics and international economics and a master's degree in statistics and econometrics from the University of New Hampshire. His bachelor's degree is in economics and mathematics from Fairleigh Dickinson University. He also served in the U.S. Army during the Vietnam conflict.

Part I

Preamble

1 Introduction

A forecast is a statement about the future and usually takes the form of a single number, which is always wrong. Another definition is that a forecast is a prognosis and forecasting is the process of identifying the future state of the business. Forecasting is an integral part of business management, and its purpose is to generate information that can be used to make intelligent decisions. It is the discipline responsible for developing projections to help decision makers make sound decisions. A broader definition of forecasting is the anticipation, expectation, and prognosis of how company operations will develop in the future.

Strategic decision forecasting (SDF) is a broad discipline focusing on the analysis, development, and validation of long-term forecasts that form the basis of major business decisions. First, it is an independent, critical, and objective assessment of the business intended to introduce discipline in decision making and the art of translating qualitative and quantitative information into usable inputs by computational models to predict future outcomes. It is the function that brings together data, analyses and evaluations, business experience, modeling techniques, and scenarios to describe the transition from the current to a future state. Stated alternatively, SDF predicts future performance based on a systematic approach using relevant data, competitive intelligence and environmental assessments, modeling tools, past performance, analogs, and industry knowledge. It is an activity that goes beyond projections in that it identifies uncertainty and risks and develops ways to manage them.

Forecasting is also a process used to identify enabling scenarios to produce more effective decision making and to create operational targets. But by its nature, whether tactical or strategic, forecasting creates conflict and is viewed as a liability by most organizations where teams are not set up, organized, and managed appropriately. Also, SDF solutions that come in various forms and shapes are misunderstood and are widely misused in the hands of amateurs. On the contrary, when the SDF function is appropriately structured, defined, and managed, it is a most useful tool in evaluating and projecting the long-term state of business.

Often, questions such as why we forecast and whether we can really predict the future many years out are raised. The simple answer is that we forecast because all decisions involve future expectations or predictions, whether they are implicit or made explicit. Every decision involves uncertainty and risks, but in the process of creating forecasts their sources and causes are understood and ways are created to reduce them and help to deal with them. But can we really forecast the future, especially for long horizons typical of strategic projects? There are two diametrically opposing viewpoints among professional forecasters, reflected in the following two perspectives:

1. "Those who have knowledge don't predict. Those who predict don't have knowledge," according to Lao Tzu—a sixth-century Chinese poet.
2. "Forecasting is the foundation of all planning and if it must be done, we are the ones who can do it better" is the opinion of many forecasters.

Wake-up calls to forecasters who believe that forecasting is the foundation of all planning and decisions have come from forecast clients and senior management in several different ways. Figure 1.1 demonstrates the point that the basis of decisions is mostly relationships and processes. However, the implications of that chart go unnoticed and even some strategic forecasters focus mostly on the substance part of forecasting.

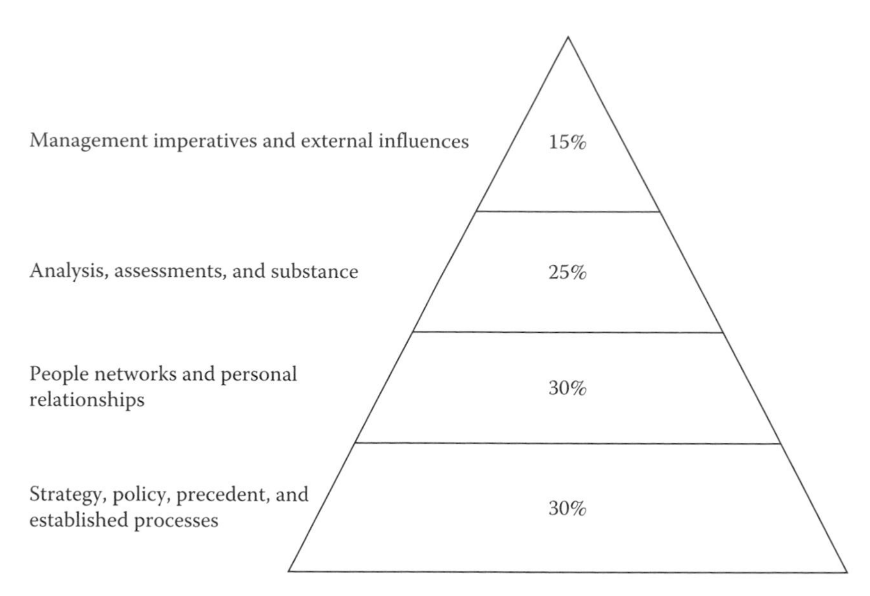

FIGURE 1.1 Key elements of decisions. (Based on Long Range Planning Associates research.)

The truth and reality of the situation is somewhere in the middle, and this has implications for how long-term forecasting is approached. Our view is a balanced perspective whereby forecasting is a tool to provide direction and guidance to reduce project uncertainty and that of surrounding decisions; and if one ignores it, they do so at their own peril.

1.1 FORECAST TYPES AND THEIR EVOLUTION

Forecasts may be classified under different criteria, but the most common classifications are those using the length of their horizon or their uses; short-term or tactical forecast, medium-term or lifecycle management forecast, and long-term or strategic management forecast. Following are the main differences along characteristic and usage lines:

1. *Tactical forecasts*. Tactical forecasts are data-intensive, short-term (six months to two years out), demand-driven, operational forecasts used for marketing and sales, production planning, and financial assessment purposes. They tend to have short turnaround timelines, and although they require familiarity with the market and may include causal factors, they do not incorporate environmental factor, industry, and SWOT (strengths, weaknesses, opportunities, and threats) analyses. Tactical forecasts tend to be subject to adjustments to facilitate client as well as senior management needs and objectives.
2. *Lifecycle management forecasts*. They are medium-term forecasts, two to four years out, with a focus on demand, competitive analysis, and future events. They may be number intensive, based on qualitative or quantitative methods, and more inventive in terms of including more causal factors and industry analysis. They are collaborative among forecast stakeholders and are used to develop product strategy, manage small investments in existing products and facilities, and are sometimes employed in product portfolio management (PPM) decisions.
3. *Strategic forecasts*. Strategic decision forecasts are long-term forecasts that are collaborative and integrative of qualitative and quantitative approaches in nature. Their emphasis is on environmental factor analysis, issue and project management, and industry and demand analysis. Importance is also attached to comprehensive modeling of company operations, scenario development and planning, and uncertainty and risk identification

and management. These forecasts are used in developing or evaluating corporate strategy, project and investment evaluations, scenario planning, and future-state realization (FSR) plans.

Over the past forty years, extensive progress has been made in many aspects of forecasting and inclusion of strategic forecasts into planning functions. The discipline of forecasting has gone from delusions of grandeur to disappointment to realism; and in that path, advances have been made in the following areas:

1. Increased data sources; improved quality of data; and better data collection, validation, and analysis techniques
2. Development of internal and external databases, database management software, and enhanced, user-friendly statistical forecasting and simulation packages
3. Applications of new approaches, methods, tools, models, quantitative and qualitative techniques, and empirical validations
4. Improved tactical forecasting processes, forecaster qualifications and training, and spreading of collaborative forecasting and consensus forecast development
5. Enterprise resource management systems designed to achieve functional integration with large forecast subsystems that include monitoring forecast performance measures

The main beneficiaries of the progress achieved have been the tactical and lifecycle management forecasting areas with SDF still being in its infancy. In the current state of strategic forecasting, failures commonly occur due to narrowly focused and restrictive approaches, inadequate methods and models, and improper tools used. Ad hoc environmental assessments, undue reliance on mathematical models and analogs, inadequate forecasting processes, and poor organizational structures magnify the challenges of strategic forecasting. Long-term forecast failures are also caused by another sort of factors that include the following:

1. Scarcity of people skills, experience, and training in the long-term forecasting discipline
2. Lack of appropriate forecast ownership, accountability, and performance measures
3. Poor project management of the strategic forecasting process, communication, delivery of forecasts, and instructions for proper use

Strategic forecasting progress has been hindered by disagreements over who owns the forecast and what constitutes a good forecast. In tactical and lifecycle management forecasting, the forecasting organization owns the forecast. The components of a good forecast are the speed and cost with which it was created, the accuracy measured by some forecast error indicator, and the ability to meet client needs. By extension, under the current forecasting paradigm, good strategic decision forecasts are judged by the same criteria even though they are substantially different creatures, supporting unique situations and needs.

A common definition of a good long-term or strategic decision forecast is that it is a forecasting solution that incorporates all of the following elements:

1. Strategic intent, constancy of purpose, and consistency of assumptions underlying forecasts in different project phases
2. Ownership of and accountability for forecasts across organizations and timelines and project managed with appropriate performance measures
3. Integrative thinking and future scenario development and evaluation along with sanity checks, uncertainty and risk assessment and management, and FSR plans
4. Inclusion of the client perspective, other stakeholder inputs, and corporate interests used to manage respective expectations

5. High levels of communication, coordination, cooperation, and collaboration throughout the forecast development process
6. Integrity, respect for the evidence, and a sense of history and learning from past experiences

Because there are misconceptions about the nature of strategic forecasting, we have advanced the following definition: It is forecasting for large impact, strategic decisions and projects using an approach that entails customization of method, analyses, evaluations, models, scenario planning, sanity checks, and uncertainty and risk assessment. SDF is a versatile tool customized to the particulars of each strategic decision and situation; and, in the minds of experienced forecasters, it is like telling a convincingly unique story that enables them to gain acceptance of the forecast. It assumes basic knowledge of SDF team members in statistical modeling, demand analysis and forecasting, and exposure to strategic planning, PPM, financial analysis, new product development (NPD) principles, and widely used marketing and general managerial tools.

Think about this: *The difference between good long-term forecasts and sound long-term forecasts: Sound forecasts are the gold standard; good forecasts are the minimum acceptable quality forecasts.*

In most tactical and lifecycle management cases, the forecaster owns the forecast and there is limited stakeholder involvement and input, and baseline forecasts do not drive big impact decisions; they are only another reference point. On the other hand, in SDF, collaboration is essential and the ownership of the forecast is shared by the three key stakeholders: (1) The client who provides information, support, and help with sanity checks; (2) senior management who shares strategic intent and insights, makes adjustments to the forecast, and provides the support needed; and (3) the forecaster who brings all the pieces together and manages the forecasting process. It is a collective effort managed by the forecaster.

1.2 WHY ARE GOOD FORECASTS AND SDF TEAMS NEEDED?

In companies where forecasting has been well established, the value of good forecasts of all types is recognized although creating good forecasts can be expensive. In companies where forecasting is viewed only as a cost center, one question still persists: Why do we need to spend so much to get good forecasts? Our experiences and research in the area of forecasting costs and benefits indicate that good forecasts are worth the investment and are needed for the following reasons:

1. Because strategic decision forecasts drive the long-range, business, and financial planning functions, it pays to do them well.
2. Errors in long-term forecasts compound and multiply in the planning horizon, give wrong project value estimates, and complicate the decision-making process.
3. It is a lot cheaper and quicker to develop good forecasts than fix problems in corporate planning areas caused by poor forecasts and ongoing revisions.
4. The impacts of controllable factors such as changes in price, competitive actions, and other business drivers on any decision variable can be predicted, and changing trends in a product's demand and revenue can be detected early on.
5. Valuable knowledge is gained from doing thorough external environment assessments, examining megatrends and subtrends, evaluating industry trends, and quantifying their impacts on company performance.
6. They help to describe future states of business and how to get there, anticipate events, create appropriate responses to ensure forecast realization, and update business and corporate strategy.

7. Good strategic decision forecasts synthesize the results of the analyses, the evaluations performed, and the intelligence and insights they create into a solid decision foundation and actionable recommendations.
8. With appropriate monitoring, they can provide early warning signals, which would give sufficient time to respond to unexpected changes, threats, or new opportunities.
9. They enable the company's Investor Relations group to provide reasonable guidance to industry analysts and enhance its credibility with the investment community.

Good forecasts are also needed for the SDF team to gain recognition, build organizational competencies, participate in the decision process, and become a trusted advisor to decision makers. From that perspective, the underlying reasons for good forecasts are to:

1. Build the broader organization's confidence in the forecast solutions it creates and their usefulness.
2. Develop all-around credibility and gain the ongoing support of clients, senior management, and project stakeholders.
3. Help manage clients' expectations by educating them about forecast processes, principles, uses, implications, limitations, and support needed to achieve forecast expectations.
4. Ensure command of resources needed to perform the analyses and evaluations the SDF team must perform to develop sound long-term forecasts.
5. Enhance communication, coordination, cooperation, and collaboration, and manage more effectively the forecasting processes, issues, and politics.
6. Provide guidance and advice to help clients plan, navigate through project future uncertainty, and manage risks more effectively.

An SDF solution can take different forms, depending on the decision or problem at hand. In one instance, it can be the modeling of a business process and simulations of scenarios to forecast and test different model input ideas. Or, it can simply be a long-term numerical projection based on quantitative, qualitative, or integrated methods. On other occasions, an SDF solution may take the form of building distinct, plausible scenarios to capture the impacts of controllable and uncontrollable events on decisions and thereby describing future states and the necessary enabling conditions. The assessment of project uncertainty, the identification of competitive threats, and the creation of plans to respond to external threats are also forecast solutions as is the evaluation of proposed strategy changes or entry new business areas. Another example of an SDF solution is the modeling of negotiation positions, creation of trade-offs, and quantification of their impact on project value, which help company negotiators to evaluate efficiently different positions and trade-offs.

Strategic forecasting is a long-term forecasting approach whose process starts with the corporate strategy and business situation; develops its own strong strategic intent statement; and uses a holistic, seasoned, mature, and broad-thinking approach to forecasting. As such, it chooses from the many forecasting methods, models, and tools currently available; borrows from other disciplines; and creates innovative ways to evaluate future state of business options that reduce project uncertainty and risks. Because of that, it prepares the organization for changes that may be required to meet the objectives that drive a strategic project decision.

A salient feature of SDF is that it links with and integrates forecasting with other key planning functions in a large organization, such as Strategic Planning, Finance, R&D, Business Development, Corporate Business Planning, and PPM. By its nature, it focuses on helping clients to explore new and different future opportunities and helps to manage not only management expectations but also stakeholder relationships and organizational politics. When executed well, it serves senior management needs because it is a trusted source of objective recommendations and advice. Also, because of the assessment complexity of strategic projects passing over a number of decision stages and gates,

SDF teams project manage the forecasting function across all processes and stages. As a result, consistent SDF solutions are created.

1.3 THE ROLE OF SDF TEAMS AND MISCONCEPTIONS

Tactical and lifecycle management forecasts are demand driven, derived from statistical models, and are mostly mechanized. In the tactical case, the role of forecasting is to be a resource to clients and provide data and reports, conduct some analysis, and deliver forecasts as needed. In the lifecycle management case, forecasting plays a bigger role as a collaborator and, in addition to services performed in tactical forecasting, helps the project team to understand the external operating environment and marketplace trends.

In SDF, the weight of other forces acting on the forecast is assessed and incorporated into the forecast solution, such as the following:

1. Understanding of megatrend and subtrend effects on the company's future business
2. Assessment of the external operating environment, the industry, the company's SWOT vis-à-vis those of key competitors
3. Evaluation of project uncertainty and risks, the effects of external changes, and occurrence of black swans
4. Assessment of the implications and risks involved in different forecasts on corporate resources

In this type of forecasting, the SDF team becomes a business partner and is called upon to perform and synthesize the results of analyses, evaluations, and intelligence gathering in order to create insights, to assess the impacts of actions or events, and to help manage decision uncertainty. Hence, its focus is on broad thinking; integrative forecasting methods; identifying relationships, feedback loops, and causality; assessing the impacts of forecasts and uncertainty and risks; conducting sanity checks; and developing probable scenarios of the future business.

The SDF discipline has made some progress in its quest for acceptance and recognition, but it is facing a difficult context in which it operates because common misunderstandings among forecast users abound. While there is some clarity of purpose and process in tactical and medium-term forecasting, in long-term forecasting and SDF, many misconceptions are still present. The confusion between tactical and strategic forecasting is exemplified in the oxymoron piece of a major forecast consulting firm's services in 2001 claiming to perform "tactical strategic forecasting services."

Some of the recurring negative perceptions about the nature and value of SDF shared in forecasting workshops and discussions with strategic project participants are the following common themes:

1. Due to the high cost of carrying inventories, a number of companies consider short- and medium-term inventory forecasting strategic forecasting.
2. There is widespread confusion that since financial long-term projections are not tactical demand forecasting, they are of the strategic forecasting kind.
3. Often, anything beyond a year or two out in the minds of clients is considered a strategic forecast. Translation: It is a dated forecast that does not affect them directly; the focus is on the latest forecast.
4. Long-term forecasts do not really enhance the company's competitive responsiveness, nor do they impact business performance because things change and people move on to new jobs.
5. Regardless of the method used, errors increase as the forecast horizon increases and long-term forecasts have much larger errors than short- and medium-term forecasts. So, simply extrapolate the future from short-term, rolling forecasts.
6. Business forecasters lack the mindset, skills, tools, and experiences to do long-term forecasting properly. That is why the SDF function is outsourced.

7. The business forecasting discipline lacks comprehensive methods to forecast company performance for new market entry, technology evolution, product licensing, and PPM adjustments.
8. SDF processes are nonexistent; incomplete; inadequate; and, in most cases, very complex. The processes and models used require extensive training and experiences.
9. Logical inconsistencies in current processes and models and lack of forecast project management are reflected in the absence of forecast reliability and the plans it drives, but are routinely ignored.
10. Strategic forecasts should ordinarily be outsourced to industry consulting firms, although they have no real ownership of the forecasting function and no forecast realization responsibility assigned to them.

1.4 UNIQUENESS OF THE BOOK AND BENEFITS TO THE READER

SDF assumes understanding of computational modeling and sufficient knowledge of quantitative and qualitative techniques. Instead, it focuses on the analysis and evaluation of results and the integration and synthesis of disjoint assessments to generate long-term forecasts. Its uniqueness lies with the actionable recommendations that flow out of the long-term forecasts it creates. They form the foundation of decisions and support projects of large impact and consequences. And what makes those recommendations reliable are the knowledge and understanding gained through the following aspects of SDF, which we address in this book:

1. Customizing the approach, models, and techniques to the specific type, situation, and needs of each project
2. Creating close functional links with all planning functions, external contacts, and client groups and emphasizing the value of strong personal relationships and their continuous nourishment
3. Making megatrend, subtrend, industry, and market trend analyses integral parts of the SDF process
4. Assessing developments in the geopolitical, economic, social and demographic, technological, and legal and regulatory environments to draw conclusions about their likely effect on the future company business
5. Incorporating the results of SWOT analysis to determine internal and competitor competencies and capabilities and the support needed to achieve forecasted levels of performance
6. Using a highly disciplined, methodical, structured, well-balanced, validated and verified, and cost–benefit-oriented approach to create SDF solutions
7. Introducing extensive checks in determining project uncertainty and risks, analyses to reduce uncertainty to residual uncertainty and develop a strategic posture, and sanity checks when assessing the implications of forecasts
8. Creating assumptions after project uncertainty and risks are identified, testing them with reality checks and against benchmarks, and updating them as conditions warrant
9. Triangulating forecasts and integrating the key principles of system dynamics modeling, real options, and scenario planning to guide the development of SDF solutions
10. Creating management dashboards to monitor strategic project performance and early warning systems (EWSs) to communicate the need for action when risks appear on the horizon
11. Performing project postmortem reviews and sharing knowledge and lessons learned to improve the chances of success of future strategic projects
12. Developing FSR plans to ensure appropriate responses to competitor reactions and realization of project goals and objectives

We have witnessed substantial performance improvements in companies that have established SDF teams due to the application of the SDF paradigm, and similar results can be obtained by the readers by applying the knowledge and experiences shared in this book. Benefits are immediate, accrue over time, and build competence in the discipline. They are reflected in the example cases of SDF team, client, and senior management acknowledgments that follow, with the most commonly cited benefits being the following:

1. Forecasters become more integrative in mindset, approach, processes, and knowledge; and, as a result, there is an increase in communication, cooperation, coordination, and collaboration, which takes place in every project.
2. Once an SDF team is established, it can create a good business definition, develop key processes, and determine the analyses and evaluations that the team will perform in different types of projects.
3. Gaps in existing competencies and capabilities are identified, how they impact project execution and forecast realization is determined, and ways to fill those gaps are recommended.
4. The holistic SDF approach is applied successfully to provide support to clients of any type of strategic decision project.
5. SDF solutions are transparent, not black box, models that enable clients to see clearly the rationale for many descriptive future states and ways to get there.
6. Strategic forecasting can fashion several future states beyond status quo projections and can identify the factors and support needed to accomplish them.
7. The use of forecast and SDF team member performance metrics, management dashboards, and EWSs to monitor progress increases significantly the chances of project value being realized.
8. SDF teams create actionable recommendations based on well-balanced considerations, which translate into solid bases and good strategic decisions.
9. Time from idea to forecast to decision is reduced while understanding of project uncertainty and risks is enhanced enough to manage them more effectively.
10. Competitive response plans and FSR programs are created, which increase the chances of project success in the implementation stages.
11. Senior management and clients are able to see, evaluate, and reward the SDF team's contributions as a business partner present at the decision table.
12. SDF teams consistently provide sound long-term forecasts supporting strategic decisions, which over time can be a source of competitive advantage.

1.5 STRUCTURE OF THE BOOK AND DISCUSSION TOPICS

Due to the many failures of long-term forecasting and the misconceptions that exist about SDF, we devote Chapter 2 to discussing such failures and presenting the results of our root cause analyses. The chapter deals with the current SDF paradigm and its failures, which are traced back to several factors: the lack of forecaster collaborative relationships with clients and other planning organizations; wrong forecasting focus, approach, and methods; and broken processes and wrong practices. The lack of clear forecast accountability and ownership, organizational issues, and inappropriate performance measures contributing to failures are also examined. Lastly, the effects of limited communication and coordination, deficiency in forecaster skills and qualifications, and absence of project and knowledge management issues in forecast failures are discussed. These, in turn, provide a blueprint of what is needed to address the shortcomings of the current strategic forecasting paradigm and develop the methods, processes, and tools needed to do justice to this important forecasting discipline.

Chapter 3 introduces the promise of the SDF paradigm, with the expectation that it will clarify issues and eliminate the confusion that surrounds it. Here, the SDF function is defined in more detail by describing its purpose, objectives, and focus; some universal process components and uses of SDF solutions; its organizational charter and structure; the role of SDF team members; and the benefits and limitations of the SDF paradigm.

Chapter 4 is devoted to the business definition of the SDF team, starting with the organization's mission, vision, values and principles, and forecasting stakeholders. A discussion of the SDF team's goals, objectives, and strategies follows and then some governance structures are sketched out. The scope, analyses, and services along with the team's roles and responsibilities are then spelled out. Because of their importance, the network links of the SDF team, its performance measures, and its communication plan are also discussed.

Chapter 5 addresses the mindset, processes, and principles that govern the operations of SDF teams. Here, the role of team's culture; the SDF team members' ways of thinking; and the approach, processes, and guidelines under which team members develop forecast solutions are captured from information in case examples and from client and other strategic project stakeholder statements shared in our discussions.

Chapters 6 and 7 are dedicated to discussing SDF techniques and implements used to enhance the team's effectiveness. It is a summary review of critical thinking concepts that form the foundation of the SDF approach. The mindset aids, anchoring analyses, and evaluations along with elements of the forecasting due diligence and forecast project management are discussed. The SDF systems used; how forecast solutions, options, and recommendations are developed; and the particulars of forecast performance measures, incentives, and the forecast realization plan elements are also presented.

To demonstrate the usefulness and practicality of SDF, Chapters 8 through 19 are devoted to the application of the methods, processes, principles, and techniques to the major types of strategic projects. Each chapter takes the reader through a review of the current state of strategic forecasting failures and the particular issues and challenges to each type of project. Then, the specific SDF approach and methods employed to develop appropriate, tailor-made solutions are articulated along with the analyses, evaluations, and techniques used. Each of these chapters ends with a review of either case examples of success factors or useful practices as well as expert testimonies and unique SDF team's contributions made through its participation in each kind of strategic project.

Chapter 8 deals with the application of SDF approach, processes, and tools in the evaluation of changes in corporate strategies and policies. This chapter is a precursor to the discussion of the role SDF teams play in supporting corporate reorganizations and turnarounds. The focus changes in Chapter 9 which shows how the SDF paradigm is applied to NPD projects. Namely, the approach, processes, models, and techniques used to improve the effectiveness of the traditional state–gate forecasting approach are shown. Here, the need for a high level of coordination, cooperation, collaboration, and communication to execute strategic projects successfully is demonstrated again.

Chapter 10 concentrates on forecasting to support a company's new market or business entry projects, be they domestic or international in market or geographic nature. The application of the SDF paradigm to merger and acquisition (M&A) and joint venture (JV) projects is treated extensively in Chapter 11, where the SDF team's participation is a crucial element to project success. The commercial evaluation aspects of product and technology licensing projects also require the application of the SDF paradigm to generate sound long-term forecasts in order to value such transactions. This is dealt with in Chapter 12.

Corporate reorganizations and turnarounds are another type of strategic projects that lend themselves to application of the SDF methods, processes, models, and techniques with good results. The analyses, evaluations, and success factors that the SDF paradigm brings to these projects are discussed in Chapter 13. Then, how the important issues of uncertainty, risk assessment, and their management in projects are executed under the SDF paradigm are illustrated in Chapter 14, while the application of the SDF framework to major capital expenditure and project financing undertakings is treated in Chapter 15.

Chapter 16 deals with the important topic of portfolio management adjustment projects and how the introduction of the new SDF paradigm in this area makes portfolio adjustments far more effective. The beneficial utilization of the SDF approach in strategic project negotiations and external relations support is articulated in Chapter 17. This is another area of valuable SDF contributions that increase project success rates. Chapter 18 deals with the area of FSR planning and management and describes how to do it successfully. The SDF applications section end with the more mundane, yet very important, subject of the SDF team support in the development and evaluation of business cases and business plans, which is the subject of discussion in Chapter 19.

The intent is to help the reader in all aspects of implementing the strategic forecasting paradigm. To that extent, an in-depth discussion of best-in-class SDF teams is presented in Chapter 20 where the SDF team governance, forecaster responsibilities and qualifications, and best-in-class processes, practices, and principles are discussed. Also, performance factors, measures, and success factors and beneficial practices are examined and illuminated. For the same reason, Chapter 21 addresses impediments and challenges to implementing SDF teams, processes, and performance measures. It also identifies the root causes of barriers and challenges and offers solutions that successful SDF teams have developed to deal with them.

The book ends with Chapter 22, which summarizes the key aspects of SDF and presents the consensus observations of different project stakeholders in the field of analysis and forecasting for large impact, strategic decisions. The main conclusion is that while the new SDF paradigm cannot solve all decision problems, when properly structured and managed by experienced professionals, SDF consistently creates value. And, over time, it can lead to creating a source of competitive advantage for the company.

2 Strategic Decision Forecasting

Forecasting for strategic decisions is the discipline that develops a set of forecast solutions to support strategic decisions or large impact projects. Scott Armstrong (2001) presented an outline of strategic forecasting, with the advice that forecasting methods should fit specific situations. Since then, not much has been published in the area of developing methodology, processes, tools, and organizations, and many misconceptions and failures in this type of forecasting prevail today. The progress made in this area has been by few progressive companies and some business consulting firms, but remains mostly proprietary.

Under the current paradigm, long-term projections needed to support strategic decisions are usually simple extrapolations of short-term statistical forecasts. For the most part, they fail to incorporate possible effects of megatrends, changes in the operating environment, future competitor reactions, and other critical factors. As a result, errors increase as the forecast horizon increases. For example, new product forecasts 10 years out have much larger errors than short-term forecasts. How much larger? They are 10, 50, and 100 times larger. Current paradigm long-term forecast performance also suffers because of lack of forecasting due diligence and sanity checks, no serious evaluation of forecast risks, and the resource implications of the forecasts. Furthermore, the absence of early warning systems (EWSs) and risk response planning make the monitoring forecast performance in the current paradigm an exercise in futility.

The failures of the current strategic forecasting paradigm consistently manifest themselves in a number of gaps, shortcomings, and problems, such as the following:

1. Confusion and misconceptions of what forecasting for strategic decisions is all about and when and how to use it
2. Forecasts that ignore the reality of changes in the external environment and the market structure that generates the data being forecasted
3. Required analyses and evaluations missing, no real intelligence and insights created, and no real contributions made to the decision-making process
4. Indiscriminant, wrong use of long-term forecasts that come with no instructions to the users and warnings of the consequences of inappropriate forecast uses
5. Isolation of forecasters and exclusion from discussions of future opportunities, assessments of project uncertainty and risk, and strategic decisions to be made
6. Low levels of forecaster credibility with senior managers who use forecasts only as another data point in their decisions

In the sections that follow, we identify failures and discuss root causes in key areas of strategic decision forecasting (SDF) using the fishbone diagram shown in Figure 2.1. These root cause failures, in turn, provide motivation for developing a better SDF alternative and a new approach that minimizes these failures. The summary analysis of this chapter deals with the following strategic forecasting problem contributing factors: forecaster relationships, approach and methods, processes and practices, forecast ownership and performance, analyses and functions supported, governance and communications, forecaster skills and qualifications, knowledge organization, and strategic forecasting project management.

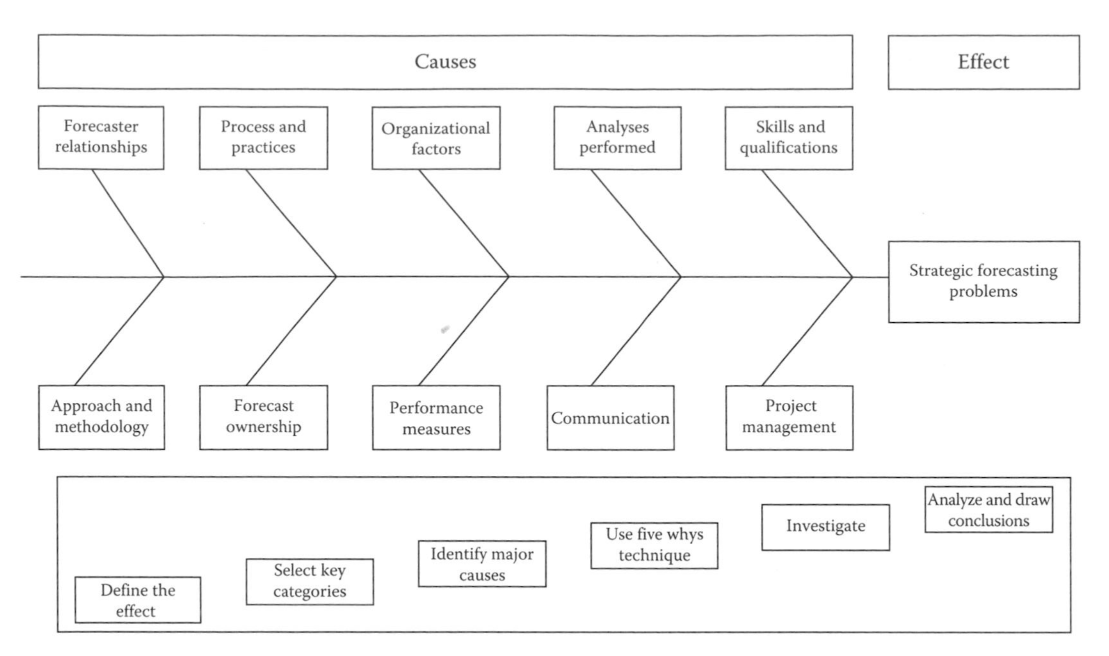

FIGURE 2.1 Strategic forecasting problems: Root cause analysis. (Adapted from Ishikawa, K., *What Is Total Quality Control? The Japanese Way*, Prentice Hall, Englewood Cliffs, NJ, 1987.)

2.1 FORECASTER RELATIONSHIPS

Forecaster participation in strategic decisions requires knowledge and appreciation of their contributions to project evaluations and acceptance of forecasting solutions created to support those decisions. It also requires the development of personal relationships and the formation of high-performing teams, which, in turn, facilitate effective and collaborative knowledge sharing. Knowledge sharing relationships involve active, ongoing interactions between forecasters and clients. Then, personal chemistry and relationships make a big difference in the influence forecasting can have on project decisions. In the current paradigm, the development of personal, knowledge sharing relationships and functional alliances is not part of forecasting job descriptions, has low priority, gets little attention by forecasters, and for the most part is nonexistent.

The purpose of modeling and forecasting is to help decision makers make decisions. But the problem with long-term forecasts is that they are always wrong and are developed without the benefit of close forecasting team functional links. They are usually created without inputs from key decision stakeholder organizations, cooperation and collaboration, critical and objective assessments and evaluations, and independent validations and verifications. As a result, the reliability of strategic forecasts suffers along with the credibility of forecasters, which are further aggravated by the absence of strong internal and external personal relationships and alliances between forecasters and clients, users of forecasts, and sources of industry and competitor data and information.

Long-term forecasting systems go wrong not only because of the poor technical quality of the forecasts, but also because insufficient attention is being paid to personal relationships. Because personal relationships are lacking, forecasts are changed through unwarranted changes in key assumptions and are used for purposes other than the original intent of the forecasts. As Jenkins (1982) points out, "Even when forecasts are done well, if insufficient attention is paid to the use of forecasting by decision makers, then the forecasts may be rendered useless, because they are ignored or manipulated by the organization's power structure."

Take heed and remember the basis of business decisions statistics of Figure 1.1 which shows that 75% of decisions are based on other factors and only about 25% on analysis, assessments, and substance.

2.2 FOCUS, APPROACH, AND METHODOLOGY

The current strategic forecasting paradigm is characterized by a narrow focus and mindset, that is, a narrow focus on quality short-term forecasts, which are then extrapolated in the medium- and long-term horizons. This is a problem and it is aggravated by the silo mindset, which is prevalent in projects involving multiple decision stakeholders with diverse interests. Lack of reflection on and observation of internal and external realities are major causes of wrong assessments, models of business operations, and, consequently, questionable long-term forecasts. Also, since the links between forecasting and other planning functions are not apparent and well understood, and evaluations of forecast implications and impacts are ignored, forecasters rarely participate in producing a consensus and consistent view of the future.

There are various accepted approaches and comprehensive methodologies to model business operations, but there is heavy dependence of forecasts on past history and industry norms. Forecasters seem focused on finding modeling and long-term forecasting capabilities in information technology systems, computer software, and new statistical techniques. While new forecasting software and systems technology are helpful, they are the center of attention at the omission of fully understanding the environmental context and underlying megatrends and subtrends, the relationships and direction of causality, the project uncertainty and forecast risks, and the impact of uncontrollable events.

Progress made in tactical forecasting areas is in terms of increased efficiency and accuracy, but it seems that expedience drives forecast processes and models and not strategic intent, project-type needs, goals and objectives, or forecast achievability. Hence, focus and effort are not in the forecaster's plan to help shape the future today through understanding that comes from approaches such as system dynamics modeling, distinct scenario development, and simulation planning and analyses. This is a lost opportunity for forecasters to demonstrate their expertise and contributions; instead, it is passed on to clients and higher level managers to create their own "flight simulators." Furthermore, lacking a value creation focus and cost–benefit orientation, forecasting teams are not invited to provide input to strategic decisions, and forecasts are routinely outsourced to external experts with no real responsibility assigned to anyone.

One of the striking deficiencies of the current state of strategic forecasting is the absence of comprehensive future-state realization planning. This important function is ordinarily reduced to contingency planning or risk management, with no active monitoring of project performance and forecast variance analysis. Project performance dashboards and EWSs are nothing more than trivial computer reports that do not monitor external environment changes. Nor do they monitor competitor reactions to the company's strategic project of the first, second, and third order. Why is that? It is because competitor response modeling, business war gaming, and tactical and strategic response planning are not in the forecasters' job descriptions.

2.3 PROCESSES AND PRACTICES

The absence of comprehensive and sound strategic forecasting processes is one of the striking features of the current paradigm, and in rare cases they exist, they are incomplete, flawed, or inadequate. What happens then is that extrapolated short-term forecasts and processes are used for strategic forecasting with little regard to the need to fit the forecasting solution to the particulars of the decision's strategic context. That is, assessments of the external environment, project uncertainty and risks, corporate needs and risk tolerance, internal competencies and capabilities, and expected level of ongoing strategic project support are lacking.

It is true in most companies around the world that a good part of business decisions are not based on facts, and because of unclear decision-making processes and forecast ownership and accountability, there is widespread inappropriate use of forecasts. For instance, one US-based company uses long-term forecasts to set high sales performance targets that remained frozen in outer future periods. In other cases, there are no comprehensive long-term forecasting processes in place. Why is that? That is due to incomplete forecaster knowledge of the company business and limited forecasting experience, lack of personal relationships, limited resource commitment to the forecasting function, and unclear policies concerning involvement in decision making. There is only late and minimal involvement of the forecasting group in strategic project decisions, and ad hoc processes are made up as needed.

Other observations in the process area of the current strategic forecasting state include the following major gaps:

1. No real project management of the long-term forecasting function across handoffs between different project phases, and much effort is expanded reconciling conflicting stakeholder objectives at decision gates.
2. Limited senior management commitment, understanding of strategic forecasting basics, involvement in facilitating the process, and resources allocated to support forecasting. Thus, strategic decisions are supported by more expedient alternatives: outsourcing the function to consultants or extrapolating short-term forecasts.
3. Because of misconceptions about strategic and business planning, currently the strategic forecasting function is often left to part-time forecasters who have no real forecasting experiences. Hence, these functions are unable to integrate external environment effects with other corporate planning functions' initiatives and incorporate forecasting tools, methods, and evaluations.

Unbalanced focus on product adoption, market shares, and reliance on analogs is common practice as opposed to understanding thoroughly the company's product attributes versus those of competitors, what drives product adoption, and the path of competitor action–reaction–action convergence. Also, time and again, we see duplication of the forecasting group's functions by Marketing and Finance organizations and waste of corporate resources. And when long-term forecasts are created by different internal groups and external experts, forecast users are more likely to pick those forecasts that better fit their current needs or provide most support for their desired type of decision. Lastly, it has been shown that in many cases, long-term forecasts are reacting to market changes; that is, they are revised frequently to keep up with those changes after they take place.

2.4 FORECAST OWNERSHIP AND ACCOUNTABILITY

In the case of short-term and lifecycle management forecasts and the current strategic forecast development paradigm, there is not much communication, coordination, cooperation, and collaboration among forecast stakeholders to speak of. Forecasters are by default accountable for the forecast and users own the forecast only because they are subsidizing the effort. In the current state of strategic forecasting paradigm, there are notably low levels of communication, cooperation, coordination, and collaboration among all stakeholders, and forecast accountability is a diffused responsibility.

In the current long-term forecasting regime, there is no expertise in project managing the forecasting function over long horizons, which include several decision gates with different needs. And as a project moves across different phases or when circumstances change, new players are introduced to projects, and assumptions are changed over time by clients, monitoring the performance of forecasts developed years earlier is not of any value. As a result, forecast clients or decision makers change forecasts to fit the current situation and forecasters are not even involved in the process any more. That is, nobody is really accountable for the forecast, no meaningful performance measures are in place, nor are in-depth variance analyses performed.

Lack of corporate policies about forecast ownership and accountability and project management of the strategic forecasting function causes finger pointing and conflict among forecast stakeholders. And a tremendous amount of energy is devoted to harmonizing different viewpoints instead of generating good long-term forecasts. Furthermore, failure to plan at the start of a project or a decision process and assign specific roles and responsibilities in all cases leads to chaos being created where the most vocal interests dominate. For this reason, forecast clients often view strategic forecasting as a liability, are glad to outsource it to external consultants, and maintain good relationships with all planning organizations.

2.5 ORGANIZATIONAL FACTORS

A major cause of strategic forecasting failures has its roots in unclear business definition of this function. There is confusion about the mission, strategy, goals and objectives, and the governance structure of strategic forecasting teams. Hence, there is diffusion of the scope and the roles and responsibilities of the strategic forecasting function—absence of a strong, central focal point. What happens then is that prevailing winds determine them arbitrarily. Examples of this problem include assignment of long-term forecast responsibility to R&D, Strategic Planning, Marketing, and Finance organizations, depending on the power structure and influence of these organizations.

Another major cause of strategic forecasting failures in the current paradigm is poor organizational structure, that is, the forecasting group's governance and management reporting structure, its composition, and links with internal planning groups and external experts. A good example of a wrong organizational structure is the case of a group responsible for forecasting for strategic decisions that reside in the Finance organization of a European public utility. This group is made up of two finance managers who rarely communicate with the Strategic Planning, Corporate Planning, and Business Development groups or the outside world, but are focused on generating and updating on a monthly basis long-term financial projections. Revised forecasts are better suited for their purposes then used by these organizations to update their plans.

Isolation of forecasting teams from mainstream decision-making activities, inappropriate organizational positioning, and few and weak links with decision makers and planning functions are key causes of failures. The causes for these problems are frequently traced back to heavy-duty internal power struggles, politics, conflicts of interest, and organizational silo mentality with divergent objectives. The end result is that strategic forecasting is a compartmentalized and often trivialized function, separate and apart from R&D, Finance, Product Management, Strategic Planning, and Product Portfolio Management (PPM) activities.

Other limitations in the prevailing paradigm occur because of weak links of forecasting teams with the external world, which includes industry associations, industry analysts, and subject matter experts. Failures also occur in cases of early organizational maturity or when integration of forecasting with planning functions begins to take place at the functional, but not at the personal relationship level. Organizational politics are present in all forecasts impacting strategic decisions and projects and since forecasting groups are usually not well positioned in large corporations, they tend to be ignored or relegated to number crunching. Lastly, when strategic forecasting is viewed only as a cost center or is treated as part of logistics planning, failures occur because strategic forecasting team contributions to decision making are considered marginal and are, basically, ignored.

Key takeaways: *The organizational factors responsible for strategic forecasting failures that must be addressed right away are as follows:*

1. *Unclear business definition of the SDF organization*
2. *Poor organizational governance structure*
3. *Missing internal and external functional and professional links*
4. *Forecaster isolation from mainstream decision-making activities*
5. *Absent or weak personal relationships and relationship development*
6. *Management of project stakeholder expectations and organizational politics*
7. *Wrong perception that forecasting is only a cost center and discounting its value*

2.6 FORECAST PERFORMANCE MEASURES

In short-term forecasting, performance is closely monitored through various measures, such as mean absolute percent error, mean squared error, or root mean squared error. As the forecast horizon increases, the business environment and structure that generated the initial forecast change and those changes affect the forecast perhaps more than the drivers included in the forecasting models. By the time the forecast goes out five years, the measures used to judge short-term forecasts become less relevant and the prevailing long-term forecasting paradigm has not developed useful measures of long-term forecast performance or measures of forecasting team success over different phases of strategic projects.

We mentioned earlier that there is a general lack of direct accountability for strategic forecasts for a number of reasons, the main one being that they involve transitioning from stage to stage over a number of years of the decision and project implementation process. In the current strategic forecasting paradigm, there is no systematic monitoring, uncertainty and risk management, risk response planning, and no end-to-end forecasting project management oversight. Why? Because of lack of experience in these areas and also because many original project stakeholders move on to new jobs and replacements are not accountable for earlier decisions and forecasts that drove them. Thus, absence of forecast accountability translates to absence of interest in addressing the lack of reasonable forecast performance measures and useful variance analysis. Furthermore, in such environments, the absence of incentives and rewards for exceptional forecast performance and the value added to decision making are striking. For these reasons, senior managers often question the need for internal strategic forecasting organizations.

The most dangerous failure in the current strategic forecasting paradigm is failure to correctly identify the driving factors of company performance, to measure causality, and to mistake the effect for the cause. This is common in strategic forecasting because of complex and changing interactions among the decision variables, drivers of a system, and environmental changes over long forecast horizons. The end product is poor technical quality and accuracy of forecasts and no insights gained from analyses and evaluations in developing the forecast. When forecasts with no meaningful performance metrics are used to drive large impact decisions, inconsistencies abound, the reliability of plans is questionable, and the basis of the decision is not well grounded.

2.7 EXPECTATIONS, ANALYSES, AND FUNCTIONS

Wrong ideas and expectations about roles and responsibilities and deliverables in strategic forecasting abound. On the one hand, there are naïve forecaster expectations that users should understand the basics of forecasting and that forecasters do not need to be concerned with learning the users' business. On the other hand, unreasonable management and client expectations concerning what it takes to develop strategic forecasts, achieve forecast effectiveness, produce superior forecast deliverables, and explain forecast variances are widely present. We have also witnessed management expectations of the strategic forecasting team to address problems that have to do with issues such as internal politics and corporate structure, strategies, policies, procedures, and corporate risk tolerance.

In organizations where the current strategic forecasting paradigm reigns, forecasters expect that historical data and industry analogs are available and reliable, decision-making problems are sufficiently defined, and their tools can easily be applied to generate long-term forecasts. This, however, is rarely the case when a multitude of decisions need to be made in strategic projects in short timelines and in the absence of historical data, precedence, or analogs. These circumstances do not afford the luxury of training forecasters in the firm's operations, decision-making process, and managing forecasts through conflicting stakeholder objectives.

There is limited support provided to strategic forecasting functions for a number of reasons, one of which is the apparent limited forecaster knowledge of and ability to use tools to create strategic forecasts simply because they are not taught in forecasting curricula. Thus, the services forecasters provide in the current paradigm are limited to data analysis, parameter estimation, some simulations, and superficial environmental, industry, and, rarely, some ad hoc strengths, weaknesses, opportunities, and threats (SWOT) analyses. Again, the weak links with Strategic Planning, R&D, PPM, Business Development, and Investor Relations groups are reflected in differences in expectations of these groups, which make it difficult to create sound and efficient consensus forecasts and to create value according to corporate expectations.

Know this: *Effective management of expectations goes a long way toward managing project stakeholder behavior, enhancing the strategic forecasting team's standing, and generating sound long-term forecasts.*

2.8 COMMUNICATIONS

In every instance, the disparity of forecaster–user perceptions concerning the forecasting group's status and needs is traced to poor communications between forecaster and stakeholder groups. Those perceptions come with many unstated and usually unreasonable expectations, which remain unspoken and are not dealt with until conflicts arise. Broken communication channels and lack of or limited interactions between forecasters and clients cause failures to achieve the high levels of coordination, cooperation, and collaboration required in creating successful projects. These are due to and/or manifested in problems in the following categories:

A. *Forecaster communication issues.* Failures, inefficiencies, and setbacks attributed to inexperienced forecasters include the following communication limitations:

1. Little understanding of the business needs, and little, if any, guidance to clients on project uncertainty and forecast risks, and how long-term forecasts should be used
2. Inadequate understanding of company operations and decision-making processes and, for that reason, forecasters being branded as introverts not interested in participating in mainstream business activities, but only in numbers
3. Inability or unwillingness to educate clients on forecast processes and to explain the implications of forecasts developed, potential risks surrounding forecasts, and ways to mitigate them
4. Allowing the scope of forecasting to be very broad and become expensive due to costs of data collection and analysis, storing and retrieving data, modeling and forecasting, performance monitoring and variance analysis, and report generation, which are labor-intensive activities
5. The concepts of stakeholder management expectations and communication plans and their value that are not familiar to forecasters and absent from their tool bag

B. *Client/stakeholder issues.* The major client and project stakeholder communication issues in the current strategic forecasting state evolve around

1. Confusion of what constitutes a strategic decision forecast, who does it, how it is updated, and what role forecasters and clients play in the development and the use of the forecast.
2. Misunderstanding and wrong impressions of project stakeholder roles and responsibilities in generating long-term forecasts, which abound up and down the line.
3. Ambiguity about the scope of involvement and confusion about their own roles and responsibilities that are present even within forecasting teams.
4. Widespread client confusion about future-state forecasts, future plans, and operational targets. They are used interchangeably and sometimes targets become forecasts, plans become targets, targets become plans, and plans become forecasts.
5. Disparity of preparer–user perceptions of company forecasting needs and costs and benefits. Forecasting is not viewed as a critical partner by senior management, while forecasters believe they are a crucial link in the value creation process.

C. *State of affairs predicament.* Communication problems attributed to the current paradigm's view of the world originate with the following factors:

1. Minimal client or decision maker exposure to how strategic forecasts are derived, what it takes to generate a strategic decision forecast, and how forecasts help decision makers and add value in a project
2. Forecasters not engaged in the decision-making process are perceived lacking strategic forecasting expertise and credibility. The prevailing view is that nobody knows what the future will bring, so why waste time forecasting it.
3. Clients, decision makers, and other forecast stakeholders are focused on their own areas, are not fully aware of their roles and responsibilities in forecasting, and are not communicating with forecasters.
4. Communication failures lead to lifecycle management forecasting viewed as strategic forecasting if it is longer than one year or to financial forecasting considered appropriate to support strategic decisions.
5. Forecasts are rendered useless, ignored, or manipulated by the organization's power structure because use of instructions, forecast implications, and risks associated with them are not well communicated.
6. Numerous implicit assumptions and disconnects on the part of the forecaster, client, and senior management go without being articulated, discussed, and settled, and their implications understood.

2.9 SKILLS AND QUALIFICATIONS

Forecasters, for the most part, have a reputation of being intelligent, well schooled, and good in working with numbers. And they are familiar with many qualitative and quantitative modeling and forecasting techniques. However, that is not enough to get them invitations to the decision table not only because of lack of relationships and skills in general management areas, but also because of lack of qualifications and training within the forecasting for strategic decisions discipline itself.

To begin with, it is now recognized that there is a lack of forecaster training in SDF that goes beyond environmental and industry analyses such as knowledge of the basic principles of R&D, strategic planning, competitive analysis, scenario development, risk assessment, and development of competitive responses. Other required strategic forecasting skills that are usually missing are training in business process assessments, new product development (NPD), technological forecasting, portfolio management, and several other business areas, which are intensive strategic forecast solution users.

The forecasters' common lack of in-depth industry knowledge and understanding of company operations is compounded by their lack of training in planning and other broad managerial areas, such as financial management, risk management, negotiations, business case development, and business planning. Consequently, there are few lateral transfer opportunities to expose forecasters to other areas of the business. And why is that? Because forecasters are not expected to and do not know, or make an effort to know, how to add value to large impact strategic decisions outside of producing numbers and talking about the latest modeling techniques.

Interpersonal skills of current paradigm strategic forecasters are weak, but since their discipline can have large impact on the future business, one would think that they would be trained to become more effective communicators. The reality is that the majority of forecasters are not very effective communicators, relationship builders, and managers. They have traditionally paid insufficient attention to alliances with decision makers and instructing them on the use of forecasts to navigate strategic decisions effectively. Thus, they have wrongly been said to have limited competence and emotional intelligence.

Key takeaway: *To be recognized as valuable business partners, strategic forecasters must first change perceptions by managing expectations, close the wide gaps in the current state, acquire and develop communication skills, develop personal relationships with clients, make all the required process changes, and demonstrate that they do add value in each and every project.*

2.10 KNOWLEDGE AND PROJECT MANAGEMENT

Knowledge management is the process of systematically acquiring, evaluating, storing, disseminating, managing, and leveraging volumes of knowledge in a company on an ongoing basis. It is related to information management, but it is much broader in that it creates intelligence and insights, which are made available to all internal users who have a need to know. In competent strategic forecasting teams, the key elements of knowledge management are the creation of an organizational memory of knowledge, knowledge sharing, and knowledge transfer.

Knowledge management is considered a corporate resource management function, and in the current strategic forecasting paradigm, it is underdeveloped and limited to report generation. Failures in this area impact strategic forecasting teams and their ability to perform their functions and to disseminate effectively the intelligence and insights they create. Thus, we observe knowledge concentrated in functional silos and not having a central repository of data, information, analyses, and experiences made available to strategic forecast stakeholders. Also, in most projects, there are no postmortem analyses performed to find out what worked and what went wrong, and to create organizational learning to be used in future projects. And so, valuable knowledge is lost, often to outside consultants who develop strategic forecasts and, eventually, to competitors.

Project management is a structured and methodical approach to plan, organize, guide, and manage all processes in a project from start to finish across all stages. Knowledge of the technical aspects of strategic forecasting is required to do it successfully, but so are the communication and relationship management skills discussed above. In every instance, forecasting function project management is crucial, yet it is a lacking competence in the current paradigm. As a consequence, we observe failures due to it, such as the following:

1. Late involvement of the forecasting group in large impact decisions and incomplete forecaster knowledge of company and competitor products
2. Forecast assumptions created and revised at different stages of the decision process by different stakeholders, forecasts updated with no one managing the process, and maintaining a record of changes and the rationale behind them

3. Cumbersome transition of the forecast from stage to stage that is fraught with disagreement and conflict and valuable resources being wasted in duplicate forecasts
4. Forecast budget overruns and delivery delays due to repetitive forecast revisions because nobody really monitors the development of forecasts and establishes the need for different views

Again, due to communication failures, decision makers do not seem concerned about the methods used and are ordinarily satisfied with financial forecasts being used as the basis of decisions. There are several important reasons for this: First, success of decisions and projects is usually judged in terms of financial value created, which is a widely accepted measure. Additionally, senior management and decision makers have a good understanding of financial terms and relationships, but know little if anything about the forecasts underlying financial projections. Also, senior managers are not convinced that strategic forecasting teams add value in managing this function other than creating intelligence out of data and insights into possible future outcomes.

First, because forecasters have not demonstrated the value of their contributions by project managing the modeling, analysis, scenario development, risk assessment and management, and the decision-making process. Second, there is limited opportunity to demonstrate the value added because financial forecasts can easily be produced with Excel spreadsheets by first-level managers very quickly and subjected to all kinds of "what-if" simulations of financial relationships. Third, Excel spreadsheet forecasts for strategic projects are easy to understand and somewhat accepted by clients and senior management, but are driven only by unsupported assumptions of relationships. This, however, does not constitute strategic forecasting or project management of this function.

Due to failures in the current strategic forecasting paradigm being so widespread, senior management and decision makers often look to outside advisors to project manage strategic forecasts. Sometimes, Finance organizations not trained in forecasting for predicting demand, which drives revenues and costs, and for overall project assessments, are assigned supervision of that task. More often than not, strategic decision forecasts are outsourced to external experts. At other times, forecasts developed by Finance groups are simple extrapolations of past performance into the future, assuming some average growth rate to prevail in the forecast horizon. In either case, the results have been far less than impressive.

Clarification note: *Why is so much effort devoted to discussing current strategic forecasting paradigm problems?*

- *To know what is currently broken and needs to be fixed*
- *To see the impacts of these problems on the success of strategic projects*
- *To avoid the same problems and issues in the new paradigm*
- *To get some ideas on how to implement the new SDF paradigm in different organizations*

3 The Promise of Strategic Decision Forecasting

In Chapter 2, we discussed the current state of forecasting for strategic decisions and its failures and touched on the root causes. We made reference to strategic decision forecasting (SDF) solutions because in addition to numerical predictions, the solutions include intelligence and insights created, evaluations, sanity checks, scenario development, simulation assessments, and various analyses performed by SDF teams to help clients make effective decisions.

3.1 DEFINITION OF SDF

SDF can be described as the set of methods, processes, models, tools, analyses, knowledge, experiences, the mindset, core values and practices, and organizational structures and links. It is a broad-thinking, methodical, highly collaborative, knowledge-intensive approach to describe, project, and assess future states. That is, to conduct sound analyses, outline plausible future states, generate credible long-term projections, plan for future-state realization (FSR), and provide actionable recommendations to clients and decision makers. These elements are brought to bear on assessing opportunities, proposals, plans, or projects and reach decisions effectively.

SDF is integrative forecasting that creates long-term forecast solutions for specific, large impact, strategic project decisions. These solutions and the nature of the business need are the first step in assessing value creation in every decision or project. However, the nature of the business problem, the impact of the decision, and the use of the SDF solution determine the type of solution developed. This concept definition of SDF is consistent with Scott Armstrong's advice that the situation should determine the methods and models used in forecasting.

A key element of strategic forecasting is to understand the project context, model the company's business operations, and incorporate the impact of distinct scenarios on the future state of business. By building the equivalent of "flight simulators," it helps decision makers to influence, alter, or shape the desired outcomes of those decisions. It is a unique approach to identify, explain, and project business drivers by the outcome of the interaction with external factors that impact a strategic decision. SDF is intended for planning to win. In other words, it helps decision makers better understand the current situation and the options to shape the future through scenario planning and use of controllable levers, mitigation of uncontrollable event impacts, making appropriate decisions, and executing effective responses to competitor reactions.

To help in planning to win and earn the right to sit at the decision table, the first step of the SDF team is to conduct a value creation audit to understand how a company is run and generates profit and then to determine how to add value to the process. To accomplish that, SDF needs to be a centralized function different than tactical forecasting for marketing, sales, operational, and financial plans. It also needs to be a function that employs modeling processes that use a focused strategic thinking to lay out a logical sequence of events that lead to future states. As such, it requires strategic thinking applied to specific issues, questions, and problems and understanding of the key success factors for different decisions.

Strategic decisions involve major risks and require a different mindset, processes, tools, skills, organizational setup, and high levels of communication, coordination, cooperation, and collaboration (4Cs). Above all, they require strategic intent and focus, critical and objective assessments, and

to influence future states and outcomes. It is also needed to provide an organized approach and process that all project stakeholders understand well and know to use it effectively and one that resolves the dilemma of "what constitutes the best forecast." When carried out appropriately, SDF solutions can successfully describe the future state of affairs for the benefit of decision makers, which includes concepts, beliefs, goals, and actions required. Thus, SDF teams clarify the need for long-term projections; identify the key elements, decisions that need to be made, and realistic options; and ensure appropriate uses of forecasts.

A major need not filled under the current long-term forecasting practices is the capability to develop, update, and monitor key assumptions effectively and analyze their implications and effect on decisions and risk profiles associated with them. The new paradigm fills that need, reduces uncertainty to its residual component, creates business process models, and develops distinct scenarios to simulate and forecast. Through these, it helps resolve competing viewpoints by pointing to the resulting outcomes and corresponding risk profiles, and providing actionable recommendations to clients. In the process of doing that, SDF identifies resources needed and drivers to leverage to ensure that SDF forecast solutions are realized.

Residual uncertainty definition: *Starting with total uncertainty surrounding a decision or project, it is the uncertainty remaining after SDF team performs analyses and evaluations to eliminate unknown factors. As unknown factors become known, this process reduces project risks and makes it easier to develop strategic posture for a decision* (Courtney et al., 1997).

In the new paradigm, the SDF team project manages the development of long-term forecasts, which necessitates sound processes and unambiguous assignment of roles and responsibilities to project stakeholders. This helps a great deal in better defining, understanding, handling, and solving decision problems. Sound processes and clear assignment of roles and responsibilities to forecast participants are missing from the current paradigm. The new SDF paradigm not only fills the gap, but it also helps clients understand the basics of analytics, the process and models, how forecast solutions are developed, and how they should be used for maximum benefit. SDF teams help to do it through determining key decision problem parameters, drivers, and controllable and uncontrollable factors and events, and anticipating future events, structural changes, technology evolution, competitor reactions, and new trends. Additionally, they identify ways to leverage controllable events and deal with uncontrollable factors; that is, they help to do forecast realization planning.

The SDF approach is a methodical and structured method that accomplishes these new paradigm crucial requirements: (a) sharing knowledge and insights and affecting closer coordination of effort, common assumptions, inputs, and assessments; (b) helping create close business and personal relationships and interactions with clients and decision makers; and (c) answering questions on forecast risks and risk management such as the following:

1. Who does what, why, when, where, how, and so what?
2. Is the forecast reasonable and consistent with historical experiences and current market realities?
3. Has the best forecast been created; can it be trusted to base decisions on it?
4. What are the forecast risks and implications; can we live with them?

Another reason the new SDF paradigm is needed is that it constructs the analytical framework and processes to perform the series of tasks required in different projects to arrive at reasonable and probable outcome scenarios. It is also needed to determine the methodology and evaluations required to develop the assumption set, create appropriate models, and assess how the assumptions affect the forecast solution. SDF teams conduct internal and external environment evaluations, develop dynamic models of business process realities, and describe what can happen when things change. They methodically move from observation, mental models, and data collection to computational models, scenario development, sanity checks, and forecast risk evaluation. Then, they arrive at a consensus forecast solution that drives the decision. The methodology employed is project and circumstance specific and helps to find sound forecasting solutions: first by defining and dissecting the decision problem and then by developing tools and criteria to determine feasible best solutions. This sets the ground for outlining a few distinct and plausible scenarios, simulating "what if-then" achievable scenarios, gaining insights into how to create impacts, determining the steps needed to make scenarios happen, and recommending the best option most likely to emerge.

Strategic decision makers have desperately been looking for reliable support to assess future business outcomes, to find ways to help them navigate through uncertainty, and to make value-creating strategic decisions. The new SDF paradigm fills this important need, which is a must in every major impact decision because

1. It specifies the forecaster qualifications required and the capabilities needed, which are missing from the current paradigm, and provides a systematic, sound way to quantify the value of opportunities.
2. Circular and broken processes are eliminated with a methodical evaluation of risk and uncertainty when SDF processes and principles are used because they are trusted by decision makers to reduce them. Thus, time to decision is reduced substantially.
3. In developing solutions, SDF teams evaluate and project environmental factors and the impacts of megatrends and subtrends, industry trends, structural changes, and marketplace shifts. They also determine the project-required support levels needed and key initiatives to undertake to manage risks.
4. It fills the need to have an assessment of competitor reaction to a strategic project and ready-to-deploy reaction plans of the first, second, and third order based on well-considered competitor response models.
5. It monitors project, forecast, and SDF team performance with appropriate measures, management dashboards, and early warning systems (EWSs) to observe progress toward project goals and objectives and communicate the need for intervention when necessary.

The average forecast horizon range for strategic decision projects, summarized in Table 3.1, is between 10 and 15 years; and during that period, many changes occur both internally and in the external operating environment. Strategy, goals, and objectives change in response to changes and shifts in the environment, which often resembles nothing like the initial state. Therefore, if

TABLE 3.1
Average Strategic Project Duration and Forecast Horizons

Strategic Decision	Project Duration (years)	Forecast Length (years)
New product development	2–5	15–20
M&A projects	0.5–2	10–15
Joint venture projects	2–4	10–15
Technology/product licensing	1–2	10–15
New business/market entry	1–3	10–15
Infrastructure projects	2–5	15–20

Note: M&A, merger and acquisition.

companies are serious about getting the strategy right, they need to implement the SDF approach so that it supports the strategic planning function.

A need senior managers have articulated in the project communications area is for the forecasting solutions to be understandable and actionable by the client. This is achieved by SDF teams sharing highlights of external and internal analyses and results as they are performed and by gaining insights into how they could impact the future of the business. To be effective, SDF teams deliver forecast solutions with adequate explanations, sanity checks, and assessments of implications, which help the client build consensus around the best solution. Then they also make clear the risks involved in the project and go two steps further and spell out the follow-up actions required to sustain forecast occurrence and outline the possible consequences of not making the decision or not executing the project. Thus, a big benefit of the new SDF paradigm is the results it produces by ongoing communication of findings and results with project stakeholders. This is very helpful in creating personal relationships and alliances.

In the current environment, long-term forecast error measures such as mean absolute error or root mean square error are monitored, but nobody looks back 10–15 years to see the accuracy of a single-number forecast that drove a major decision. This happens because nobody is fully accountable for the forecast; there is no discipline and no insightful forecasting. The new paradigm addresses these problems by requiring focus and attention to its methods, processes, and principles; assigning specific roles and responsibilities to stakeholders; and project managing the forecasting process. In the areas of strategy development and evaluation, the focus is not on plans, but on insights into how to create value and make SDF solutions part of the process of developing insights to be turned into strategic actions. For that reason, SDF models contain all drivers relevant to the decision problems, and their projections are key inputs to the Strategic Planning function.

Why SDF is needed—key takeaways:

1. *Address the current paradigm problems and eliminate its failures*
2. *Increase confidence in and reliability of long-term forecasts*
3. *Manage the development, updates, revisions, and recordkeeping of a common assumption set, which drives all project planning*
4. *Create complete, sound, and effective processes communicated to all stakeholders*
5. *Identify project uncertainty and risks and help manage them*
6. *Develop techniques and tools to deal with strategic project issues*
7. *Create the basis to get the corporate strategy right and evaluate strategy changes*
8. *Reduce time to decision significantly and enhance forecast realization*
9. *Help decision makers navigate through project decision uncertainty*
10. *Enhance the chances of strategic project success and value creation*

3.4 THE FOCUS OF SDF

The foundation of the new SDF paradigm is a framework developed to determine the elements needed to address complex strategic decision issues and arrive at reliable solutions. It includes the methodology, which is the set of processes, practices, methods, and the means used in a systematic way to develop SDF solutions. The overarching SDF framework shown in Figure 3.1 has three components. The first one includes the organizational culture, the mindset of associates, and the approach and methodology used to develop solutions. The second component includes the SDF team's goals, objectives, and strategies; the plan of attack; the

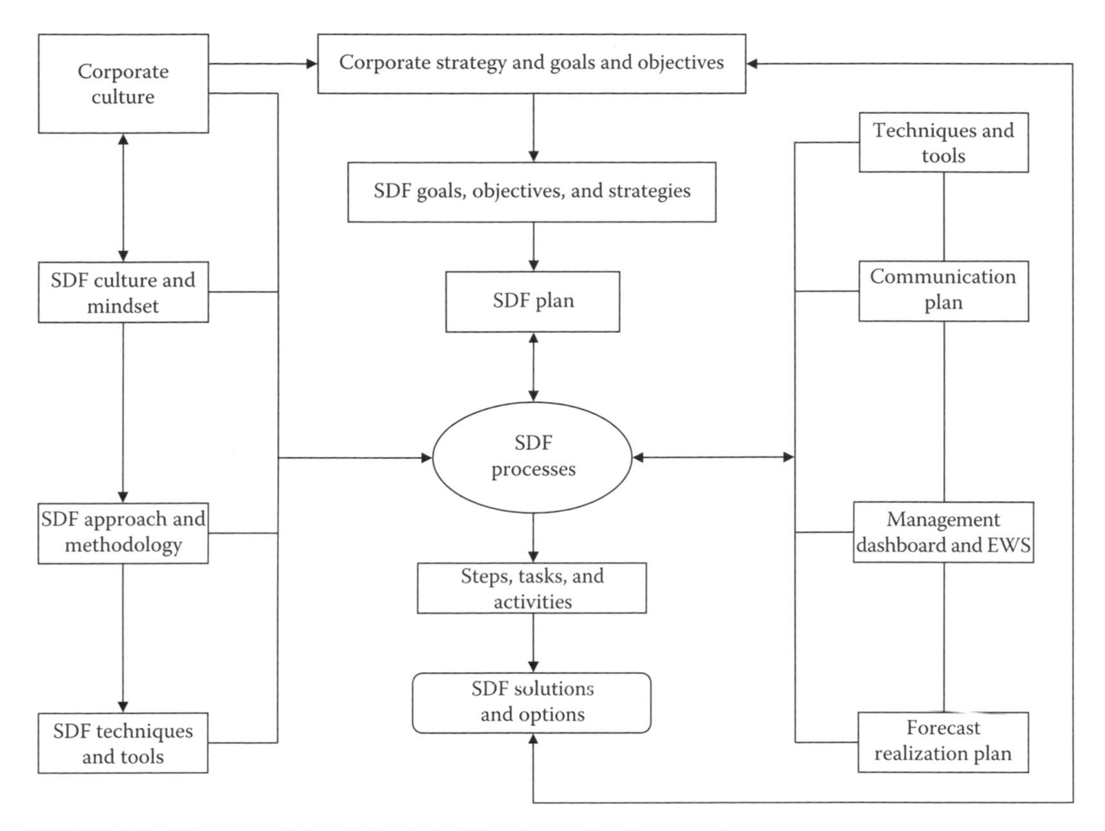

FIGURE 3.1 The overarching strategic decision forecasting framework.

key processes; and the steps, tasks, and activities to develop solutions. The third component contains the techniques and tools used, the management dashboard, the EWS, the competitive response plan, and the SDF solution communication plan.

> ***Management dashboard:*** *It is usually a computer interface that receives, defines, and communicates the status of a project; that is, it shows the big picture in key performance indicators (KPIs) that clients use to manage projects effectively. It is a management reporting tool that shows a snapshot of KPIs at a point in time in easy-to-read and understand format. Dashboard elements may include forecast performance, budgetary metrics, planned versus actual implementation schedule progress, milestones achieved, deliverables completed, and targets met.*

The focus of SDF teams is on a number of interrelated elements of the process, but special emphasis, concentration of energy, and attention are attached on clarity of purpose of the decision and the key factors in developing sound SDF solutions and recommendations. The most important of these factors are the following:

1. A value creation audit to better understand the needs surrounding a decision, create a specific plan for creating a solution for that decision, and perform a quick cost–benefit analysis to determine the value of different approaches
2. Determining what needs to be known, what is known and unknown, knowable or unknowable, what factors and events are controllable and uncontrollable, and figuring out how to leverage the controllable factors for forecasts to be realized

3. Strong alliances and personal relationships with clients to affect open communication, broad-based cooperation, and close coordination of effort, which help establish close links with other planning functions in the company, reduce the silo mindset-related conflicts, and reach decisions effectively
4. Development of SDF team strategic intent, which is focus and ambition out of all proportion to its resources and capabilities to create compelling targets for corporate commitment and effort
5. Assignment of clear roles and responsibilities to stakeholders in the SDF plan and development of performance measures that encourage collaboration
6. Thorough understanding of megatrends, subtrends, and industry and market trends; correct interpretation of market signals; and evaluation of their implications
7. Accurate reading of the operating environment and prevailing conditions in the industry, perceptive and objective evaluation of internal processes and capabilities to develop good scenarios and ascertain the company's ability to execute the project successfully
8. Openness to and ability to work with conflicting viewpoints and different types of decision makers, consider all available methodologies, techniques, and tools; and employ the most suitable models for the current market realities with minimum biases
9. Conducting a forecasting due diligence to verify the quality of data, information, intelligence, and insights; reviewing knowledge gained in past projects; determining the validity of analyses, processes, models, and techniques used; and verifying relationship measures, causality, feedback loops, and lags in responses
10. Continuous testing of the commonly held views, perceptions, assumptions, methods, models, scenarios, estimates and results, and impacts and their implications, which helps to identify the sources and causes of errors and manage project uncertainty and risks from SDF solutions
11. Striving to obtain certainty in an SDF solution and then mistrusting it because black swans always occur at some point in time, and the best policy is to prepare the organization to be in a position to respond effectively

Black swans: *They are events or phenomena that occur beyond what is expected in a given situation and are extremely difficult to predict. Black swan events are random and unexpected and thought to be impossible. These events have three characteristics: They are surprise; they have a major impact; and, after the fact, people assert that they expected them to take place. Examples of black swan are the invention of the Internet, the collapse of USSR, and 9/11.*

12. Common forecaster–stakeholder learning and institutionalized knowledge from internal and competitor successes, mistakes, and the results of postmortem analyses distilled in key learning points
13. In the presence of rapid technology changes, product changes, market instability, and high risks, focusing on erring on the safe side and selecting scenarios and forecasts consistent with senior management risk tolerance
14. Sharing forecasts in a story format of things that need to be in place, of what events happen in the future, and why in such a way that project stakeholders can understand and preparing guidance for senior management, clients, key project stakeholders, and the Investor Relations group
15. Monitoring what takes place after a decision that is made through a management dashboard, reassessing risks through an EWS, and determining how well different parts of the scenario used in the decision are performing
16. Developing a competitive response and forecast realization plan to course correct and shape required support to realize the projected state in the next few periods

The basic premise and starting point of the new SDF paradigm is reason. Then come the economic laws of supply and demand, the laws of physics (gravity, action–reaction), and the motto "believe, but check; trust, but verify." Due to the importance of being right in strategic decisions, the focus of the SDF team is guided by several dynamics. To create valuable intelligence and insights, the SDF due diligence is of the environmental, commercial, and business type and not just of the financial type that looks for confirming and disconfirming evidence. Second, all strategic intelligence and insights created are safeguarded in-house and used to validate and integrate diverse perspectives and all-around evaluations.

The more the qualifications and the broader and deeper the knowledge of forecast stakeholders, the better the SDF team members' ability to assess and represent the business reality, model it, simulate it, and forecast it. And, because of its position in the company and its competencies, the SDF team concentrates on assessing and validating proposals, projects, and decisions and supporting proposed courses of action. However, since SDF models of the company's operations are based on the team members' own experiences, it is beneficial to look at driving factors that impact forecasts objectively from different angles and figure out how things fit together. Realizing that no system is perfect, SDF solutions are intended to be guides to strategic decision making and not to serve as operational targets because each has a different underlying set of objectives and support requirements.

Recapping the areas of SDF team concentration:

- *Developing an SDF team strategic intent, which is an ambition out of proportion to resources*
- *Creating strong alliances and personal relationships with clients*
- *Conducting a value creation audit to determine key drivers and how to add value*
- *Determining what needs to be known versus what is known or knowable, what is controllable and what is not*
- *Assigning roles and responsibilities to project stakeholders and obtaining their support*
- *Accurate interpretation of external environment and market developments*
- *Ability to work with conflicting interests and points of view to reach consensus forecasts*
- *A thorough forecasting due diligence to reduce uncertainty and risks and strive for forecast certainty*
- *Continuous testing of perceptions, assumptions, methods, models, and scenarios*
- *Ongoing learning and institutionalizing knowledge to apply to all strategic projects*
- *Selecting realistic, balanced scenarios consistent with senior management risk tolerance*
- *Communicating findings and forecasts in easy-to-understand terms to all stakeholders*
- *Monitoring risks through management dashboards and EWSs*
- *Developing FSR plans with specific actions to take as threats appear on the horizon*

3.5 UNIVERSAL SDF PROCESS COMPONENTS

What goes into developing SDF solutions and the extent of the effort involved in each project varies according to the corporate goals and objectives, the position of the SDF team in the firm, the needs of the situation, the project objectives, and the purpose and intended use of the forecasts. The first common element in all SDF applications is learning well the clients' business; understanding their responsibilities; and identifying their goals and objectives, their true needs, and possible conflicts of interest with other stakeholders. Detecting what each party expects to get from an SDF solution and the decision or project and determining the feasibility, scope of involvement, and level of stakeholder commitment to the project are important common elements as well.

Discovering the exact use to which the SDF solution will be put into; instructing clients on forecast implications, risks, limitations, proper use, and caveats; and finding out the expectations of senior management, their intended involvement in, and the support to be provided for an SDF solution are key activities performed in all projects. Other common steps are listing the information needs and requirements for a given decision, establishing what needs to be known and is an essential input to the SDF solution and the decision, finding out what is known and unknown, and determining what is knowable and unknowable, as well as identifying controllable and uncontrollable current and future events that bear on the SDF solution and are leveraged to offset adverse impacts on the predicted future state and affect forecast realization.

Creating the SDF plan that outlines the steps and requirements, roles and responsibilities, knowledge and resource requirements, timetables, and specification of deliverables and follow-up actions is a universal activity. It is followed by ongoing forecaster–client and other stakeholder interaction and relationship building, all of which make SDF team members better business partners. Developing the SDF team's strategic intent for the project is another common element, which is part of the team's standard operating procedures. Subsequent to that is the need to bring a flexible forecaster mindset and the right SDF framework to the project. That is, the mental attitude and approach that determine the responses to business needs; the interpretations of information and events; and the set of processes, practices, and rules used to build SDF solutions. Furthermore, managing client and senior management expectations, project managing the forecasting process through the various stages and gates of the company's decision-making process and stages, monitoring and recording assumption changes, and updating forecasts are all pieces of the common SDF approach.

Every strategic decision forecast requires that the SDF team conducts a forecasting due diligence, which is an in-depth, methodical investigation to sort through the maze of data sources in order to uncover adequate data and information and determine their validity. At times, it is prudent to conduct data analyses, detect and understand outliers, and make graphical representations of data and the intelligence created. In this step, megatrends and subtrends impacting the industry are identified; particular market trends are reassessed; and prior knowledge, analogs, and assumptions are confirmed. In parallel to that step, models, techniques, and experiences that may be applicable to the problem on hand are determined, and the internal strengths, weaknesses, opportunities, and threats (SWOT) and capabilities related to a decision or a project implementation are evaluated.

SDF teams identify and use best practices needed to create their processes and procedures; define the format, medium, and means of delivery of the SDF solution; determine analyses and evaluations to be performed; and select or develop the tools to address specific issues in each project. Next, they build a representation of the business process reality; that is, the business operations modeling, which starts by identifying the problem and then

1. Observing closely how the business works, confirming, and portraying that understanding.
2. Translating the conceptual model into mathematical equations linked by reasonable assumptions and supported by evidence.
3. Developing, testing, and updating scenarios as sanity checks dictate.

In the scenario development phase, SDF team members lay out a plausible sequence of events that lead to a certain outcome in the future. This helps identify the true driving factors and sources of risks, assess the impact of changes in these factors; and identify conditions and support that make the forecast achievable.

Evaluating uncertainty and business risk surrounding the decision is done on a universal basis. Sensitivity and impact thresholds are assessed, company risk tolerance and adaptability are tested, and scenario planning is initiated. Also, evaluating the implications of the SDF solution on the decision objectives and on company operations and resources and determining the appropriate scaling of decision objectives and the required support to affect forecast realization are done in every project. Sanity checks are conducted again to evaluate the validity of the models used, confirm that the SDF solutions make sense, and ensure that the model representations of the future are logically consistent. Once convinced of their validity, part of the common process is communicating the SDF solution in a manner easily understood to senior management and forecast stakeholders, especially the forecast implications and risks when things go wrong and obtaining concurrence, buy-in, and the required support.

Every strategic decision forecast includes the creation of market change and competitor reaction models and helping clients prepare first-, second-, and third-order response plans to put into effect once certain threats become apparent. To make them easy to implement, response plans list all the follow-up actions to sustain forecast occurrence and monitor the evolution of the scenario used in the decision, understand the causes of deviations from what was expected, and suggest course correction actions. Performing project postmortem analysis and managing the creation, storing, and dissemination of intelligence, insights, and knowledge accumulated are other common elements in the new SDF paradigm. Lastly, sharing forecasting experiences and knowledge, coaching and helping clients organize intelligence for a decision, prioritizing alternatives and actions, and managing resources effectively are part of every SDF project plan.

Summary of SDF elements common in strategic projects:

- *Knowing the clients' business, their responsibilities, and their goals and objectives*
- *Determining expectations, the scope of involvement, and stakeholder involvement*
- *Knowing how the SDF solution is going to be used and has senior management support*
- *Listing the information needs and requirements and determining what is knowable*
- *Creating processes, procedures, and computational models of the company's operations*
- *Assessing decision or project uncertainty and risks and getting to residual uncertainty*
- *Developing the SDF strategic intent and SDF plan with deliverables and timetables*
- *Conducting external environment, internal assessments, and forecasting due diligence*
- *Generating scenarios, long-term forecasts, and assessing their implications and risks*
- *Developing a competitor response and forecast realization plan with specific actions*
- *Monitoring the evolution of the scenario selected and updating the forecast*
- *Conducting a project postmortem analysis and sharing the knowledge and experience gained*

3.6 COMMON USES OF SDF SOLUTIONS

There are as many and varied SDF solution uses as are proposals, business ideas, decisions, projects, evaluations, and decisions to be made. Applications of SDF solutions and how they are actually carried out are discussed in Chapters 7 through 18. The following are examples of common motivations for using SDF solutions:

1. Reducing uncertainty to residual levels in a dynamic business environment by conducting environmental studies, industry analyses, and competitor and internal assessments of strengths and weaknesses, and identifying the risks associated with decisions and projects
2. Providing input, forecast support, benchmarks and reality checks, and guidance to the Strategic Planning, R&D, NPD, Product Portfolio Management (PPM), Business Development, and Investor Relations groups

3. Using management dashboards and EWSs and performing evaluations and analyses to separate market signals from opinions in an independent, critical, and objective manner
4. Creating and testing strategic forecasting models for future states through scenario development, planning, and assessment, which identify the support and conditions that need to be satisfied in order for the predicted states to materialize
5. Extracting insights from the analysis of scenarios, simulations, and options created; developing a better understanding of processes and interrelationships; and defining, measuring, and managing uncertainty and risk, and value-creation volatility
6. Using SDF solutions as benchmarks for independent confirmation of senior management expectations and for IVV of earlier business assessments and forecasts

For more immediate purposes, clients and decision makers are looking for SDF solutions to help them understand market relationships, what drives business performance, and quantify the impact of various decisions and actions. SDF team members help them with that through determining the market needs for new products or technologies, projecting technological advances, identifying and evaluating new technology-based products and services, and projecting market adoption rates for new technologies. Decision makers also expect to have alternatives of distinct scenarios created, model simulations performed, future options with evaluations of implications and risks involved completed, and actionable plans to deal with risks provided to them. These are the functions SDF team members deliver in the new paradigm.

Evaluating proposals and estimating economic value added for investment opportunities and developing inputs for financial forecasts that lead to project valuations is a major responsibility of SDF teams as is identifying and quantifying the value of different trade-offs used for more effective negotiations. These are areas of invaluable help to clients and decision makers as is developing the needed support for the rationale and evidence for feasibility studies, business cases, and business plans. Also, because of the systematic way in which SDF solutions are developed, they help reduce time to decisions, reconcile conflicts of interests, and help manage risk. And, the need for monitoring progress on the predicted state of business is met through the development of a management dashboard, an EWS, and competitive response and FSR plans for each project.

3.7 SDF TEAM CHARTER AND STRUCTURE

The SDF business definition and governance structures are discussed in detail in Chapter 4, but this is a brief introduction to the topic. There are only a few leading companies that use the type of new SDF paradigm described here. In the cases where it exists, it resides in the Strategic Planning or the Portfolio Management organizations. In other cases, it reports to the R&D or the NPD group. In a rare case, the SDF team reports to the CFO organization where project financing needs dominate other considerations and fully control the outcome of a strategic project.

The charter of the SDF team is to provide sound and reliable support to strategic decisions. Hence, it defines the scope and functions it performs broadly enough to cover the entire spectrum of activities in support of strategic decisions, projects, and growth opportunities. It also defines its function as being the provider of

1. Information, intelligence, and insights.
2. Analysis, modeling, evaluations, forecasts, and scenario planning.
3. Risk assessment and management and forecast realization planning.
4. Specialized assessments and solutions.

These support elements make SDF teams a trusted source of recommendations and advice for strategic decisions.

The client groups supported by SDF teams often include the Office of the Chairman, strategic decision-making senior managers, business unit heads, large project sponsors, and sometimes project managers of large impact projects. The functional groups they are closely connected with and are given support are the corporate-wide and business unit planning functions and key external experts, namely, the Strategic Planning, R&D, NPD, Portfolio Management, Business Development, Investor Relations, and Finance organizations, and, in a few instances, the external funding sources engaged to finance large company infrastructure projects.

The SDF team's responsibility spans across all areas of its charter and scope of operations, and it is accountable for project managing the entire SDF function from start to finish, the knowledge and insight creation, and the forecast monitoring process. Again, the nature of the SDF team support provided to the various clients and functional areas depends on the nature of the decision to be made, the business problem, the project type, the investment proposal advanced, the growth opportunity presented, and the resource constraints present. The functions, activities, and tasks required to carry out its charter mandates are described in Chapter 6 dealing with methodology, processes, analyses and evaluations, and tools, and in Chapters 8 through 19 dealing with applications of SDF solutions to specific types of projects.

Professional SDF teams are a relatively new development, but reliable long-term forecasts come out of teams that have a structure similar to the ones seen in operation. That organizational structure includes the following:

1. A competitive analysis function with thorough knowledge of the industry, the main competitors, and the own and competitor products and services. This subteam has a good understanding of the internal SWOT, competencies, and capabilities and handles all questions and issues that pertain to competitors.
2. A market research function that engages in the collection and evaluation of data about customer preferences, usually for new products or services, and information about the effectiveness of marketing programs. This function is responsible for profiling trends in consumer or user markets, alerting the organization of changing consumer behavior, working with market research vendors, and evaluating their findings.
3. A demand analysis and tactical forecasting function composed of qualified quantitative modeling and forecasting experts who handle all short- and medium-term forecasts. This function handles supply chain and logistics-related issues and, occasionally, lends statistical modeling support to strategic decision forecasters.
4. An external environment assessment function that deals with ongoing monitoring and assessment of megatrends, subtrends, industry positions and developments, and changes in the current SWOT assessment. A key output of this function is the evaluation of the company's ability to execute a strategic project successfully with existing competencies and capabilities.
5. An SDF function made up of seasoned and broad-thinking managers with a good understanding of the industry, long-term forecasting applications, and decision support experience. This function works closely with and leverages the expertise of the other functions to create SDF solutions and perform project evaluations. In some cases, assessments normally performed by the other functions are performed by SDF team members.

3.8 THE ROLE OF STRATEGIC DECISION FORECASTERS

To avoid the shortcomings of the current state, the first activities of strategic forecasters are to understand the company's business, the decision maker's needs, and the project rationale. Understanding the current corporate strategy, the risk tolerance of the firm, and the perspectives of decision stakeholders before attempting to model the operations reality is another prerequisite. Since SDF teams project manage the SDF solution creation process, they are the keepers of assumptions, internal and external data, competitive intelligence, business insights, and the

Intranet database and knowledge center. Sometimes, they are responsible for identifying but are always called upon to assess megatrends and key business drivers, external versus internal influences on the forecast, and what is needed to create a plausible future state.

The forecaster–client interaction is concentrated on collaborating and jointly developing assumptions and information inputs, evaluating project uncertainty and risks, assessing forecast implications, creating recommendations, and managing forecast realization planning. After modeling business operations, a major responsibility of strategic decision forecasters is to describe future states, events and their timing, and the paths to get there and develop actionable recommendations. This requires that their main task is to understand the client business and personal needs, their expectations, those of senior managers, and other forecast stakeholders. In silo organizational structures, it also needs to manage expectations and organizational issues impacting the forecasting function as soon as they arise. A key role SDF teams play is to be business partners, sounding boards, trusted advisors, and credible resources to clients and other project stakeholders. Part of this role is to be a coach and instructor on issues relating to principles of analytics, modeling and forecasting, the methods and tools used, and how they can be applied to increase the success rates of strategic projects.

In addition to SDF team members observing and assessing the external environment, the industry, megatrends and subtrends, and company operations, they obtain a good understanding of problems surrounding the decision to be made in order to help clients clarify priorities and identify alternative approaches. Only then can they create the analytical framework and models to develop a solution and provide answers to questions raised, using best practices and benchmarks. SDF team members also develop distinct plausible scenarios describing future business operations and states using sound methodology, expert analysis, and evaluations. Namely, they build "flight simulators," which enable clients to create future states by varying drivers of the modeled scenarios and which lead to good business solutions. This process provides an increased degree of comfort to decision makers about the range of outcomes derived from SDF solutions and their ability to manage future uncertainty using controllable levers and having environmental, market changes and competitor response plans ready to deploy.

Again, to be effective in these roles, SDF team members establish close working and personal relationships with counterparts in the internal planning organizations, with external experts and industry analysts. They leverage these relationships to create intelligence and insights and they share knowledge and methods, data, analyses, models, results, forecasts, evaluations, scenarios, and results with clients in a simple and easy-to-understand manner. They also communicate, cooperate, collaborate, and coordinate with all forecast stakeholders in order to resolve conflicting interests and perspectives and arrive at a consensus forecast effectively. In their capacity as trusted advisors, experienced SDF team members provide solutions, which come with considerable amount of confidence, and make clients aware that black swans do occur sometimes; and while there are no ways to predict when they will appear, one must always be prepared for such eventuality.

3.9 THE UNIQUENESS OF SDF

A commonly asked question about SDF is how it is different from the current paradigm and what makes it unique. In this section, we examine the differences between them, which make the new SDF paradigm unique, starting with what functions each approach supports. In the current paradigm, the most common applications are to produce rolling, static forecasts based on historical data, statistical models, and industry norms and analogs. These forecasts are used primarily to support supply chain management, create stretch objectives, and establish performance targets. The motivation for SDF solutions, however, is to develop dynamic, collaborative forecasts based on scenarios using all the environmental and market intelligence and insights; perform in-depth company analyses; evaluate internal SWOT; assess uncertainty and risks; and develop plans and actions to manage decision uncertainty and risk.

In the current paradigm, integrative forecasting means combining different forecasts or pooling time series and cross-sectional data. In the new SDF paradigm, it means employing integrative thinking to create a blueprint to go from goals and objectives, to create alternative scenarios, and to select the best alternative after it passes a number of sanity checks. Yes, it also means combining forecasts from different sources. In the current paradigm, forecasts are single perspective and deterministic; in the new, they are multiperspective and involve decision models where the projected future-state outcome falls within a certain range.

The unique contributions of SDF have to do with overcoming the limitations of statistical forecasts, which are based on numbers, past patterns, and order. The new SDF paradigm is about the art of assessing trends and revealing patterns and identifying relationships, drivers, and levers. It involves testing hypotheses and measuring the impact of events and interrelationships; creating reasonable assumptions, models, and scenarios; and extrapolating them into the future. It also involves planning for forecasting support to make sound decisions using a value chain-based methodology. It does that by employing a methodology parallel to the stage-gate to validate data, models, and assumptions, by managing the creation of forecast solutions, and by performing value audits in each project stage.

Another distinctive element of the new SDF paradigm is that it utilizes a flat organizational structure that links the SDF team with all planning groups and helps to align objectives and interests. Its uniqueness comes from the totality of components used by the new SDF approach, which are the following:

1. Monitoring, evaluating, and interpreting the impacts of megatrends and subtrends, the operating environment, the industry, the company's capabilities and its products and services, and its customer and competitor actions and reactions
2. Evaluating first-, second-, and third-order competitive reactions and preparation of competitive response plans to be implemented as threats begin to arise
3. Identifying and managing the underpinnings and support requirements of forecast realization, which include resource allocation, project value relationship drivers, product attribute and marketing levers, controllable events and external influence programs, uncertainty and risk management programs, and forecast realization plans
4. Using complete and consistent processes based on the strategic forecasting value blocks, which are a forward-thinking mindset, IVVs, critical and objective assessment of the business, and future scenario development and planning
5. Selecting only appropriate data sources, conducting reliable evaluations and analyses, creating intelligence and insights, picking suitable methodology and models, and using appropriate techniques and tools to provide the necessary support in each strategic project
6. Producing forecast solutions that become common key inputs to the strategic decision and surrounding planning functions because of their acceptance due to the confluence of unique experiences in SDF teams
7. Sustaining a culture of innovation and process improvement, borrowing from Six Sigma and using a Janus approach, leveraging an extensive network of internal and external contacts, benchmarking of best practices, and continuous development of team members

Experiences in forecasting due diligence, demand analysis, modeling of operations, statistical forecasting and environmental, industry, market, and competitive analyses are part of the strategic forecasters' job description qualification requirements. Familiarity with the strategic planning, business development, technology and product licensing, and project and PPM functions is a requirement, and skills in creating management dashboards and EWSs are assumed as is experience in risk management. These experiences enable team members to provide a sound, logical structure and the tools to benefit decision makers. These tools are extensive hypothesis testing, sanity checks, and evaluation of forecast risks. Also, the culture of innovation helps not only in identifying the critical success factors in each strategic decision but also in adopting performance benchmarks and best practices, and knowing what goes into creating and using SDF solutions effectively.

Knowing that no system is perfect, the new SDF paradigm recognizes the need for appropriate forecast performance measures, management dashboards, and adequate EWSs. However, institutionalizing and managing learning and knowledge and project managing the strategic forecasting process from start to finish ensure consistency of methodology, assumptions, and scenarios across decision stages and gates. Also by incorporating total quality management (TQM) elements that make maximum use of organizational intelligence and collaboration, SDF teams deliver a more timely and effective development of solutions and higher forecast acceptance by clients and senior management.

3.10 BENEFITS AND LIMITATIONS OF SDF

The new SDF approach fills many of the gaps present in the current long-term forecasting paradigm and benefits users in ways that become apparent in the specific applications of SDF solutions. The main benefits from using the new SDF paradigm solutions include the following:

1. A team structure with close links and ties with planning functions, which produces better communication, collaboration, and coordination, and leads to less conflicts and reaching agreement on forecasts quicker
2. SDF teams serving as a conduit of both internally and externally residing expertise, studies, findings, and best practices
3. Answering the basic question of what is the best way for decision makers and forecast stakeholders to understand and make sense of the world around them more effectively
4. Prioritizing the requirements and inputs needed for a decision, creating a plan that lays out the steps to arrive at an SDF solution, assigning responsibilities to stakeholders at each step of the decision process, and explaining how to use it
5. Determining the levels of uncertainty associated with different future outcomes, identifying which uncertainty matters most, reducing it to residual uncertainty, and developing strategic posture
6. Employing processes that, with appropriate modification, can be adopted to any long-term forecast application need and strategic decision situation
7. Using a set of validated techniques, tools, knowledge, and forecasting function project management approach that produces consistent and reliable results, which leads to a more complete description of future states than other approaches and shortens time to decision
8. Eliminating process problems by using tested practices, which make stakeholders accountable for their parts and by measuring and improving forecast, SDF team, and overall project performance
9. Introducing discipline in the decision-making process through adherence to preestablished processes, evaluating the execution of assigned specific roles and responsibilities, and transitioning between project phases effectively
10. Making project stakeholders knowledgeable about issues surrounding SDF solutions, sharing insightful findings, bringing together value created by its support teams, and synthesizing the results of all analyses and evaluations

SDF yields benefits to clients, senior management, and other stakeholders, but its implementation involves a number of challenges and limitations, of concern to all project participants. Examples of these challenges include the following:

1. Stringent forecaster training, experience, and qualifications in the midst of limited availability of suitable candidates for SDF positions. Even when such candidates are available, there are training periods involved to understand company operations and how its customized SDF solutions are developed.
2. Creating SDF solutions can be expensive because they are human resource intensive, with many of its activities and tasks not lending themselves to being automated.

3. The high levels of 4Cs prevailing in SDF projects are time consuming and not as quick to achieve as clients may wish them to be, even though that is what makes them effective.
4. Each SDF solution necessitates matching the method, analyses, processes, and tools to the specifics of the decision need, the interpretation of the implications of results, and the development of recommendations and advice to clients. This requires a confluence of many skills and experiences acquired over years of participation in the strategic decision-making area, which is not a common occurrence.

To date, software needed to integrate the SDF function with the Strategic Planning, R&D and NPD, Business Development, and PPM functions is not available. Large enterprise management systems such as SAP and Hyperion do not lend themselves to the customization required to affect true integration of these functions with SDF. Again, the more significant constraint is that the filtering of emerging and incomplete information, the analyses and evaluations needed, the creation of intelligence and insights, and the synthesis of all inputs in developing SDF solutions take a high degree of skill and competence. These are not present to required levels in the existing enterprise management systems. While conflicting stakeholder perspectives and corporate politics are lessened under the new paradigm, when SDF solutions are outside the decision makers' expectations set, they resurface and diminish the effectiveness of any type of solution. Furthermore, the state of newness of SDF and the fact that relatively few companies have adopted it so far are responsible for the less than expected acceptance and credibility needed for other companies to embrace SDF teams in the same way they have recognized the value of tactical forecasting teams.

Word to the wise: *The new SDF paradigm solves many of the problems and issues related to forecasts supporting strategic decisions, and it produces more reliable long-term forecasts. However, it cannot be used to address systemic problems and give answers to all the questions project stakeholders may have. Additionally, it can only be effective if it enjoys the support of senior management and demonstrates value added on a continuous basis.*

Part II

A New Paradigm for Strategic Decision Forecasting

4 SDF Organization Business Definition

Strategic forecasting is an essential support function to strategic business decision making, and for that reason, articulating a good business definition for the strategic decision forecasting (SDF) team is vital. To make SDF teams effective, a clearly defined and well-considered charter, scope, and support functions are needed along with well-balanced vision, mission, values and principles, goals, objectives, and strategies, and a sound organizational structure. Ordinarily, a business definition begins with the organization's charter and scope of activities, but we start with a description of the SDF team's vision, mission, and values, and later we discuss its charter and scope. Figure 4.1 shows the elements of the SDF team's business definition, which is useful for articulating its charter and scope of its activities.

Developing an appropriate business definition, which meets company needs requires answering truthfully several questions prior to establishing the SDF team, such as the following:

1. Where are we today? Specifically, what are the client needs, what is the current state of the company's strategic forecasting, what functions does it perform, how does it meet the needs of clients, what needs go unmet, and how can the current state be improved?
2. What do we have to work with? Namely, what is the corporate organizational and cultural context, where does strategic forecasting reside, what is its governance structure, how qualified are the incumbent forecasters, and what resources are available to support this function?
3. How credible is strategic forecasting today and how satisfied are senior managers and clients today? Where does the company want to be in, say two to three years, in terms of an organization and its clients base; its processes, capabilities, and competencies; the types of solutions and support it provides; the skills and qualifications of team members; and their roles and responsibilities?
4. Is the company able and willing to fund a team of strategic forecasting experts? Is it willing to accommodate this team and make changes needed to ensure its success? Do the other planning groups support the development of such capability?
5. How do we get to the desired state of SDF? What are the gaps between the current and the desired state and how do we close those gaps? What strategies, methods, and means should we use to get to where we want to be and meet our objectives?
6. How would we know that we have been successful in arriving at the desired state? What criteria would tell us that we indeed have arrived? How would we measure the performance of the SDF team in meeting client needs? What have we learned in the process and how can we improve it?

4.1 THE MISSION OF THE SDF TEAM

In every company, each organization has a reason for existing, filling a need that is met by its work, and a purpose that it strives to achieve. This is the mission of the organization. SDF teams have recently come into existence because of the failures of the current strategic forecasting paradigm and the need for better long-term forecasting support. Their purpose and intent is to address those shortfalls; provide better solutions, recommendations, and advice; and support decision needs of clients effectively. The key elements required to ensure that the SDF team's mission is fulfilled

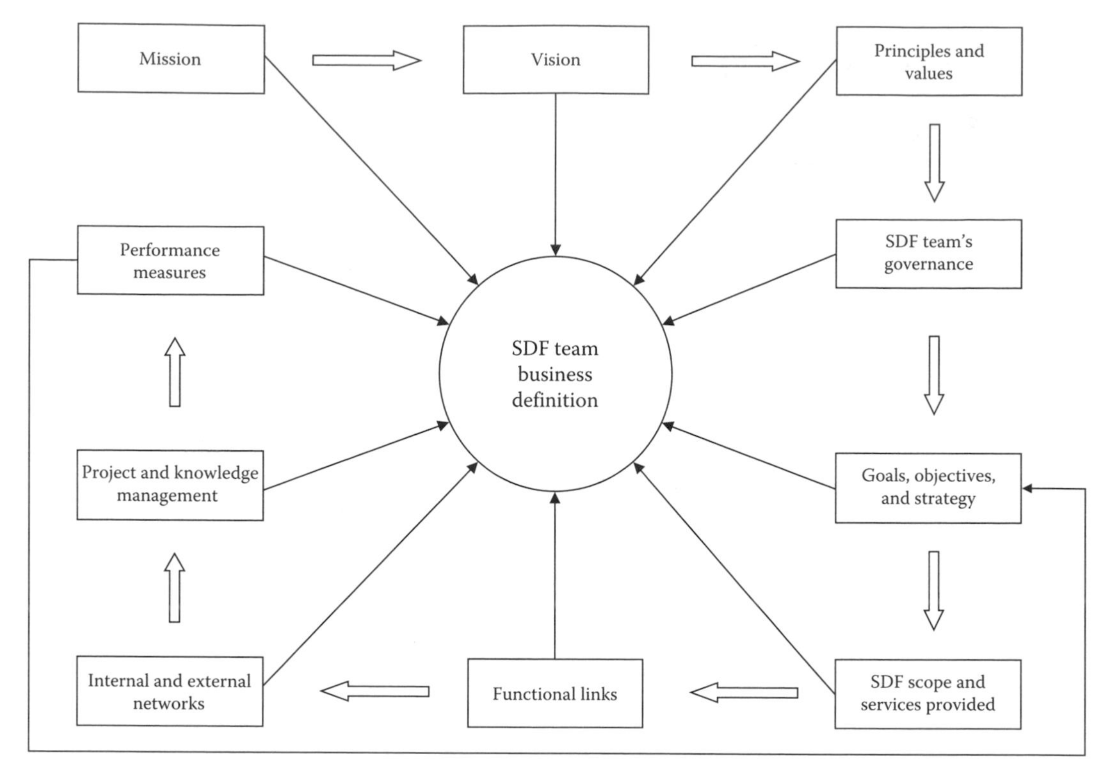

FIGURE 4.1 Components of the SDF team's business definition.

are to provide intelligence, insights, methods, models, plausible scenarios, and evaluations. These elements are needed to develop long-term forecasts, measures of uncertainty and risks, and ways to minimize them so that expected future states materialize. They are elements of SDF solutions to decision problems affecting the future state of the business several years out.

A typical example of an SDF team's mission statement in a large company is as follows: The mission of the SDF team is to enable its clients, decision makers, and senior management to do their jobs better by improving the processes and their effectiveness, shortening the time to decision, and creating intelligence and insights to enhance the likelihood of decision success, to provide reliable assessments and descriptions of future states, and to deliver superior solutions and services to forecast stakeholders. Another example is the following mission statement of the SDF team of a European company: Our mission is to produce superior analyses, evaluations, forecasts, intelligence, and insights for strategic decisions and provide reliable recommendations and advice to our clients and senior management in the most effective manner through building strong relationships and alliances with all forecast stakeholders.

4.2 THE VISION OF THE SDF TEAM

The vision of the SDF team is often an idealized statement that creates a clear picture of the desired state of the group. It is the first step in establishing its ultimate destination and determining ways by which the SDF team creates value and performs its functions successfully. The vision defines the purpose, often in terms of values, and communicates it to team members and clients. For team members, it provides a clear direction and inspiration to strive for excellence, and for senior management and clients, it helps their understanding of what the SDF function is all about and what they can expect from it. The quality of its vision statement determines the innovation, creativity, and value of the team by stretching expectations and setting high aspirations. And it is most effective when it is consistent with the senior management's view of the world.

The SDF team's vision statement also needs to resonate with clients and project team members and help them feel confident and proud to be affiliated with the group. Once an SDF team is in place and begins to make progress in its acceptance, the vision statement is updated to aim for even higher levels of service to clients. The vision statement of an SDF team of a US-based firm reads as follows: The SDF team's vision is to become a trusted advisor and highly valued partner to clients, decision makers, senior management, and teams we support with high professional standards and in accordance with our values.

A less ambitious but practical SDF team vision statement of an Asian firm is expressed in these words: to create a professional SDF team that provides superior services and meets client needs consistently and reliably and presents a viable alternative to external experts currently used to develop forecasts to support strategic business decisions. We do so by continuous learning, innovation, and improvement in all aspects of our business consistent with the corporate value system.

A third example of an SDF team vision statement is one of a Canadian company's where relationships come first: Client needs are the reason we exist, and to fully satisfy them, the SDF team will build strong functional links, business alliances, and personal relationships to provide credible, reliable, and effective strategic forecast solutions for decision making, our goal being to be a team easy to do business with that makes the clients' jobs easier.

4.3 SDF TEAM PRINCIPLES AND VALUES

The SDF team's values are the strong beliefs, qualities, and guiding principles of the organization shared by all team members and espoused by senior management and clients. They influence the group's culture, determine its priorities, and drive the behavior of team members. The importance of these values is apparent when defining clear and acceptable standards, which govern team member behavior in dealing with clients, senior management, and other forecast stakeholders.

Core values are crucial to the success of the SDF function and require that SDF team members share the qualities, traits, and principles considered worthwhile. Clearly articulated SDF team core values are also a very powerful communication tool both within the team and for clients, senior management, and other stakeholders because they give a concise overview of the group's style and way of conducting business. They are viewed as crafting the SDF purpose-driven team identity and are considered worthwhile. The reason so much attention is paid to values is that when communicated effectively, they guide team member behavior and tell clients how to deal with the SDF team. This, in turn, forms a solid foundation for developing forecaster–client and senior management relationships. Values compiled from statements of experts are listed by major category.

A. *Mindset principles*. The core principles related to the SDF team's mindset include the following:

1. Ambition out of all proportion to SDF team resources and capabilities is essential to create value in the client decision process.
2. Broad thinking is required in selecting among distinct options, developing solutions, assessing scenarios, and creating practical recommendations.
3. Clarity of purpose drives the approach and actions of team members and is a key factor in producing effective SDF solutions.
4. Empowerment is a highly valued capacity of SDF team members to make choices as they see appropriate, to transform them to desired actions and solutions, to build the knowledge base of the group, and to improve the efficiency of processes.
5. Flexibility is prescribed in the application of methods, approaches, tools, models, and innovations, and in dealings with other project team members and clients.
6. Innovation is expected, encouraged, supported, promoted, and highly valued by the team and clients, especially when focused on increasing efficiency and productivity of processes and shortening time to decision making.

7. Insistence is on high professional and quality standards in pursuing better understanding, innovation, productivity enhancements, and better client service.
8. Persistence, balanced with relationship and cost management, is a core value apparent in team member behavior manifested in tenacity in pursuing leads, resoluteness in getting to the evidence, and determination in validating information, intelligence, and solutions.
9. Resourcefulness is crucial in the SDF cost–benefit balance approach and refers to the ability of team members to cope with constraints through imagination and ingenuity, inventing new means, ways, and approaches to get things done.
10. Understanding and reason are the qualities of comprehension and assimilation of knowledge, the ability to fully grasp the meaning of information, and the ability to make inferences. It is the capacity to observe business operations and classify, describe, explain, recognize, report, and distill patterns, trends, and insights.

B. *Behavior standards*. The behavior standards that guide SDF team member transactions include the following precepts:

1. Challenging frozen mindsets, the status quo, and long-held beliefs is expected of SDF team members and is part of integrative thinking where conflicting perspectives are reconciled into a better alternative.
2. Commitment to provide exceptional client support consistent with corporate risk tolerance, strategic goals and objectives, and the SDF team's core values
3. Courage to stand up to wrong and unreasonable positions, expectations, and assumptions; face dilemmas and difficult situations with confidence; and pursue SDF solutions with conviction, evidence, and empirical proof
4. Dignity, respect, honoring and being honored and esteemed, and ethical behavior must be evident in SDF team interactions with internal stakeholders and the outside world.
5. Fear of the new or unknown has no place in the SDF team's activities and this is particularly true for ideas originating externally, different approaches to problem solutions, unusual thinking patterns, and out of the known and ordinary processes and practices.
6. Honesty in seeking and speaking the truth in all situations, admitting ignorance, and conducting business without cheating, lying, or any other form of deception is a hallmark SDF core value. Without it, the SDF team has no credibility.
7. Integrity extends across all activities and is reflected in the quality of SDF team members' conduct in terms of consistency of values, principles, actions, methods, performance measures, and walking the talk.
8. Positive thinking, behavior, and reactions constitute a required core value. In every situation, SDF team members are expected to be a voice of reason, affirm positive client perceptions, and strengthen business and personal relationships.
9. Results and not excuses orientation refer to SDF team members obtaining expected outcomes and effects as a consequence of team member actions, relationships, and clarity of focus regardless of the intensity of effort required.
10. Trustworthy and owning up to its conduct and obligations is a mature SDF team's core value, which is differentiated by being able to be relied on as honest, dependable, and reliable partner, and one characterized by responsible, truthful, and consistently exemplary behavior.
11. Versatility is a core value of team members, which requires competence in different business disciplines, being able to deal with ease from one discipline to another. This requires having a wide variety of skills and being capable of doing many things well, being able to move in different directions in creating solutions, and serving many functions for clients and senior management.

C. *Communication premises.* The premises that govern how SDF team members communicate inside and outside their team encompass the following elements:

1. Articulation of plans, goals and objectives, ideas, opinions, findings, and results in clear terms and in a consistent and coordinated manner with the project team and clients is essential.
2. Communication of the continuous, open, direct, accurate, and 360-degree type characterizes all transactions with clients, senior management, and other project stakeholders.
3. Constructive exchange, dialog, and criticism are encouraged among SDF team members, clients, and outside experts, always in the spirit of cooperation to arrive at well-founded decisions.
4. Empathy and being able to appreciate the emotions and feelings of decision stakeholders are essential, especially when managing expectations and situations of conflict of interest.
5. Listening attentively to senior management and client needs, wishes, concerns, and objections, and paying close attention to the voice of customers, is required to determine what must be accomplished through SDF solutions.
6. Resolution of differences, opposing viewpoints, and conflicting interests in an amicable manner is an important part of developing SDF solutions and is essential in enduring relationship building. It is also indicative of the maturity of the team and a must-have capability if the SDF team is to survive organizational politics.
7. Respect of individual is a core value that refers to SDF team members accepting, understanding, valuing, and honoring the contributions of others. The second respect component is respect of the evidence, which means giving it attention as being worthy of and not dismiss it without careful consideration, and the third component is deference to client and senior management prerogatives.
8. Sensitivity is awareness of the needs and emotions of team members, clients, and other stakeholders and the ability to respond to changes in the business environment and personal relationships. Sensitivity and ability to respond are core values in satisfying client needs, handling proprietary information, assessing the internal and external environment, and discerning market signals and changing trends.

D. *Relationship imperatives.* The relationships of SDF team members with clients and senior management are governed by the following imperatives:

1. Accomplishment of agreed-to goals and objectives and meeting client expectations in each and every project
2. Coordination, cooperation, and collaboration with key stakeholders to exchange information and ideas, obtain input and feedback, resolve issues, reach agreement on forecast assumptions, and enhance functional and personal relationships
3. Courtesy, respect, and valuing the perspective of others characterize dealings within the team, but especially exchanges with clients and senior management.
4. Focus of effort on meeting client needs; identifying the crucial, essential, and practical solutions; and balancing project stakeholder with corporate perspectives
5. Openness to input, feedback, and guidance from clients and other stakeholders; to new methods, approaches, and tools; and to emerging evidence, intelligence, and evaluations
6. Personalized touch in the delivery and presentation to clients and senior management of customized SDF solutions that fit the situation goes a long way in strengthening relationships.
7. Responsibility, a core value of the SDF team, is the functional, mental, and moral accountability of team members for the trust placed in them to help guide the company decision through various phases and the functions it performs for clients.
8. Service to clients and senior management to make their jobs easier is the SDF team's priority one by responding in a positive manner to their requests, understanding and satisfying current and anticipating future client needs, and performing all activities in a professional and helpful manner. Quality service also includes increasing, clarifying, and

augmenting the client's knowledge base and understanding; delivering timely, professional, quality solutions to problems or decisions; and being easy to do business with.

9. Supportive team members. It is another distinctive core value of SDF teams characterized by a deep sense of responsibility for the outcome of decisions based on SDF solutions and it is manifested in the following behaviors:
   a. Listening attentively to understand needs, problems, and constraints
   b. Having a positive, affirming, approving, respectful, and balanced attitude toward stakeholder perspectives and contributions
   c. Working together with clients and senior management, being an advocate, and strengthening their decisions with objective, sound, and actionable advice
   d. Trusting the ability of clients to implement SDF recommendations and standing ready to assist and prevent decision errors
10. Teamwork in this age of cooperation where even cats and dogs know the value of teamwork is the ability to leverage SDF resources to its maximum. How? By creating an environment where solutions are a result of collaborative, combined efforts. Each participant contributes different skills, but all are committed to the project mission and, by channeling diverse expertise, achieve a common objective and attain high levels of team performance and satisfaction.
11. Yielding to reasonable client and senior management wishes and decisions, but stating clearly the SDF team position and the implications of their wishes or decisions is a key element of relationship development.

E. *Process rules.* These are essential guidelines that SDF team members follow in order to develop successful solutions and incorporate the following characteristics:

1. Ownership of and accountability for managing the SDF process, intelligence creation, and dissemination of usable insights means that SDF team members project manage the development of long-term forecasts in a trustworthy manner and keep their promises, commitments, and timelines at all times.
2. Coordination of effort in all phases, stages, and gates of the process is necessary both within the project team and with other project stakeholders.
3. Discipline, order, adherence to guidelines and processes, focus on the crucial not just important, and scope control is the preferred approach and type of behavior.
4. Judgment, reason, consideration, and balance are required at every step of developing forecasts for strategic decisions, but are crucial in the drafting of recommendations and advice to senior management.
5. Objectivity, independence, and criticality of assessments are core SDF values, which require striving to eliminate biases and preconceived notions and perform evaluations supported by evidence and verification of inputs to long-term forecasts.
6. Quality of SDF solutions requires excellence in delivering solutions to decision problems, which are free of biases and deficiencies. This is achieved by assessing, anticipating, and fulfilling all the analyses and evaluations required by the SDF process.
7. Sanity checks are tests performed on every step in the SDF process to verify that inputs to the development of solutions are free of errors, conform to specified criteria, pass the test of reasonableness, and produce expected results. Reality checks are also performed by SDF team members to determine if evaluations and solutions conform to the reality of the world around them.
8. Stewardship of strategic forecasting is the personal responsibility of associates related to collaboration in and end-to-end management of developing solutions. Stewardship extends beyond process project management and includes management of learning, discoveries, intelligence and knowledge creation, organizational assets, corporate resources, and intellectual property.

F. *Knowledge management principles*. These are basic principles that dictate how SDF team members handle knowledge creation and management when developing SDF solutions and consist of the following elements:

1. Contribution and sharing of team knowledge, tools, evaluations, insights, and solutions to fill gaps in the client's decision support set are expected of all SDF team members.
2. Fairness in assessing evidence and sources of information; in treating different perspectives, methods, and approaches; and in evaluating existing forecasts, the impact of risks, and proposals requiring strategic forecasts. This enhances the SDF team's credibility.
3. Influencing and helping shape the decision foundation based on evidence, knowledge, insights, benchmarks, best practices, sound reasoning, and objective evaluations are appropriate. Outside of that context, interference with client decisions constitutes inappropriate behavior.
4. Intelligent and intelligence refer to team member traits and the creation of valuable and usable input to SDF solutions and decisions, respectively. Both are valued, nourished, and channeled toward effective decision making, innovation, and productivity enhancements.
5. Learning of the continuous type from primary and secondary market research, competitive analysis, publically available business cases, Marketing and Sales organizations, the groups the SDF team supports, and external contacts and experts augments the project team's capabilities.
6. Observation and reflection are important in understanding the business environment, the industry and market, and the operations of competitors and the company in order to create images of reality and then translate them into computational models. Reality checks, however, are required in all steps of the SDF process, including the observation and reflection stage.
7. Verification and validation is an evaluation to check that the forecasting system and solution meet the requirements and that they fulfill their intended purpose, that is, to determine compliance with the conditions described at the start of forecasting plan's requirements, its processes, and best practices. Verification refers to evaluating if the strategic decision's forecasting plan is being correctly implemented and validation refers to determining if the resulting SDF solution meets the client needs.

G. *Technical and professional tenets*. These are canons that deal with the area of technical and professional values of the SDF team, which include the following:

1. Ability to manage uncertainty in the context of changing industry and market trends, conditions, uncontrollable events, and competitor initiatives is a required SDF team member skill.
2. Competency in key areas that strategic forecasting impacts is a requirement, and development of additional competencies is expected and encouraged.
3. Dependability of client support is essential to ensure accuracy and reliability of data and competitive intelligence; the selection of methods, models, and plausible scenarios; and the completeness of evaluations and sanity checks performed.
4. Efficiency in developing SDF solutions and reducing time to decision are two valuable targets, but the real objectives are effectiveness, reliability, credibility, and client satisfaction.
5. Priority setting to manage the demands on team members is essential in planning the development of SDF solutions and the principle used to do it is to always ask the question: Which are the three most important things at any moment? Another priority setting principle is to anticipate problems, trends, and diversions from plans and prepare alternative priorities, always taking into account business and personal relationships.
6. Reason, reasonableness, and balance are three key elements incorporated in SDF solutions. They refer to scope of engagement, processes, activities, methods, models, tools, hypothesis testing, results, forecasts, or solutions, and in their execution and relationship building and management.

7. Reliability of SDF solutions is the strength and the dependability of the inferences, conclusions, solutions, and recommendations arrived at with minimum failures, under stated conditions and within a required time period.
8. Strategic intent is about clarity, focus, and inspiration that create a compelling target for SDF team commitment and effort. This, in turn, creates long-term, stable objectives and leverages resources and resourcefulness to reach stretch targets rather than fitting objectives to limited resources.
9. Wisdom of SDF solutions means sound judgments derived from common sense, insight, good judgment, and ability to judge what is true and right. That wisdom begins with doubting the solution, comparing it with previous experiences, testing it, introducing emerging evidence and new perspectives and assumptions, creating new hypothesis, and discarding some and keeping others.

H. *Skills and competencies.* These SDF team qualifications and principles are capabilities that lead to creating sound and unique solutions. They consist of the following elements:

1. Creativity in developing tools, models, approaches, and scenarios; compiling intelligence and evidence; presenting new ideas and unexpected results; and overcoming objections are required strategic forecaster qualifications.
2. Diversity of educational, industry, and business function backgrounds, training, skills, experiences, perspectives, opinions, thinking, mindset, and problem-solving approaches are highly valued and promoted.
3. Excellence and highest levels of technical expertise are required as is demonstrating character and integrity in all business transactions and personal relationships with clients and senior management.
4. Independence of thought and action in all aspects of strategic forecasting provides opportunities for personal growth and professional learning, and it is a highly regarded core value and privilege of SDF team members.
5. Insightful is a distinct, valued trait of SDF team members who show the ability to understand the total decision context and an attribute of solutions, which synthesize distilled knowledge with balanced reasoning.
6. High-level performance in carrying out activities and monitoring the performance measures of strategic forecasts are valued functions of SDF teams. Performance measures are intended to evaluate the manner in which activities are carried out and the extent to which projected states materialize.
7. Professionalism is a highly prized value and refers to views, standards, qualities, and conduct of SDF team members. It is marked by adherence to courtesy, honesty, and responsibility in dealing with clients and excellence that goes well above and beyond job description or contractual requirements.
8. Proficiency of associates is the knowledge, skills, expertise, and competence in developing long-term forecasts for strategic decisions. They include analytics and decision support, knowledge and project management, forecasting due diligence, modeling and forecasting, scenario development and planning, summarizing findings, preparing written reports, and effective communications and relationship management.

SDF team core values are the source of differentiation: *The major categories of principles and values governing SDF team behavior are mindset attributes, behavior standards, communication premises, relationship imperatives, process rules, knowledge creation and management, and skills and organizational competencies.*

4.4 PROJECT STAKEHOLDERS AND FUNCTIONAL LINKS

Project stakeholders are individuals, groups, and organizations that have a significant direct stake in a given project's or a decision's forecasting solution because they can impact it but can also be affected by it. They are part of the broad project team, play a role in the process, and have a vested interest in the SDF team's ability to develop sound solutions. Therefore, they need to pay attention to SDF goals and objectives, follow established processes and plans, and be active, collaborative participants. The creation of an SDF solution may involve any and all of the following stakeholders: the client; senior management; representatives of planning organizations; providers of data, information, and analysis; and business units impacted by the strategic decision. From time to time, the Marketing and Sales, Finance, and Corporate Planning organizations have stakes in SDF solutions and serve as advisors, business partners, and providers of valuable intelligence, insights, and sanity checks.

We mentioned earlier that for an SDF team to function effectively, it must have strong functional links, business exchanges, and personal relationships with all the organizations it supports. Stakeholder parties supported, besides the client or decision maker organization, are the senior management team, the Office of the Chairman, Business Unit heads, and Corporate and Business Unit Strategic Planning groups. Other project stakeholder parties may include the R&D and New Product Development (NPD) organizations; corporate and subsidiary Business Development entities; the Product Portfolio Management (PPM) group; and external experts, consultants, or advisors.

The SDF team considers essential the active involvement and cooperation of stakeholders in all phases of getting to a solution and building open communication and feedback loops. This builds confidence in the solution and increases the acceptance of recommendations flowing from it. Another reason for active stakeholder participation is that they get to better understand the SDF goals and objectives, processes, and roles and responsibilities of each participant. This leads to enhanced communication, coordination, cooperation, and collaboration (4Cs) and management of stakeholder expectations. For these reasons, key stakeholders are involved in the creation of the SDF solution from the start of the process.

4.5 SDF GOALS, OBJECTIVES, AND STRATEGIES REVISITED

Goals are broad, general intentions that relate to purpose, vision, and aspirations, while objectives are narrow, more specific attitudinal or behavioral propositions. Goals are to set direction for the organization; objectives are the steps on the path to reaching the goals. They are powerful ways to motivate SDF team members because they focus effort to goal-relevant activities and increase the effort expanded and the likelihood of accomplishing them. To produce the desired results, goals and objectives must be clear, challenging, and achievable, and have SDF team member involvement and the benefit of client and senior management feedback to measure their effectiveness.

The key goals and objectives that flow out of the organization's core values, lead to sound solutions, and are embraced by SDF team members, clients, and senior management include the following:

1. Forming lasting relationships and alliances and being a source of reliable and trusted advice on all strategic decisions
2. Helping clients and senior management navigate project uncertainty and make better decisions
3. Making the decision process less stressful and the jobs of clients and senior managers easier

A secondary set of SDF team objectives includes anticipating and meeting stated and implied client and senior management needs, making deep commitments for big contributions to the success of a decision, and reducing time to decision and decision uncertainty while increasing decision-making effectiveness company wide.

SDF team success is achieved by sharing knowledge and insights, providing professional support of unmatched quality, and elevating the role of SDF team members and making them business partners committed to the success of strategic decisions. In addition to these, three more SDF team objectives are to displace all external experts as the source of strategic forecasts and analyses for strategic decisions, to increase client satisfaction beyond expectations, and to ensure resource availability and support to the SDF team throughout the project.

The means used to achieve the stated SDF team goals and objectives include several of the earlier discussed elements common in the development of SDF solutions, such as the following:

1. Creating an effective SDF team and governance structure founded on the stated vision and mission statements and on core values and principles
2. Establishing close business links with all functional and stakeholder organizations and personal relationships with counterparts in these groups
3. Customizing the approaches, methods, applications, techniques, and tools used in each project to fit the particulars of the situation and current needs
4. Benchmarking and adopting best strategic forecasting processes, practices, and learning from the best in the business and from competitors
5. Building effective processes and procedures to ensure consistency, discipline, and transparency, and to increase the productivity of SDF team members, clients, and other project stakeholders
6. Providing extensive training and learning opportunities for team members in broad business disciplines and in forecasting basics for clients and other stakeholders
7. Performing sanity checks throughout the process to ensure that solutions display logical consistency and pass the common sense test
8. Identifying project uncertainty and forecast risks and their sources and causes, practical options to offset the impact of uncertainty and risks, and options to manage risks as they appear on the horizon
9. Using all available internal and external expertise, insights, analyses, evaluations, and scenario development and planning to develop sound SDF solutions
10. Conducting extensive hypothesis testing, performing sensitivity analyses, and providing implementable recommendations based on reasonable scenarios and sanity checks and backed up with concrete and reliable evidence
11. Encouraging continuous team member relationship building and expecting significant ongoing productivity improvements in all aspects of SDF
12. Living up to SDF team values and principles at all times, in all cases, under any circumstances, and with no exceptions

The unique implementation strategies that set SDF teams apart from the current strategic forecasting groups are contained in the practices that enable it to communicate effectively the SDF mission, vision, values, and principles to stakeholders so they know what they are dealing with and what to expect. By creating extensive networks of internal and external contacts, SDF teams are able to obtain and validate data, information, intelligence, and insights in the most effective manner. They also project manage the entire forecasting process end to end, the creation of and updates to assumption set, and the forecast revisions at each stage and gate of decision making. Furthermore, they ensure that project team members perform the assigned roles and responsibilities based on the SDF plan that was jointly created with clients at the start of the project. That plan is a blueprint telling stakeholders what has to be done, when, by what date, and who is responsible for what.

Part of the SDF implementation strategy is to customize the approach, models, and tools to fit the situation and decision needs. In SDF, one size does not fit all and ingenuity and innovation are required to come up with sound solutions. That strategy also requires conducting a forecasting due diligence to uncover sources of data and intelligence, determining the validity of data and intelligence, identifying internal process and method weaknesses and problem areas, and correcting them or devising options to compensate for them. In addition, sharing experiences, intelligence, knowledge, and insights with key stakeholders are done with no fear that the position of the SDF team is weakened by such a gesture.

Building and managing knowledge is done by reviewing earlier cases; previous associate experiences; industry practices and norms, averages, and analogs; and stakeholder input and feedback. Making knowledge and information available through an Intranet repository of knowledge is another element of its strategy. This approach is useful in applying learning to developing distinct, plausible, and easy-to-monitor scenarios; to test hypotheses, ideas, plans, and interventions; and to measure impacts and related risks. All this amounts to enabling clients to create the company's future state today. Other SDF team activities related to the implementation strategy include the following:

1. Developing early warning systems (EWSs) and competitive response models and assisting clients to prepare plans to implement them in the event that threats materialize
2. Performing postmortem analysis after decisions are implemented, distilling key lessons, and institutionalizing knowledge gained
3. Providing cross-training and job transfer opportunities to broaden team member understanding of the client's business and wider business principles

4.6 SCOPE OF SDF TEAM ACTIVITIES AND SUPPORT PROVIDED

The scope of the SDF team activities is determined by its vision, mission, and value statements and by its goals, objectives, and strategies, and is validated by all the requirements crucial to developing dependable forecasts. The result is the set of functions and activities performed by the SDF team that defines its scope: Starting with taking on the end-to-end forecasting function project management responsibility and transitioning through different stages or phases, the SDF team implements its strategic intent. Its scope is influenced by the alliances formed and by client help needed to understand models developed, the projected states and their implications, and how to use and not to use forecasts. It also includes creating alternative options, determining enabling factors to achieve decision or project objectives, and addressing human resource requirements, timelines, budgets, and other support requirements.

Major corporate strategic projects are characterized by complexity of decisions and, sometimes, conflicting or changing corporate objectives and strategy. The duration of these projects requires long forecast horizons that involve changes in the environment, the industry, the technology, the competitive landscape, and the company itself. As a result, strategic projects are done in phases, which involve multiple handoffs in forecast ownership. SDF solutions are the confluence of diverse bodies of knowledge and experiences brought to bear on creating customized projections of future states to be created by strategic decisions. For these reasons, the SDF team must provide services that make it a trusted source of advice and help to navigate decision uncertainty, which, in turn, determine the scope and services provided by SDF teams.

Services SDF team members provide take many different forms depending on the type of project and the needs of the decision to be made. Following are examples of the forms these services often take:

1. Lending help in problem scoping, definition, and analysis and reconciliation of project participant views and interests based on successful past experiences
2. Sharing the team's conclusions about the state of competitor positioning, possible reactions, and ways to counter their responses
3. Conducting independent, critical, and objective assessments of the market and business segments impacting the strategic decision

4. Providing intelligence, insights, and advice in clear statements about megatrends, potential impacts, the future of technology, customer needs and changing tastes, and other environmental changes
5. Creating strategic intent for the SDF team and executing it through research, modeling, and evaluations performed, which provide sharp focus on project implementation activities
6. Evaluating the impacts of restructuring and reorganization on business performance through evaluation of impacts on demand and competitive positioning
7. Setting the direction for analysis and evaluations, outsourcing the strategic forecasts when necessary, and managing the work of vendors
8. Evaluating market research results and external forecasts, making appropriate adjustments, and assessing the validity of client forecast adjustments and their implications
9. Outlining plausible and reasonable future decision options for senior management to understand how projected states can be realized and helping develop operational performance targets
10. Providing analytical support and recommendations to senior management and to Investor Relations groups concerning the key features of a forecast, the implications and impacts of a decision on the future of the business, and the guidance to share with the external world
11. Creating a future view of the business today by identifying means and levers to affect it, determining what is required to get there, and using them as inputs to long-range plans
12. Helping develop negotiation elements, positions, and trade-offs and evaluating counterpositions and offers based on models of the business and scenario evaluations
13. Challenging current thinking and the status quo of the forecasting function and providing alternative, value-enhancing perspectives to clients
14. Evaluating corporate strategy and policy through assessment of the underlying forecasts and their implications, and determining support requirements needed to make them happen
15. Creating management dashboards and EWSs to monitor project performance, detecting and communicating forecast risks as they appear on the horizon, and helping manage them through forecast realization planning

Establishing what is needed to make a decision, what is known and unknown, and what is knowable and unknowable internally and externally, and identifying controllable and uncontrollable factors, is a key part of the SDF team's scope and first activity. Administering internal, competitor, and industry data, and information collection, validation, and verification and distilling usable intelligence and insights are also its responsibility. However, its scope is incomplete without performing the following functions:

1. Identifying the business and client needs, determining the client requirements for the SDF solution, and agreeing on its form, content, and delivery timelines
2. Observing company operations and creating flow models, which are then translated into computational models and option scenarios
3. Developing assumptions jointly with clients and managing the assumption set through the different decision stages and gates
4. Gathering competitive data, information, intelligence, insights, and benchmarks, and conducting the forecasting due diligence
5. Creating, maintaining, and updating a centralized SDF database containing project data, information, findings, analyses, and conclusions

Creating a collaborative forecasting environment by coordinating, evaluating, and managing stakeholder inputs and feedback; by sharing knowledge; and by coaching clients through the process is in the range of SDF activities. Collaboration is most useful in conducting an in-depth forecasting due diligence and using the conclusions of the findings to address specific decision needs. It is also a prerequisite in conducting a critical company and competitor strengths, weaknesses, opportunities, and threats (SWOT) analysis, identifying the degree of corporate and client risk appetite, deciding on approach and scenarios to entertain, and determining enabling factors to make them achievable.

Another part of the SDF team's span of activities is performing a number of essential analyses required for all strategic projects. Monitoring technological trends and evolution, producing technological forecasts, identifying market drivers, determining causality and feedback loops, and providing a sense of history and respect for the evidence and the intelligence created are important elements as well. Determining industry standards, practices, averages, analogs, and benchmarks relevant to the SDF solution; measuring the strength of market relationships; and assessing variability over geographies and through time are also common SDF team activities. Determining future events impacting the forecast and the timing of their occurrence and identifying black swans are necessary activities. Also, determining the implications for human, financial, and other resource constraints on forecast realization is an additional element of the SDF team's scope.

Selecting the appropriate methodology and models to use in projections; developing distinct, plausible, and reasonable scenarios; and performing model simulations and sensitivity analysis to test hypotheses and identify different options to achieve the projections are important elements of the SDF team's scope. Monitoring conditions surrounding the forecast and performing value creation audits between stages to ensure smooth transition of assumptions and forecast updates between stages are distinctive SDF team activities as are obtaining, assessing, and combining forecasts from different sources, approaches, and models.

Clients rely on the SDF team to conduct a number of reality and sanity checks to establish the validity of the forecast solution, identify forecast risks and their sources, and measure potential impacts on the business. This is useful in ensuring forecast acceptance. Managing revisions of assumptions, stakeholder inputs, and forecasts, and maintaining records are other activities managed by the SDF team. Assumptions are then used in developing and testing scenarios, picking the best suited for the project needs, creating a consensus view, and assessing the resource implications and possible constraints. All these SDF team activities help increase the likelihood of projected states and plans being realized.

Depending on the nature of the project, in addition to the essential analyses and services mentioned above, SDF teams provide other services for specific types of projects discussed in Chapters 8 through 19 dealing with applications, such as the following:

1. Developing testable hypotheses for different ideas and viewpoints, building models to assess management and client hypotheses, testing them, and providing expert opinion concerning their value in defining the future state
2. Conducting elasticity, customer response, and empirical investigations to assess impacts of individual drivers, responses, and impacts on resources needed to achieve the expected results
3. Determining methods, tools, and means needed in special situations identifying the required project support to ensure that projections indeed materialize, and helping clients with risk management and contingency planning
4. Creating decision options based on selected scenarios, communicating the solution and potential impacts along with appropriate caveats, instructions, and cautions on its use, and preparing a set of recommendations
5. Building competitive response models, helping clients evaluate remedial actions, and preparing response plans to put into effect when threats begin to materialize
6. Providing support to ad hoc projects, such as project proposals, major capital expenditures, growth investments, and strategy or policy development and evaluations
7. Conducting special studies to confirm or disprove ideas and beliefs, and scrutinize the effectiveness of various initiatives and programs
8. Addressing project performance measurement needs of clients and senior management and helping to develop appropriate project performance targets
9. Serving as a coach, instructor, and trusted advisor to clients, senior management, and decision makers in each project

Determinants of SDF team scope—Key takeaways:

- *Obtaining senior management approval of team mission, vision, values, goals, and objectives*
- *Providing the needed services that make the team a trusted advisor*
- *Establishing what needs to be known and what is knowable and going after it*
- *Creating a collaborative strategic forecasting environment*
- *Adapting analyses and evaluations to fit the type and needs of each project*
- *Selecting appropriate benchmarks, methods, models, best practices, and tools to use*
- *Creating, validating, updating, communicating, and managing the assumption set*
- *Preparing recommendations on the use and implementation of forecasts*
- *Creating management dashboards, EWSs, and monitoring performance*
- *Conducting project reviews, report findings, and managing knowledge gained*

4.7 SDF TEAM GOVERNANCE

The SDF team governance structure outlines the organization, purpose, goals, and functions of the team. It is the system by which the team is managed and a framework by which stakeholder relationships and interests are balanced. Under a broad governance definition, the SDF team's governance structure is the set of relationships between the team members, their leader, and forecast stakeholders, which define the way objectives are set, the means of reaching those objectives, and how monitoring performance is performed. Another way of looking at the SDF team's governance is as being the composite of functions around client relationships; knowledge and project management; and the policies, rules, and procedures governing relationships among key project stakeholders.

In light of the failures of the current strategic forecasting paradigm, the authority to establish and fund a centralized SDF team is derived from a clear need for sound and reliable long-term forecasts to support strategic decisions corporate wide. The SDF team's charter assigns responsibility for its activities to its supervisory body, which sets the rules of engagement that include the following:

1. Specifying the distribution of rights and responsibilities among participants in the SDF process and laying down the rules and procedures for forecast decision making
2. Specifying the reporting arrangements, practices, policies, and rules affecting the way the SDF team is oriented and managed, and its goals and strategies
3. Providing a structure to set objectives and the means of attaining those objectives and focusing the group's efforts to major impact activities
4. Enhancing the team's productivity on an ongoing basis and increasing the likelihood of accomplishing its goals and objectives in each strategic project
5. Reviewing, validating, and updating the SDF team governance, policies, processes, rules, and relationships periodically
6. Evaluating the effectiveness of SDF team strategies, processes, tools, and performance measures, and producing progress reports to clients and senior management
7. Linking the SDF team to functions, initiatives, and decision processes it impacts and is impacted by and participating in broader decision-making corporate activities
8. Establishing performance measures, monitoring performance, and influencing how team members perceive themselves and how they communicate forecasts across the entire corporation
9. Avoiding duplication of effort by leveraging existing capabilities in other groups, operating the SDF team effectively, and monitoring its acceptance progress
10. Sharpening the skills of team members and focusing innovation on developing a source of competitive advantage for the company

When creating the SDF team's governance structure and desired scope and capabilities, attention is given to structuring, operating, and managing the team to achieve its objectives. This begins with creating a culture of sound business ethics and professional excellence that advances the mission of the team and setting priorities for engagements aligned with team objectives. Other goals include meeting immediate client needs and satisfying company long-term forecast-related objectives; balancing the interests of the SDF team, the clients and stakeholders, and the company; and building and nourishing alliances and relationships with clients and senior management. Also, aligning functions with business strategy; ensuring responsible and best use of human resources and corporate investments; delivering sound, reliable, and acceptable SDF solutions; and identifying forecast risks effectively and responsibly are influenced by the governance structure.

The SDF team governance requires incorporating performance measures in the forecast plan; defining functional boundaries, responsibilities, and tasks; and including forecast, SDF team member, and project operational performance indicators. These elements require establishing specific SDF deliverables; instructing clients on the use of its outputs and solutions; and building checks and balances to ensure that SDF team members pursue goals and objectives consistent with the corporate mission. Additional elements of effective SDF team governance include stating clearly the activities that make the SDF team accountable to organizations it reports to or supports, its clients, and senior management, and being operated effectively. Furthermore, SDF team accountability encompasses providing a clear understanding of project objectives, expectations, and issues and priorities; assigning joint responsibility to forecasters and clients for executing the forecast plans; and managing forecast risks through forecast realization planning.

Governance processes are critical in managing SDF solutions and a sound structure helps the SDF team supervisor to guide the work and make critical decisions. However, corporate culture, structure, and strategy are the real determining factors. Other important factors are the maturity in the design of the SDF team and how stakeholders communicate, cooperate, coordinate activities, and work together in developing and applying SDF solutions. Therefore, effective SDF team governance structures are purposely designed and are revisited as SDF teams mature. They have a simple configuration, which is easily understood and minimizes confusion as to who does what, how, and when. Also, sound SDF team governance structures produce clear, challenging, and achievable objectives; team member involvement in setting goals and objectives; and ongoing client input and feedback.

An SDF team organizational chart is a visual display of its structure that depicts formal lines of reporting, communication, and interrelationships of positions within the company in terms of authority and responsibility. Two types of SDF teams encountered are the synergistic and standalone teams, the charts of which are shown in Figures 4.2 and 4.3 or 4.4, respectively. In most synergistic cases, the SDF team resides in the Strategic Planning or PPM organization, and it is a self-contained team, separate from but closely linked with its support teams. Occasionally, the SDF team resides in the R&D or NPD group under the Business Research and Analytics group alongside the External Assessments, Tactical Forecasting, Competitive Analysis, and Market Research groups. In some rare, standalone cases, the SDF team resides in the Corporate Planning or Finance organization; in one case, it is part of the Office of the Chairman staff, but in the latter case, a good part of its work is outsourced.

In the synergistic case, the SDF team reports directly to the vice president of the organization it resides in and dotted line to the client or a senior manager, who is responsible for a project or a decision, from one of the organizations it supports. There are some instances, however, where the SDF team reports dotted line to more than one senior manager from organizations impacted by the SDF solution. For example, in the case of an NPD project, the SDF team may reside in the PPM group, but reports dotted line to the vice presidents of Strategic Planning, R&D, and NPD organizations.

The synergistic SDF team structure shown in Figure 4.2 is particularly effective for several reasons, such as better alignment of functions, minimal duplication of effort in supporting strategic projects, and increased quality and reliability of service to clients. Under this structure, reduced time

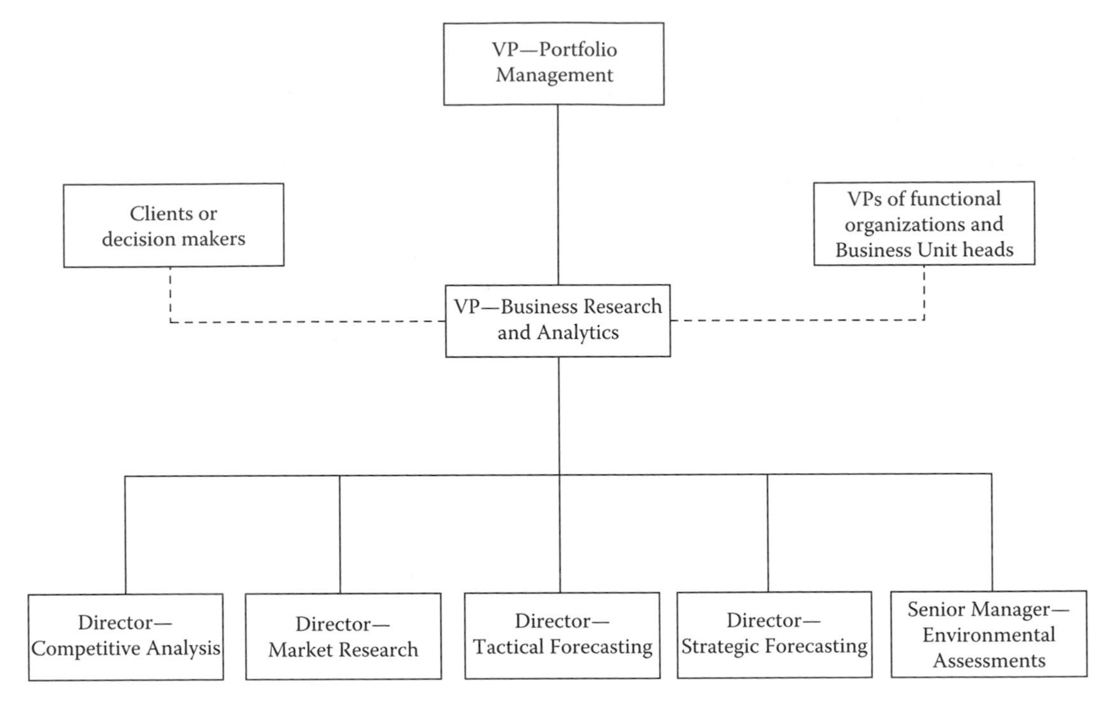

FIGURE 4.2 Sample of a synergistic SDF team organizational structure.

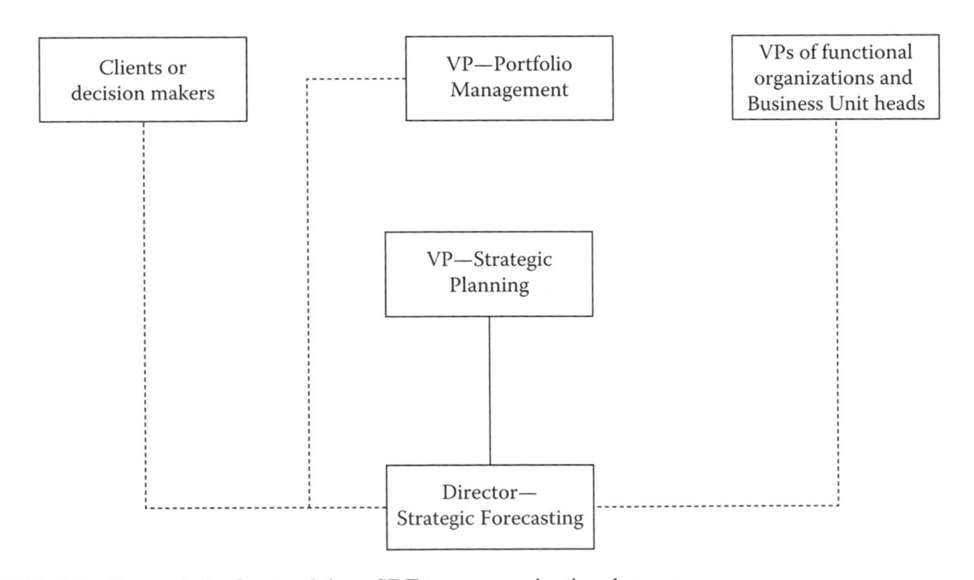

FIGURE 4.3 Example 1 of a standalone SDF team organizational structure.

to decision through effective processes and navigation through uncertainties inherent in different future scenarios and reduced decision risks, effective sanity checks, and creation of response plans are typical results. A side benefit of this structure is lower cost arrangement for developing SDF solutions and recommendations and obtaining continuous productivity gains.

Examples of a standalone SDF teams are shown in Figures 4.3 and 4.4, where the SDF team does not benefit from residing in the same organization as the other disciplines that could lend resources, valuable analyses, and insights into the development of solutions.

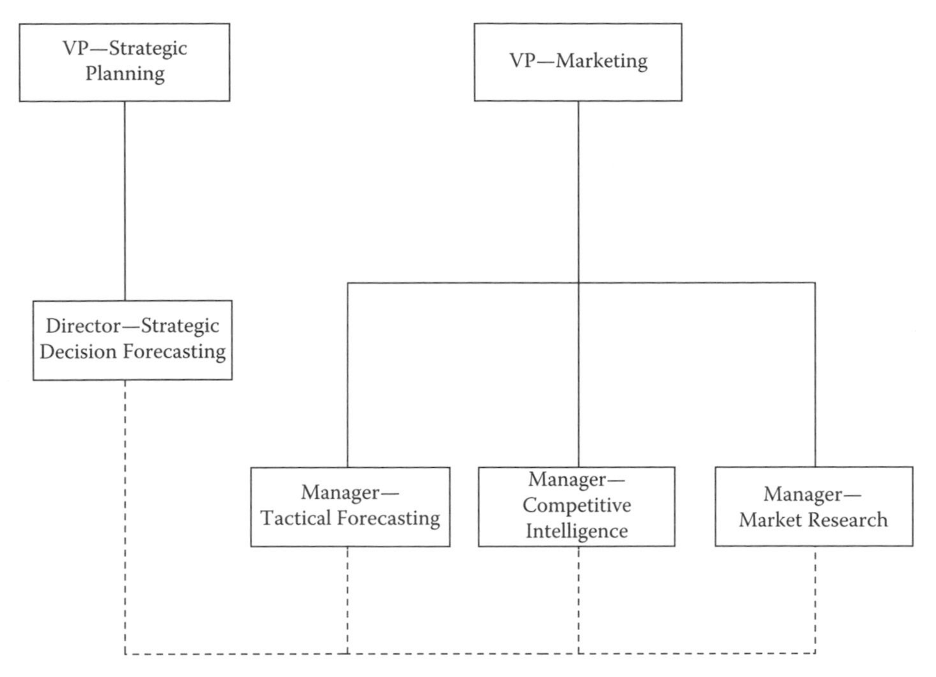

FIGURE 4.4 Example 2 of a standalone SDF team organizational structure.

These types of SDF team structures have three major shortcomings, which are as follows:

1. They make it difficult to leverage any synergies present in the example shown earlier where the SDF team is supported by other business research and analytics functions.
2. The supervisor of the SDF team has ownership of the SDF solution, but only an indirect reporting relationship and involvement in the project.
3. They are associated with lower levels of acceptance, longer consensus forecast negotiations, and inadequate senior management support.

Variant reporting structures of the ones shown here exist and the main differences are the organization to which the SDF team reports to and the help it receives from that organization's support structure.

4.8 INTERNAL AND EXTERNAL SDF NETWORKS

For the SDF team to live up to its core values and principles, the team must be well connected with all groups it supports, internal project stakeholders, and external sources of knowledge. A high level of 4Cs is a core value of the SDF team, which necessitates development of strong business linkages and personal relationships internally and externally. The primary internal SDF team functional links are with the organizations indicated in Figure 4.5.

1. *Strategic Planning.* This functional link is vital because no matter where the SDF team resides, it must understand well the corporate strategy, the underlying rationale, goals and objectives, as well as periodic updates to it. It is also needed for the SDF team assessment of strategy changes or new strategic initiatives. Why? Because all elements of strategy impact SDF solutions, which, in turn, impact the realization of predicted future outcomes. For this reason, Strategic Planning is often the organization the SDF team reports to directly and creates a lot of value added.
2. *R&D and NPD.* Sometimes these two functions reside in different organizations, but more often than not they are part of the same group. When the SDF team reports to this

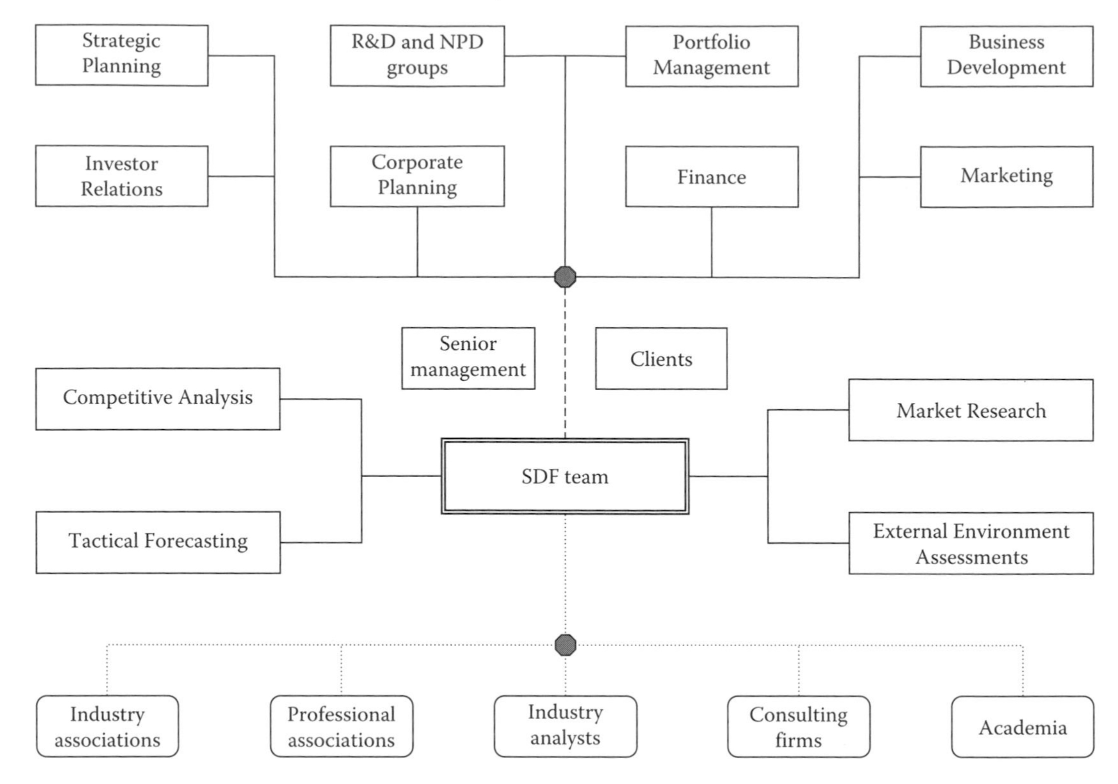

FIGURE 4.5 Internal and external SDF team networks.

organization, it is brought in the project evaluation early on, which is crucial to its ability to obtain the required level of support. It is also important because it helps to develop sound assumptions and reliable SDF solutions through close working and personal relationships, and two-way information exchanges, input, and feedback.

3. *Portfolio Management.* This organization's charter is to manage the company's portfolio of products from the idea stage to product launch to product phase out within the bounds defined by corporate strategy. This is an important function closely linked with and has overlapping responsibilities with Strategic Planning, R&D, and NPD. For these reasons, it is considered one of the best-suited homes for the SDF team because it is an especially synergistic arrangement.
4. *Business Development.* The traditional functions of this organization are merger and acquisition (M&A) projects, new business formations, product licensing, and strategic alliances. Joint ventures (JVs) and technology and product in licensing projects are considered strategic alliances. SDF team evaluations and forecasts, as inputs to decisions for projects of this nature, are crucial. Close relationships and cooperation with this organization make all the difference in the success of these projects because of the support the SDF team provides in project valuations and negotiations.
5. *Investor Relations.* This is a client organization that uses SDF forecasts and recommendations to relate to the outside world, and the investment community in particular, and to influence expectations about the future state of the business. Support needed by the Investor Relations group includes understanding how SDF solutions were developed, key assumptions used, and forecast risks, caveats, and qualifications. This organization, in turn, is a good source of industry intelligence and competitive insights.
6. *Competitive Analysis.* Whether Competitive Analysis resides in the Business Research and Analytics group, there is close collaboration between the SDF team and this group,

which produces useful intelligence, sanity checks, and insights resulting in more reliable SDF solutions. This group is best qualified to answer internal SWOT and competency and capability questions. Scenario development and the evaluations of possible competitive impacts by SDF models often raise new questions and provide direction for investigation by the Competitive Analysis group.

7. *Market Research.* As with the case of Competitive Analysis, the SDF team is a beneficiary of the knowledge, understanding, and the feel this group has for the market place, trends, and customer preferences. Market Research and the SDF team benefit from the inputs, feedback, idea exchanges, and evaluations of each other's findings and provide reality and sanity checks for work produced by the other group.
8. *External Environment Assessment.* This group has developed knowledge and expertise in identifying megatrends and subtrends, industry structure changes, and developments in the political, economic, social, technological, legal, educational, and demographic (PESTLED) environment. However, most of its value comes from the interpretation of external environment factors and their impacts on the company's business.
9. *Tactical Forecasting.* The technical skills and expertise of quantitative and qualitative demand analysis and forecasting are the domains of this group. Some of the methods, models, and tools used by Tactical Forecasting are used in generating and synthesizing SDF solutions. This group is often the source of talent to develop into SDF team members, and close cooperation with it is important for the success of the SDF team.

Interactions with senior management and clients are not functional links, but they are a crucial source of guidance and insights. They are needed to understand their vision, business and personal needs, corporate strategic intent, risk tolerance, and their decision styles and preferences. These are elements for building personal relationships and better SDF solutions, which form the basis of guidance and advice to key project participants. Secondary, yet important, internal SDF team links are with the Marketing and Sales, Finance, Human Resources, and Corporate Planning organizations. These groups are beneficiaries of many of the outputs, solutions, analyses, intelligence, insights, and services SDF provides to the primary functional links. They, in turn, offer their own insights, judgment, and evaluations to the SDF team. In fact, good relationships with these organizations are beneficial all around because they can be the best source of benchmarks, sanity checks, data and information about company policies, and initiatives that could impact options developed and the project's success.

The question of the need for external relationships and information exchanges comes up because of fear of improper disclosure of proprietary information. This is a real concern, but it is adequately addressed by monitoring adherence to SDF team core values and principles and training of highly skilled and intelligent team members in nondisclosure issues. In addition to being valuable sources of information, intelligence, and insights, external links and relationships present opportunities to recruit SDF team members from. The external links of considerable value and benefit to the SDF team are the following:

1. *Industry associations.* These associations represent the industry to the rest of the world and are rich sources of other contacts, the history of the industry, data, information, news, analyses, evaluations, and special studies. They also provide opportunities to network and create external relationships and alliances, obtain industry training, and attend seminars. Additional value to the SDF team from industry association relationships comes from getting independent industry-sponsored market research, validated industry averages, benchmarks, and best practices and processes.
2. *Professional associations.* These are groups formed to bring together people with similar backgrounds and interests and advance their status. Close relationship with these groups is advantageous to the SDF team because they provide access to industry luminaries,

accreditation, events and conferences, and networking opportunities with other SDF practitioners. Professional forecasting associations, such as the Institute of Business Forecasting and the International Institute of Forecasters, are also a potential source of recruiting experienced and trainable future SDF team members.

3. *Industry analysts*. These are individuals who follow the conditions that impact an industry and specific company performance and write about current news, events, and trends. They are ordinarily employed by financial institutions or consulting firms, and in addition to concentrating in one industry, they specialize in a subset of it and develop in-depth knowledge of a few companies. As such, they can be a tremendous source of competitive insights, assessments of megatrends and subtrends, investor expectations, and knowledge of future initiatives and events planned by other companies in the industry.
4. *Investment banks*. These are financial institutions that trade in securities, raise capital, manage corporate M&A projects, and provide independent investment assessments and opinions to buyers and sellers. Investment bankers are well connected with senior management teams of companies in the industry they specialize and have a good understanding of those companies' strengths, weaknesses, and potential. Because of that, SDF team members could obtain valuable statistics, insights, and assessments from investment bankers, which can help them to separate market signals from the distractions of massive and rapid data updates.
5. *Consulting firms*. They are companies that specialize in one discipline or one industry or one discipline within an industry. In addition to being potential partners in developing solutions, experienced SDF consultants can be a source of knowledge and learning from their previous experiences. Their feedback, critique, and independent assessments can sometimes make all the difference in creating effective and reliable SDF solutions. Good relationships with consulting firms are also needed when strategic forecasts are outsourced.

Benefits of SDF team's networks and relationships: *The benefits flowing out of SDF team functional linkages and personal relationships are as follows:*

- *Better understanding of the environment, the industry, the company operations, the corporate strategy, and the client needs and expectations*
- *Easier access to the expertise, capabilities, and resources of the supporting or partnering organizations and contacts*
- *Enhanced stakeholder understanding of the SDF team goals and objectives and familiarity with processes and the basics of forecasting*
- *Expanded and more reliable sources of data, information, intelligence, insights, feedback, evaluation, and sanity checks*
- *Helping in developing a sound, common assumption set and updating the underlying SDF solutions in different stages and gates of a project*
- *Increased participation in client decisions and in broader business issues analysis and evaluations, more balanced SDF solutions, and reduced time to decisions*
- *More effective management of stakeholder expectations, different interests, and conflicting viewpoints and developing decision options*
- *Enhanced client and senior management decision-making productivity and more fulfilling jobs for SDF team members*
- *Quicker agreement on SDF solutions, their implications and risks, and increased acceptance of SDF team recommendations*

4.9 SDF PROJECT AND KNOWLEDGE MANAGEMENT

Project management is a structured approach to planning and guiding work processes end to end and is used to control the complex task of developing SDF solutions. It brings discipline to the SDF processes in planning, organizing, prioritizing, assigning roles and responsibilities, allocating resources, budgeting, setting timelines, and monitoring progress to bring about completion of project and SDF team goals and objectives successfully. In the context of SDF, knowledge management is a holistic approach to manage the company's knowledge assets. These assets include practices, processes, and procedures; data, information, methods, models, techniques, and tools; and skills, experiences, and learning residing in the SDF team for the benefit of the entire corporation.

The primary SDF team knowledge management activities center on collecting, creating, applying, and sharing knowledge and expertise to affect ongoing SDF innovation, productivity improvements, forecast performance, and sound decisions. What is also needed is an Intranet database to house the knowledge accumulated and which is updated periodically. Knowledge contained in the SDF database is accessible to all who have a need to know.

Project management and knowledge management are different but parallel activities the SDF team engages in, with several common elements such as creating an outline of the project plan and populating it as progress occurs; setting the goals, objectives, and focus of the project; articulating project policies and delegations of authority; and making clear the success criteria to be judged against. Identifying special competencies and expertise in the team assigned to the project, assigning roles and responsibilities, and ensuring that participants understand what is required of them are other common elements as is ensuring that a project planning and monitoring system are in place.

Identifying specific activities and milestones, determining critical paths and estimating time to perform the specified activities, and managing to it are key components of project management as is ensuring that all human resources and financial wherewithal are available when needed. Spelling out the form of the deliverables, ensuring that quality standards are met, and developing alternative contingencies to the initial plan are part of the project management plan. Also, reporting the progress achieved, missed deadlines, issues, problems, and cost overruns are common project and knowledge management activities as is the practice of conducting a postmortem review and analysis and sharing with all stakeholders.

4.10 SDF PERFORMANCE MEASURES

In tactical and lifecycle management as well as in the current state of long-term forecasting, the focus is on forecast accuracy and the performance measure is some measure of forecast error. Sometimes, forecast turnaround time is included in judging how well forecasters did their jobs. In the new SDF paradigm, however, performance is multidimensional and includes forecast, SDF team, and project performance criteria as well as client and senior management evaluation of the following elements:

1. Adherence to SDF team core values and principles and commitment to meeting all promises and current needs and anticipating future client needs
2. Continuous SDF team innovations and contributions to enhance decision options and practicality of implementing SDF solutions, recommendations, and options
3. Understandable and effective forecasting processes, methods, and tools, and minimum time to decision from start to delivery of solution and recommendation
4. Collaboration in identifying forecast-enabling factors and creating response plans to be put into effect when risks materialize and project management of all forecast components and processes end to end through all stages and gates
5. Effective communication of forecast assumptions, models, intelligence, methods, scenarios, sanity checks, risk assessments, decision options, and ownership and accountability of SDF team members in every step of the process

6. Implementation of the SDF forecast plan on time and budget with all expected deliverables, alternative scenarios and future outcomes developed, sanity checks performed, insights provided, and decision options created
7. Reliability and dependability of SDF solutions and the decision options created and institutionalizing learning and knowledge management
8. Management of forecast stakeholder expectations and resolution of conflicting interests and professionalism, courtesy, and equal respect for the client and the evidence
9. Cost of developing SDF solutions and options versus perceived value of management dashboards and EWSs and SDF tools used
10. Quality of project and forecast performance monitoring and the variance analysis performed
11. Extent and quality of SDF team support provided in creating competitive response and forecast realization plans
12. Overall customer satisfaction and improvement over time and rating of the SDF team on balanced scorecard elements and trends
13. Effectiveness of the project dashboard, the EWS, and the future-state realization plan
14. Monitoring economic value added by the project, such as net present value (NPV), versus its projections based on the SDF solution and evaluating the extent to which project objectives are met

Ordinarily, the performance evaluation of the SDF team is done by rating the performance of the team member dedicated to a project on each of the elements on a scale of 1–10, assigning weight by the client or the senior manager to the elements, and calculating a weighted average. The last step in assessing performance is to assign weighted score ranges for below average, average, good, and exceptional performance categories. Notice that the weights assigned by clients or senior managers are subjective measures and so is the specification of performance categories. However, this is necessary because it reflects company culture, policies, and rules, and allows forecast users to express the strength of their emotions and feelings about the SDF experience and the solution itself. A lot more about performance measures in SDF is presented in Chapters 20 and 21.

Key point to remember: *Critics question the value of forecasts beyond two to three years out because they are always wrong; therefore, don't bother creating elaborate forecasts because they are useless. It is important, however, to remember that strategic forecasts are long-term forecasts developed by incorporating all relevant information and intelligence available to help make a business decision. As such, they are the best foundations to base strategic decisions on. And, as is the case with a global positioning system (GPS), when road reality is different than what the system shows, the sensible thing to do is to course correct. In the case of the GPS by taking a detour and in the case of a long-term forecast by incorporating updated information in the original model and scenario used. Without those course corrections, no one gets to the final destination, and blaming the accuracy of the GPS or the long-term forecast is unproductive.*

5 Culture, Mindset, Processes, and Principles

Mindset is a set of mental attitudes that predetermine a person's interpretation of situations and reaction to them, and business management experts believe that the largest cost and quality driver across all industries is the mindset of senior management. Because that mindset permeates corporate culture, the resulting corporate policies and strategies, goals and objectives, organizational structures, collaboration among internal entities, and performance incentives are reflective of that mindset. Hence, there is a need to understand the senior management mindset, especially on issues affecting the strategic decision forecasting (SDF) team's standing, and to shape the project team structure and the attitude of project participants.

The purpose of SDF is to translate client objectives into SDF management objectives and, subsequently, into sound and actionable forecast solutions, options, and recommendations. As these objectives are internalized in the organization, the focus is on creating an SDF team mindset that thinks of competitive advantage. To accomplish this, it requires emphasis on cultivating the right environment, approaches, processes, and principles. One of the tools recommended for doing this is the quality function deployment (QFD) method, which translates user requirements into design quality of processes. This is one of the differentiators of the new from the current strategic forecasting paradigm; and for that reason, the discussion of this chapter centers around the subjects of culture, mindset, processes, principles, and guidelines.

Definition of QDF: *It is a method that helps transform needs into process characteristics, prioritizing them, and developing appropriate targets. It focuses on examining key elements from different perspectives to transform customer needs and user demands into design quality, perform the functions that lead to quality, and deploy appropriate methods and specific elements to develop a process* (Chan and Wu 2002).

5.1 SDF TEAM CULTURE

SDF team culture is the personality of the team, which is defined by the way things get done and which is the product of the interaction of underlying values, beliefs, and codes of practice that make a group what it is. The SDF team's culture is multidimensional, is shaped to a large degree by the corporate culture, and reflects the understandings that the SDF team attributes to situations and the solutions that it develops to address decision problems.

SDF teams have elements of task cultures, where teams are formed to solve particular forecasting problems in a matrix organization format, which facilitates horizontal flow of skills and knowledge. This culture is developed and shaped to suit the strategic goals and objectives of the company. SDF teams also have some strong cultural elements, meaning that team members respond to client needs because of alignment with organizational values. We observe that SDF team cultures are a blend of a club culture with elements of collegial culture in that they preserve a strong sense of their mission and that teamwork is the basis on which the jobs of strategic forecasters are designed.

The SDF team's culture is important because it shapes the mindset of team members, which, in turn, influences the approaches used to develop solutions and determine the processes used. We will discuss SDF teams further in Chapter 20, but there are five key elements involved in assessing culture:

1. What the organization is about and what it does, its mission, and goals and objectives
2. Its core values, beliefs, principles, and approaches that characterize its dealings
3. The client–forecaster power structure and relationships, who makes what decisions, how power is shared, and what is power based on
4. The governance structure, reporting lines, hierarchies, and how projects flow through the organization
5. The mechanisms in place to monitor activities and progress in groups where there is reliance on innovation, teamwork, and professionalism

5.2 SDF MINDSET

A mindset is a philosophy of life, a way of thinking, the attitude, opinion, and mentality of a person or a group; it is a habitual or characteristic mental state that determines how SDF team members perceive, interpret, and respond to situations. It is an established set of notions, methods, and assumptions that provides the basis for accepting and adopting certain tools, behaviors, and choices. Mindset is congruence between beliefs, feelings, thoughts, and performance; and as such, it provides a frame of reference that, for all practical purposes, is fixed. The organizational mindset is so strong in a specific outlook that it cannot perceive other views, ideas, and perspectives, even though they are considered.

Mindset can also be thought of as the set of beliefs, preconditioning, and faith in prior experiences that prevent one from looking at new options in a realistic sense. The importance of a positive strategic forecaster mindset is well recognized, but few really understand how it works, what defines it, and how to put it into practice. In this section, we attempt to do that and we begin by introducing a chart on the determinants of SDF processes and solutions in Figure 5.1.

Corporate culture is the driving force that defines the SDF team's mindset and culture and influences corporate risk tolerance, strategy, and goals and objectives. The key determinant of the SDF team's mindset is corporate culture. There are, however, other factors that determine its mindset, such as exceptional senior management capabilities and exemplary behavior and expectations. The SDF team's positioning and internal and external links, the team members' training and prior experiences, and the team's strategic intent also play a role. On some occasions, the SDF team's mindset may have some positive feedback effects, but in all cases it influences a wide range of SDF team success factors. For example, the group's business definition and its core values, its project goals, objectives, and strategies and the SDF team's approaches and methods to problem solving are impacted by that mindset. Once the corporate culture and SDF team mindset are established, they determine the processes, procedures, and activities pursued in a project, the techniques and tools it develops and uses, the forecast project management function, and the project implementation plans.

The effect of established processes, tools and implements, and forecast project management and implementation plans cascades down to tailoring the forecasting methodology to fit the particulars of a project, namely the analyses performed and services rendered by SDF team members, the quality of forecasting solutions and decision options, and the SDF team's contributions and value created in that project. The elements reflected in the right SDF team mindset which are valued most are the following qualities:

1. SDF team members have a strategic management orientation and core values driving behavior and goals and objectives, which, in turn, drive processes and solutions.
2. Team members are flexible and supportive in dealing with clients and are easy to do business with.
3. They are broad-minded and open to new ideas, approaches, and techniques that can be applied to come up with better SDF solutions.

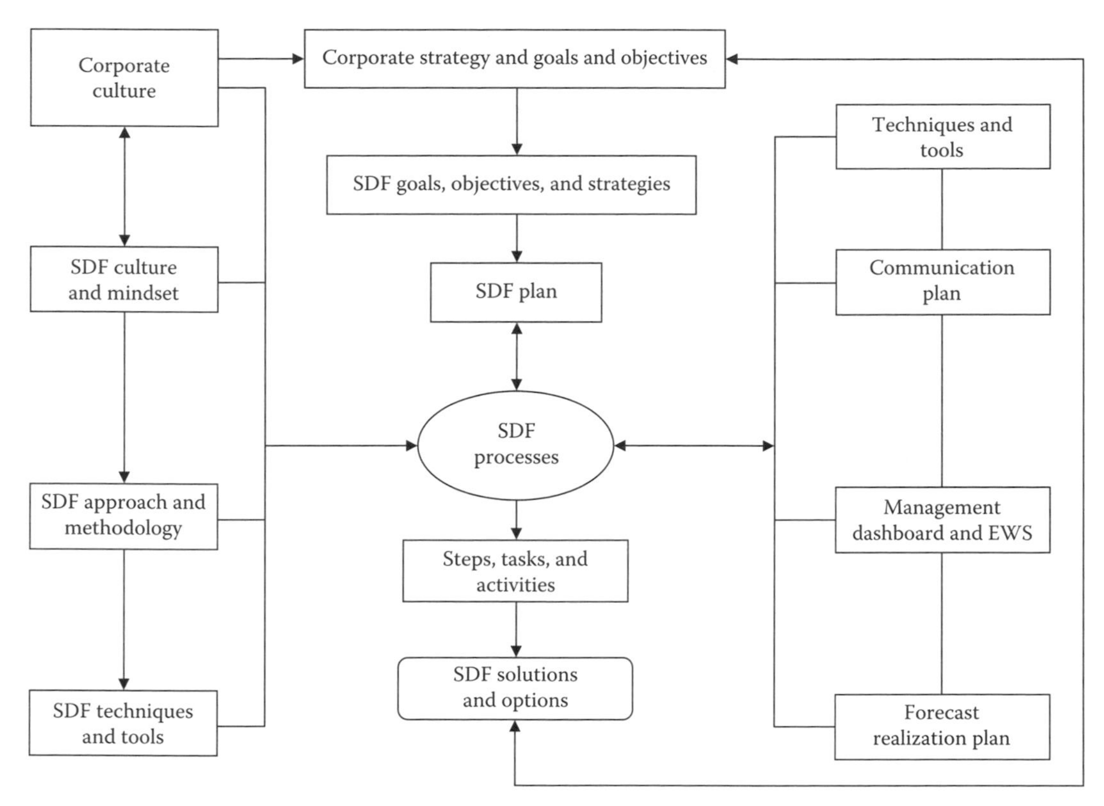

FIGURE 5.1 Determinants of SDF processes, solutions, and options.

4. They think of and have a competitive advantage attitude when presenting their recommendations to clients and senior management.
5. High levels of communication, coordination, cooperation, and collaboration (4Cs) are present in their problem-solving approach.
6. Design for Six Sigma, continuous improvement, and total quality management (TQM) principles and practices are embraced and applied to enhance SDF processes.
7. Forecast ownership, accountability, responsibility, and delivery on agreements are paramount and honored throughout the process.
8. SDF team members create systematic, disciplined, and auditable approaches and processes and no black box solutions are ever created.
9. They come up with reality-grounded evaluations based on sanity checks they perform at each turn of the process; and recognizing that no one has a monopoly on truth, they validate and verify assessments and triangulate long-term forecasts.

Definition of forecast triangulation: *It is a method of using different approaches and routes to arrive at a forecast by comparing and combining the results to get a better informed forecast. Elements of this method include combining the following:*

- *The top-down, build-up, and benchmark forecasts*
- *Forecasts driven by different factors such as own product customer demand, competitor products, company operational capacity, and strategy and project objectives*
- *Forecasts based on different patterns and market share estimates over time*

10. SDF processes are characterized by transparency; consistency; flexibility; and balance of activities, models, evaluations, and conclusions.
11. Scalability of successful prior experiences and customization of approach, process, and solution to fit the situation ensure the creation of sound support.
12. Having adopted a "what-if, then, and so what" method of evaluation, they are able to distill most useful insights.
13. Team members are well informed, reliable, and able to generate trustworthy solutions and present them in terms clients and project stakeholders can easily understand.
14. In every solution, they create, seek forecast certainty, and then distrust it because they know that unexpected and uncontrolled events can always come up in long forecast periods.
15. They are most concerned with management of uncertainty and risk and making certain that a minimum number of surprises occur in strategic decision projects.
16. Well-understood and agreed-upon performance measures are employed with corresponding incentives in order to motivate achievement of desired results.
17. SDF team members consider knowledge and learning a corporate-wide property and share their knowledge freely with project participants. Unlike tactical forecasters, SDF team members' first loyalty is to the corporate needs, then to their career path, and last to the forecasting profession.
18. In each and every project, SDF team participation reduces time to decision and makes the decision makers' jobs easier.

The value of the right SDF team mindset is evident in the improvements obtained by eliminating the factors that underlie the mindset prevailing in the current paradigm. These factors include the following wrong attitudes and practices:

1. Thinking and focusing on historical data, analysis, techniques, and analogs; everything else is not as important.
2. Be concerned about the privacy of forecasts and limit cooperation and communication with other groups to required transactions only.
3. Staying with tried-and-true approaches and new ideas are not tested because they are costly.
4. TQM issues are of secondary importance; focus on the present and let the next generation of forecasters learn from your experiences.
5. Responsibility in long-term forecasts shifts across organizations according to decision phases, and project participants move on to new jobs. So, why worry?
6. In the presence of time constraints, use any ad hoc, varying approaches and processes. Document things later.
7. Sanity checks of long-term forecasts are to be performed against past history and industry averages and practices.
8. Due to complexity, the models of forecasting operations have to be kept hidden and fixed because they are too difficult to explain to nonforecasters.
9. One-size-fits-all and customizing processes and models are not necessary; tailoring the approach means unnecessary delays in meeting client needs.
10. "What-if" analyses are best- and worst-case scenarios flowing out of deterministic models. Validation of underlying assumptions and understanding of corresponding risks is not our responsibility.
11. Forecasters are well informed on forecast issues, but they cannot know all the ins and outs of business operations and decision needs and processes.
12. The responsibility of forecasters ends with the delivery of the forecast. When actuals come in, we know how well the forecast is performed.
13. Knowledge and learning reside in the forecasting group and they stay there. Forecaster adherence to forecasting techniques first, then come client needs.

14. Forecast-related assessments are limited to areas of senior management and client directives; other requests have low priority.
15. Uncertainty and risk management are not part of the forecasting discipline and the forecaster's job description.
16. Effort is directed to making the forecasters jobs easier, not necessarily that of the client or the decision maker.
17. There is narrow definition of forecasting performance: some measure of forecast deviation from actuals such as root-mean-square error.
18. The forecasting staff's professional conduct is judged by forecast managers, not the clients or other project stakeholders.
19. Forecast knowledge and experience are key drivers of behavior and forecast approach.
20. Forecasters believe that their discipline has a unique perspective on the future state of business that others do not have.

There are several indicators used to determine the right SDF mindset. First, SDF team members invest time to understand the company business, the industry, the operating environment, megatrends and subtrends, market conditions, and what drives company performance. Second, accurate assessment of the decision makers' problems, issues, needs, expectations, biases, and conflicting objectives takes place in each project as well as appropriate escalation of detracting factors when necessary. Third, a practical SDF team structure is established, which facilitates open and effective all-around 4Cs, and sharing of information and knowledge. Fourth, SDF team members own the forecasting process and project manage it effectively end to end across all phases and decision gates.

The fifth indicator of a right SDF team mindset is that SDF solutions and options are influenced by cost–benefit considerations, and team member performance is tied not only to accuracy of projected states but also to client satisfaction results and ongoing SDF team productivity improvements. The sixth indicator is delivery of well-grounded, tested, and sound solutions and options on time and on budget along with assessment of uncertainty and forecast risk and risk management recommendations. Progress reports with actionable advice on dealing with deviations from expected business states and offsetting the impacts of adverse developments is another indicator that the right mindset is guiding the SDF team.

Another sign of a supportive mindset is expanding SDF team responsibility in decision support to include creating options, competitive responses, and forecast realization tactics and actions. When SDF team members become full business partners and trusted advisors and earn a place on the decision table next to clients, it is affirmation that they have indeed demonstrated the correct mindset. Last, a practical indicator commonly used is whether SDF solutions result in reduced time to forecast consensus development, time to decision, and project uncertainty and risk impacts.

5.3 SDF APPROACH

In the context of SDF, approach is the broad set of ideas, methods, and actions used to develop solutions, options, and recommendations for a strategic decision. It is a broad framework that defines the activities and tasks to be performed in order to achieve the objectives of a given project; that is, it is what defines the outline of the SDF process. The mindset of the SDF team influences the approach to developing solutions, which, in turn, determines informational needs and also influences the selection of methodology and specific processes used. These elements play a major role in determining the SDF team goals and objectives for a specific decision or project support and the customization of SDF processes. Once the SDF processes are outlined, procedures and activities are laid out and particular models, tools, implements, and artifacts are created. Then, detail tasks and steps to be taken in developing the SDF solution are spelled out.

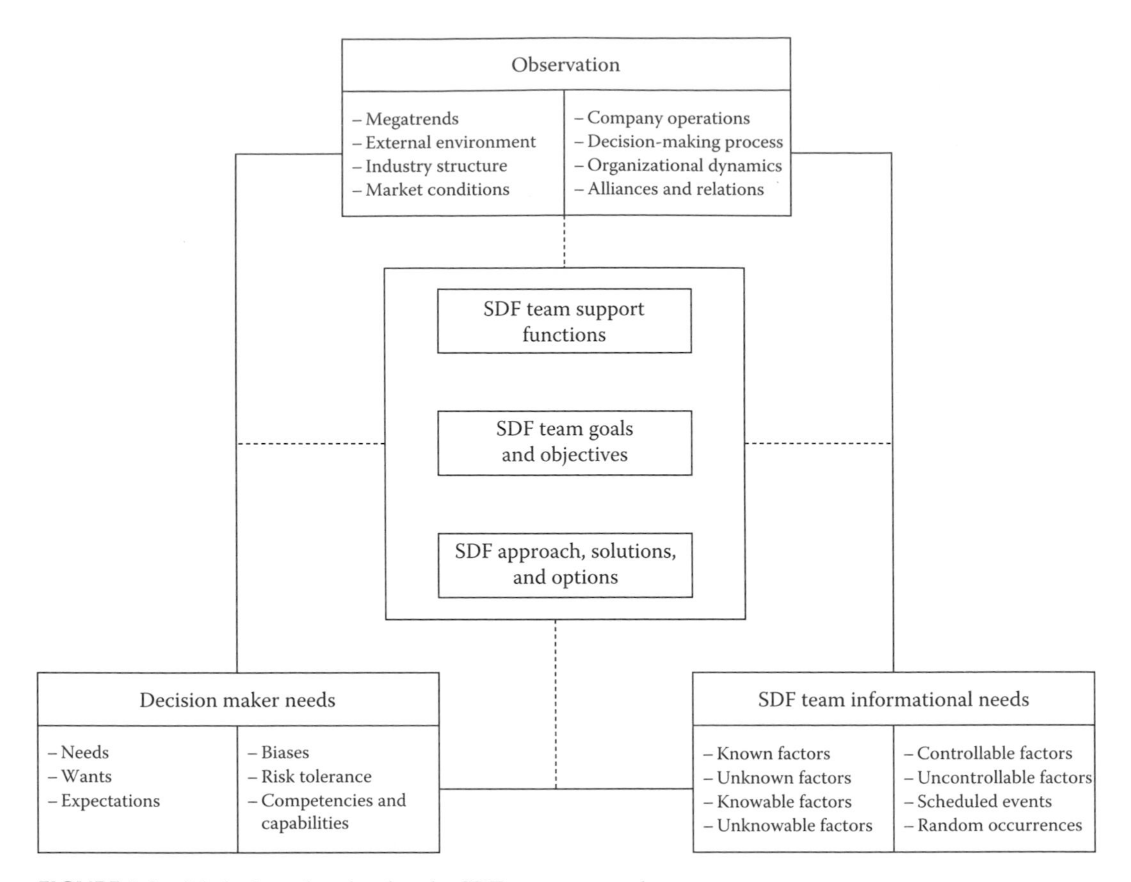

FIGURE 5.2 Method used to develop the SDF team approach.

The SDF approach builds on the broader decision under uncertainty model shown in Figure 5.2, which involves a sequence of steps, the most significant being surveying the external environment, the industry, and the market in question and observing the workings of the company to understand underlying dynamics. Then, the forecaster determines what needs to be known and what are the known, knowable and unknowable, and controllable and uncontrollable factors around a strategic decision. Identifying the real decision maker's and key stakeholder needs, biases, and expectations is also needed to create the right approach. The step of assessing consistency of corporate risk appetite with project uncertainty is next, followed by effort to reduce uncertainty through analyses, reasoning, and evaluations to residual uncertainty.

Collecting relevant data and intelligence on the important aspects of a decision is essential as is verifying the use of long-term forecasts developed and providing usage caveats that go with them. Selecting or developing methods and techniques needed to address key issues surrounding a decision is influenced by the specifics around a project, but identifying market drivers, factors, and events that can be controlled is central to modeling correctly company operations and creating plans to affect forecast realization. Performing an independent, critical, and objective verification and validation; modeling reality as recognized by the project team; and creating plausible, distinct scenarios of future states are probably the most valued SDF team contributions. Also, testing senior management and client hypotheses via simulations of alternative scenarios and ranges of the driving factors enables the project team to see what future states look like and what it takes to create them.

Assessing SDF solutions and their implications and conducting validations are key elements of the process prior to their becoming consensus forecasts as is developing decision options, recommendations, and actions to be taken in the implementation phase. Monitoring, evaluating, and reporting performance and rewarding it accordingly is another SDF process attribute because it is complemented with the future-state realization (FSR) plan. The last element of the approach is the project review that takes place to determine what was done right, what went wrong, and what lessons can be learned. No SDF project is complete without the project review step.

The SDF approach has a Design for Six Sigma and TQM orientation, and it is the starting point that enables associates to achieve a number of objectives, namely the efficient development of SDF solutions and decision options, a well-balanced approach, and a cost and benefit orientation at every step. Other objectives to be accomplished include suitable positioning of the SDF team to enhance its standing and its value-adding capability, minimizing uncertainty and risk surrounding projected future states and decision options, and increasing the chances of project goals and objectives being realized. Managing project stakeholder expectations to minimize conflicts and reduce the time from idea to decision is a major TQM objective SDF team members strive to achieve. However, the implementation of the SDF approach requires maximum stakeholder participation and 4Cs.

The SDF approach mirrors the team's mindset and is characterized by disciplined, methodical, collaborative, balanced analyses and evaluations, knowledge sharing, and open communication. We observe that when the outline of the SDF approach presented earlier is supplemented by clear SDF team goals and objectives, plans, and core values, it facilitates

1. Creating processes that help build efficient organizations.
2. Developing a firm grounding of analysis, scenario creation, and assessments.
3. Conducting a thorough forecasting due diligence.
4. Selecting appropriate methods, techniques, tools, and implements.
5. Project managing effectively the development of SDF solutions and options.

In the presence of the right SDF team mindset and appropriate approach, a number of advantages over the current state of strategic forecasting are realized. First, the ability of SDF team members to model business operations, identify key drivers and perceive possible scenarios, integrate diverse points of view, and get to consensus forecasts efficiently is enhanced. The proficiency of SDF teams in creating intelligence and valuable insights and separating warning signs from a multitude of data reaches levels of excellence as their expertise in identifying forecast enablers and describing future-state scenarios increases. Also, the right SDF team mindset promotes its members' capabilities in selecting appropriate solutions and options to control decision outcomes and their competence in going from analysis to the right choice in a highly effective manner. Furthermore, the right mindset plays a key part in raising the role of SDF team members to that of a valued business partner.

5.4 SDF PROCESSES

To simplify the discussion, the forecasting function management, operational, and support processes are lumped together under SDF processes. These processes are a collection of several sequential, structured, interdependent, and linked operations and actions, which are designed to take the client and project needs and requirements, assemble or build up all the inputs to the process, and develop the solution to support a project decision. SDF processes begin with the client and senior management needs and end up with solutions and options that satisfy those needs in the most effective manner.

The first determinant of the SDF process for a project is the specific need or application; the second is establishing the boundaries of the process and its beginning and terminal points; and the third one is ensuring there are no unnecessary activities. The conditions that apply are that there is only minimal duplication of effort, and each SDF process includes no more than five or six major sets of activities, each of which can be broken down to simpler functions and tasks. They are usually shown in a flowchart format, and we find that swim lane charts are the most effective in representing them. And because SDF solutions require process with high levels of 4Cs, they avoid many of the

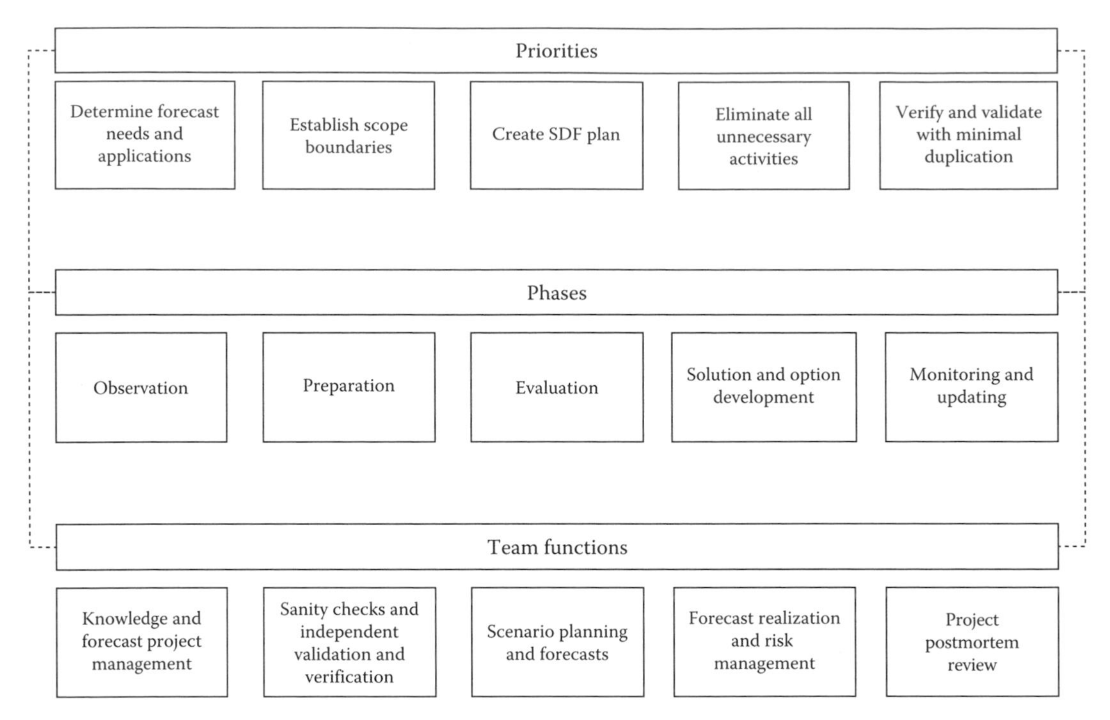

FIGURE 5.3 SDF team priorities, project phases, and key functions.

difficulties common in silo functional arrangements. The key phases common to SDF processes are depicted in Figure 5.3, but the treatment of the forecast realization, the knowledge management, and the postmortem activities is deferred to later chapters.

When developing the forecast plan and processes, SDF team members ensure that, at minimum, the following qualities are present to make them pass the acceptability test:

1. Clarity of purpose, roles and responsibilities, timelines, expected inputs and outputs, and form of deliverables
2. Comprehensiveness of plan and processes to address all key points and issues that need to be dealt with and not let things fall through the cracks
3. Efficacy that here refers to the absence of duplication of effort and minimum resources expanded to complete all activities in a timely manner
4. Replication of its key components in other situations, decisions, and projects; that is, standardization of components once they have been tested several times
5. Coordination, cooperation, and collaboration in completing evaluation and feedback tasks or validating the inputs and assessments of other participants
6. Transparency of process meaning knowing the present activity or step and ease of determining what follows and when
7. Adherence to core values and principles, established guidelines, and best practices
8. Flexibility when faced with dead-end situations and having contingency options and authority to pursue alternative courses of action

What follows is a checklist of the steps by major process phase found in companies that have dedicated SDF teams. Depending on the complexity of the end result the project aims to attain, the way customized SDF processes are portrayed varies a lot. A simple way to define an SDF process is through a process map, which is a matrix of key activities across the top and responsible stakeholders listed vertically. For parsimony of discussion, we divide the SDF process into four parts: the preparation, the evaluation and modeling, the solution and option development, and the monitoring and updating phases. Notice, however, that the functions and activities listed are not necessarily performed sequentially, but as required.

A. *Preparation phase.* The common elements in the preparation phase include the following activities SDF team members go through:

1. Initiating contacts and developing internal relationships with potential clients, senior management, and forecast stakeholders in client organizations
2. Developing a good understanding of the corporate culture, policies, strategy, goals and objectives, and risk appetite
3. Checking the company's stock performance in response to prior strategic projects and determining key investment community expectations, issues, and concerns
4. Identifying corporate and client needs vis-à-vis the expectations, issues, and concerns of the external world and understanding discrepancies, biases, and their sources
5. Identifying preferences, predispositions, and expectations of strategic forecast stakeholders and inclinations of senior managers and clients
6. Obtaining a good understanding of how the company operates and makes money; that is, examining closely the products or services offered, the different markets they are sold in, and how they stand next to competitors
7. Getting to know the industry and the major competitors well, how the company is positioned in it, and observing the industry dynamics and how they affect the company's operations
8. Identifying, monitoring, and assessing megatrends, subtrends, and market trends impacting the industry and the company's business, and determining likely effects
9. Determining what needs to be known to generate sound SDF solutions for alternative decision options, what is known and unknown, and what is knowable and unknowable
10. Verifying and validating the decision or project rationale, the need for and use of the SDF solutions, and the options desired to make a decision
11. Making preparations for project managing the SDF solution development and creating the project management plan
12. Assembling the project team and assigning and agreeing on participant roles and responsibilities

B. *Evaluation phase.* The evaluation and modeling part of the SDF process contains more elements because of the importance of accurate assessments in creating SDF solutions. The SDF team addresses process requirements with the following activities:

1. Examining key markets and identifying relationships, causality, strength of association, how stable they are over time, and what may cause shifts in them
2. Creating a mental model of the company's operations and visualizing the flows and how the different parts work together to produce revenue and cost streams
3. Developing and articulating the SDF team's strategic intent statement about the project decision in question
4. Communicating to stakeholders the SDF plan, goals and objectives, strategies, processes to be followed, the expected deliverables in required formats from each participant, and the corresponding timelines
5. Reviewing with the project team earlier strategic forecasting case studies, competitor business cases, prior SDF projects, past strategic project performance, and lessons learned
6. Performing a political, economic, social, technological, legal, educational, and demographic (PESTLED) environment analysis and a company and key competitor strengths, weaknesses, opportunities, and threats (SWOT) analysis to understand key influencers and ground the evaluation, and reviewing with the project team key findings and conclusions

7. Conducting a megatrend, subtrend, industry trend, market trend, and major competitor assessment and distilling insights about possible impacts on future company performance
8. Determining what events and key drivers of company performance are controllable, the extent to which they are, and those which are determined by factors outside the company's control
9. Clarifying the project rationale, goals, and objectives and assessing the project uncertainty and risks and reducing project uncertainty to residual uncertainty through reasoning, analyses, and evaluations
10. Developing the assumption set jointly with the stakeholder team and testing it against previous experience, industry benchmarks, norms and analogs, and internal and external expert opinions
11. Identifying reliable sources of data and proceeding to data collection, validation, adjustment, and data analysis activities with the intent to create intelligence and distill practical insights
12. Deciding on the methodology and techniques to be used in developing the SDF solutions and decision options
13. Building a computational model of the current state of affairs to identify associations and causality, measure the strength of key relationships, and conduct sanity checks
14. Conducting a comprehensive forecasting due diligence and review important information and findings with the project team
15. Developing jointly with the stakeholder team the criteria that determine to what extent the SDF solution meets the project team's quality standards and industry benchmarks
16. Generating a comprehensive set of hypotheses to be tested about future states resulting from different decision options, support levels, and industry and competitor reaction
17. Using a diverse scenario model development and selection framework, in addition to the status quo scenario, create distinct scenarios that take the company from the current to alternative future states because they provide diverse insights and reflect different influences
18. Adopting an action–reaction–action–reaction mindset and thinking through several distinct and plausible future scenarios, major events and triggers, changes in drivers, and possible market shifts and shocks
19. Conceptualizing a few catastrophic future events or eventualities (black swans), building scenarios that could lead to such states, and determining possible company responses to such threats
20. Obtaining stakeholder team input and reaction to identified scenarios, soliciting new ideas and other possible scenarios, and agreeing on the set of scenarios to test to the point of understanding how future states can be created
21. Considering outsourcing future-state forecasts to different experts and approaches and then combining them to increase client and senior management faith in the future-state projections

C. *Option development phase*. The solution and option development part of the SDF process is a continuation of the evaluation and modeling effort and includes the following activities that the SDF team members perform:

1. Selecting a set of future-state scenarios to be included in the SDF solution and identifying the enabling factors and resource requirements (e.g., human and financial resources and technological advances)
2. Conducting additional sensitivity analyses and scenario simulations along with reality and sanity checks on the future-state scenario outcomes and assessing their implications

3. Revisiting project residual uncertainty; identifying the risks inherent in the forecast future state, their sources, and likelihood of occurrence; and quantifying their impacts
4. Communicating to project stakeholders the resource constraints and the risks due to uncontrollable factors in each future-state scenario and evaluating the costs and benefits of over- and underforecasting future-state scenarios
5. Incorporating project team inputs, obtaining agreement on the future-state scenarios to base the SDF solution on, and identifying decision options that pass the test of reasonableness, quality standards, and likelihood of occurrence
6. Creating a forecast realization plan that consists of support requirements, contingencies, and competitive response preparation plans in the event risks materialize
7. Helping the project team in the preparation of a risk management plan and actions to offset major risk impacts
8. Preparing the SDF solution and decision option baseline, presenting it in appropriate format and with supporting documentation for senior management review, and soliciting client and senior management input and comments
9. Evaluating the input and adjustments of senior management, incorporating any new requirements to make the adjusted future states realizable, and pointing out the resource implications and risks associated with their forecast adjustments
10. Assisting the client in presenting the final forecast to the stakeholder team and senior management and obtaining consensus on the SDF solution and decision option
11. Creating an Intranet-based SDF database intended to record key elements from developing the decision solution and options (e.g., the stakeholder team's roles and responsibilities, assumptions made, data and information collected, analyses and evaluations performed, key findings, scenarios used, and insights created)
12. Posting the particulars of the SDF solution and options created on the SDF database and making them available to the project stakeholders and those organizations with a need to know

D. *Monitoring and updating phase.* This part of the SDF process deals with what happens after a strategic decision is made and the project begins to get implemented. When the SDF team is involved in this phase, the needs of the process require team members to

1. Define strategic forecast performance measures and indicators and develop a dasboard to monitor not only progress against projected states but also events, triggers, impacting factors, and major external developments.
2. Create an early warning system (EWS) to monitor actual forecast performance and report discrepancies, potential risks being realized, and their impacts on the business.
3. Perform variance analysis that delves into the root causes of deviations of actual from forecast states and communicate the need for action when it becomes necessary for project dashboard indicators and EWS signals.
4. Assess the impact of the project on the industry, competitive reactions, and progress made toward achievement of project goals and objectives.
5. Assist the project team to deploy the FSR program and deal with contingency planning and management issues.
6. Update the assumption set in the postdecision period as warranted and generate new scenarios and future-state projections.
7. Conduct a client and senior management satisfaction survey on the SDF process, team member performance, the extent to which their needs were met, and aspects of the SDF team's participation that need improvement.
8. Communicate the performance survey results to the stakeholder team and invite feedback, input, and comments on how to enhance performance going forward.
9. Research current industry association reports and subject matter expert views, seek alternative approaches, benchmark strategic forecasting best practices, and create improvement targets.

10. Undertake an ongoing process restructuring, team member development and practice improvement effort and report progress via the SDF balanced scorecard.
11. Update the SDF database in conjunction with performing a postmortem analysis to determine what was done right, what was done wrong, what worked and did not, and what was learned.

5.5 SDF PRINCIPLES AND GUIDELINES

Principles, on the one hand, are rules and standards, which are used as a basis for conduct. Guidelines, on the other hand, are ways to plan and determine a course of action and facilitate implementation of policies. As such, principles and guidelines provide guidance for appropriate behavior of team members and the operation of the SDF team. Sometimes they can be best practices, but often they are instruments used to ensure that critical success factors are in place. They are quality and efficiency considerations and effective habits, which are incorporated and observed in the SDF team's processes and practices and the selection of SDF solution and decision options.

The source of principles and guidelines shown below are unedited quotes from discussions with SDF experts. The reason for using this approach is that the SDF discipline is new, and not enough history exists to do a meaningful benchmarking of best practices, although several of the statements are considered best practices.

A. *SDF team member activities and behavior.* The following are considered principles and guidelines in the area of the SDF team's activities and behavior:

1. Ability to bridge the forecasting–decision making divide with broad context awareness and boldness within collaboration
2. Close ties with clients to understand their needs and the purpose and uses of SDF forecasts, develop better internal and external networks and relationships, and form alliances with other stakeholders are as important as functional expertise.
3. Creating the right SDF governance structure at the start and a solid foundation by selecting seasoned forecasters with a winning mindset and fostering a learning organizational climate, which is an effective means to operationalize knowledge
4. SDF team authority with external experts and industry analysts and credibility based on well-disciplined and well-considered approaches and well-balanced perspective
5. Effective management of forecast politics to permit mobilization of resources and commitment to realize the forecast
6. Forecast ownership and accountability at each stage of the decision process; a feel for the jugular and forecasting excellence
7. Getting immersed in the company business and getting a good handle on the client needs and other stakeholder needs and objectives
8. Hiring an SDF team manager with the right kind of intelligence, communication, and motivational skills to be able to sell the forecasts and get the best of the forecasters' capabilities
9. Independent, critical, and objective thinking and an integrative SDF team with a learning culture, a comprehensive approach, orientation and focus, and values and beliefs
10. Knowing the corporate strategy and second-tier strategies well enough to evaluate their validity and effectiveness
11. Long-term focus of the organization, linkages with all planning functions and organizations, and development of SDF team members
12. Picking and developing the best forecasters to help manage client expectations and different agendas

13. Making management of politics, expectations, relationships, and the forecast function part of the SDF team members' job
14. Starting with realistic strategy and reasonable objectives and client project expectations
15. SDF team leadership coupled with forecast owner accountability
16. Respect for the evidence by all project participants and associates seeking model, scenario, and forecast certainty and then mistrusting it
17. Sending all forecasters to charm school to learn how to deal with unreasonable clients and difficult project teams
18. A sense of purpose and direction, a sense of history, and respect for competitors
19. Sound project management, hands-on leadership, and senior management support
20. SDF managers thinking outside the traditional forecasting box and having a passion for practical innovation

B. *SDF processes*. There are fewer principles and guidelines in the area of SDF processes primarily because they are straightforward and come after SDF team member behavior has been addressed. However, SDF processes must deal with project uncertainty and risks at the outset and then drive to residual uncertainty, forecast risk assessment, and risk management. Adequate project planning, preparation, and support at all stages is crucial in coming up with appropriate processes, practices, methodologies, data, models, and scenarios. The next principles in order of importance for process creation are creating complete and effective processes, making them known to the project team, and determining forecast support requirements before hand, monitoring, and learning how to adjust support levels in the forecast period to make the forecast materialize.

Key elements of the SDF process include the following activities: a value audit, forecasting due diligence, independent validation and verification, stakeholder education, monitoring of implementation performance, and knowledge and project management. Other important principles include identifying and measuring structural and resource constraints and inclusion of sanity checks, benchmarks, and industry norms in evaluations. Also, knowledge, logic, science, creativity, practical experience, and, above all, common sense and a balanced approach are crucial. Making, recording, and updating assumptions as changes occur in the course of a project so that all relevant information are readily available about rationale, time, and magnitude is a central activity that impacts SDF performance measures, processes, steps, and activities and how they are understood. Since strategy drives planning and strategic intent drives SDF team activities, corresponding periodic reviews and assessments are also part of validating the assumptions.

C. *Forecast application principles*. On the subject of practices related to SDF applications, the following principles and guidelines are vital:

1. Clarity of purpose, project definition, team member roles and responsibilities, and common understanding of project goals and objectives
2. Application of the right methodology and best practices suited for the type of the project to be evaluated and decisions to be made
3. Assessment of project uncertainty, forecast risks, and the potential impacts of all scenarios, including the status quo scenario of affairs
4. Collaboration with all project stakeholders, free exchange of data and information, close coordination of activities, cooperation in getting to consensus, and continuous communication and information updates
5. Consistency of SDF solutions with company history, earlier project results, judgment of recognized experts, and triangulation of diverse forecasts
6. Developing the right performance metrics for the SDF solution, for the SDF team, and for project goal and objective progress realization

7. Exploiting distilled lessons learned from past company projects, external experiences, benchmarking information, and industry norms to increase project team productivity
8. Having a mechanism to find out what is happening in the implementation phase, what is not working according to expectations, and what works well and why
9. Other concurrent corporate initiatives with nonconflicting objectives with the project supported by the SDF team
10. Well-defined product strategy, senior management commitment and involvement, high-quality stage–gate kind of assessments, necessary resources in place and dedicated to the project, and high quality of execution
11. Forecast accuracy supplemented by cost–benefit analysis and Six Sigma measures
12. Getting buy-in for the SDF process at the start of the project to make the choice of forecast alternatives more efficient
13. Validation of all information, data, competitive intelligence, and environmental assessments as well as verification of the identification, testing, and understanding of controllable and uncontrollable events and ability to leverage the forecasts
14. Evaluation and inclusion of appropriate input of clients and key stakeholders concerning assumptions, future plans, future events, controllable factors, and other elements impacting forecast realization; that is, making maximum use of organizational intelligence and collaboration when measuring the impact of decisions
15. Consideration to inputs to and evaluations of forecasts by internal and external experts, triangulating forecasts from diverse sources, and use of more than one technique to develop SDF solutions
16. Paying more attention to research, data, and the analysis aspect than to newspaper headlines to do a proper identification of underlying megatrends and subtrends
17. Reasonable methodology and modeling of the company's business and testing; use of both relevant qualitative and quantitative inputs; reviewing and challenging the assumptions, starting points, data, analogs, and models; and timely development of consensus forecasts and minimum turnaround cycle of updates
18. Sound industry analysis and appropriate use of SWOT findings, good understanding of key drivers, paths of influence, and feedback effects and appreciation of forecast implications
19. Sound SDF project management characterized by a structured, accurate, and complete record keeping of past forecast project experiences and enhanced by storing major project-related research, competitive intelligence, evaluations, and results on the Intranet database and sharing them with all who need to know

D. *Project team interactions.* The following principles and guidelines apply on the issue of project team operations and behavior:

1. Proactive, risk-taking, and results-oriented SDF team members performing independent, critical, and objective assessments
2. Integrity and project team credibility are essential throughout the duration of the project as is including all stakeholders and defining clearly the roles and responsibilities of team members, clients, and senior management.
3. Understanding the decision that needs to be made, the purpose of the forecast, and how the forecast will be used, coupled with high levels of 4Cs throughout the project
4. Clients and management are educated about the forecast process, what drives it, and what has to happen for the forecast to materialize, which makes it easier to manage client expectations and conflicts among team members arising from internal politics.
5. Well-understood linkages between sales and product quality, attributes, pricing and market dynamics, and the external environment by clients and other project stakeholders

6. Proficient SDF team members, sound forecasts, competent implementation team, proper use of SDF solutions, and effective project execution
7. Selecting and preparing project team members and ensuring that the right team is in place that has not only the talent but also the right chemistry in order to win
8. Necessary forecaster and other team member qualifications and training supplemented with the right mindset and attitude
9. Evidence-based identification of opportunities and tested expert judgment to determine support requirements for projects to achieve stated goals and objectives
10. Roles and responsibilities assigned to strategic forecasters and project stakeholders commensurate with their capabilities to carry out successfully.
11. Ensuring that senior management commitment and support are well established and resources are allocated to support successful project implementation
12. Proper handoffs of responsibilities between project phases and documentation to help the implementation team understand what needs to be done for the projected states to be realized

E. *Solutions and options.* On the topic of SDF solutions and options, experts offer the following principles and guidelines:

1. In every project, the SDF solution should fit the problem, not the other way around, and that particular SDF solutions should serve as communication vehicles across all stakeholders and project phases.
2. The decision context should tell forecasters which techniques to use and the forecasters should recommend to clients the options to be considered.
3. Inputs to SDF solutions come from the extended project team, from objective sources, and from impartial investigators.
4. Innovation and application of all organizational intelligence to reduce project uncertainty
5. Consistency of SDF solutions with corporate strategy, company history, and prior project experiences is crucial.
6. Criteria need to be developed for the difficult task of selecting the method to assess SDF solutions and picking the preferred decision option.
7. Complex SDF solutions do not have to look that way to the client and senior management; they should be easily understood and delivered effectively.
8. Performance measures need to account for the interactions among decision components and should include indicators of forecast, the SDF team, and project performance measures.
9. Project dashboards, EWSs, FSR plans, and SDF team balanced scorecards are part of the SDF package.
10. Involving senior managers to oversee in the selection of SDF solutions brings about efficient consensus agreements.

F. *Competitive advantage.* Asked about whether SDF excellence can yield a competitive advantage for the company, SDF experts state that it is possible only when a number of practices are followed closely, namely, the following:

1. Focusing on strategic intent, creating a good SDF plan for each project, strong functional links and personal relationships, coordination of processes and activities, concentration of effort on key objectives, directed project team commitment, and SDF and project team determination to create value
2. Making the SDF team part of the Strategic Planning or Portfolio Management organizations; creating a suitable governance structure that includes the right mission and business definition; adopting best practices; and creating a center of knowledge, learning, training, and professional excellence

3. Capitalizing on the right mindset characterized by SDF team members being driven by strategy, proactive, entrepreneurial and risk taking, relationship oriented and results focused, and participative and collaborative, yet independent, objective, and critical in their assessments
4. Creating superior economic value by accomplishing objectives related to a long-term perspective, providing exceptional support to implement value-creating projects, and contributing to project management, implementation, and operational quality
5. Making it difficult to imitate the core competencies and capabilities of the SDF team with respect to unique knowledge databases, high levels of 4Cs, effective SDF evaluation processes, and customized techniques and tools
6. Building effective processes with functional linkages and knowledge sharing with other disciplines, a Six Sigma orientation, sound project planning, validation, verification, triangulation, and continuous process improvements
7. Eclectic approaches, techniques and implements, and application of tools to fit the project circumstances, which include market research and competitive analysis, thorough review of paradigms, analogs, past projects, and earlier learning; forecasting due diligence
8. Focused data analyses to create intelligence and insights, statistical analysis, modeling capabilities, and introducing concepts from other disciplines
9. Scenario development, planning, and sanity checks along with FSR planning

Key observation: *By now, the reader has noticed a number of items that keep coming up, the most noteworthy being the following:*

- *Consistency of SDF values with principles and guidelines*
- *Commonality of focus and activities in process creation, analyses, and evaluations*
- *Similarity of process elements for reasonable assumption creation, thorough analyses and assessments, and development of sound SDF solutions*
- *Repetition of themes and factors considered important in different areas of SDF*

What may not be obvious is that these factors permeate the development of SDF solutions for any type of strategic projects and result in higher project success rates. Understanding these factors helps to create the conditions for sound forecasts and increased chances of realizing project objectives.

6 SDF Analytical Techniques and Tools

The techniques and tools used in strategic decision forecasting (SDF) are applications of widely known, systematic, effective, and practical methods. They are procedures used to accomplish the SDF team's specific goals and objectives through activities that require specialized knowledge, experience, and skills. Tools on the other hand are instruments that have been found to work well in getting tasks completed and produce reliable SDF solutions when applied correctly on a consistent basis. They are valuable instruments used routinely to carry out the responsibilities of the SDF team in the most effective manner.

Analytical tools can be structures to evaluate a situation and systematic processes to investigate issues. They can also be tools to help visualize the operating environment and company operations or provide a structured way to organize ideas and information. In some instances, they can be procedures used to enhance the SDF team members' understanding of the nature and the causes of problems or they can be efficient ways to assess a problem and come up with alternative solutions. Often, SDF techniques and tools are aids used to make sense of huge volumes of data and information, critical thinking methods, and sound analytical tools to evaluate information and create intelligence, insights, and solutions.

SDF team members make use of these tools and instruments as prototypes of the world to represent their understanding of business realities in a clear and concise manner and to portray their findings and results of their evaluations. They can also be indicators used to assess performance or measures that show a need for certain actions. In all cases, however, they are effective decision-guiding and decision-making tools. We discuss the various techniques and tools used by SDF teams under six major headings: SDF analytical aids; anchoring of the analysis, assessments, and evaluations; forecasting due diligence; SDF project plan and processes; and methods and systems.

The discussion of this chapter centers on the toolkit of SDF team members, with a particular emphasis on the following key takeaway points for strategic forecasters:

1. Know how to use the analytical aids in strategic forecasting; they make the forecaster's contributions more valuable and noticeable.
2. Without strong anchoring of analyses and evaluations, forecasts and decisions based on them are on shaky grounds.
3. Performed well, the forecasting due diligence helps arrive at consensus quickly and enhances the credibility of forecasts and the SDF team.
4. Effective SDF plans and processes are created when using benchmarks and tested techniques.
5. Receptiveness to new ideas and approaches and making use of all SDF methods and tools creates reliable solutions.

6.1 SDF ANALYTICAL CONCEPTS AND AIDS

The SDF analytical aids are a set of cognitive tools that determine how SDF team members think, observe, interpret, model company operations, and plan responses to different situations. Following are concepts and aids useful in creating or conditioning the right way of thinking of the SDF and the project teams.

Application instructions: *How to use the techniques and tools discussed in this chapter in SDF projects:*

- *Agree on the issue to address or question to answer, research it, and understand it well.*
- *Examine how the techniques and tools discussed here are relevant or, for specific project application techniques and tools, look in Chapters 8 through 19.*
- *Select as many appropriate and likely to be fitting techniques and tools to address specific issues and question raised, apply them, and evaluate the results of each.*
- *Make an interpretation and conclusion on the best way to address the issue or answer the question raised.*

A. *Collaborative forecasting.* It is a concept originally developed to enhance supply chain integration by developing and supporting common planning practices with customers or distributors. In the SDF context, it refers to forecast solutions developed through committed involvement of forecast stakeholders; high levels of internal communication, free exchange of ideas; sharing of data and information; and intensive cooperation, coordination, and collaboration of effort between SDF team members and other project participants.

B. *Cause-and-effect diagrams.* They are also known as fishbone diagrams, which help to think through not only relationships but causes of a problem solution thoroughly and consider possible causes, not just the most obvious. Their development supports and enhances one's ability to decipher structures and identify and understand the drivers of a resulting outcome. They are also useful in recording facts and information, and show how pieces of information fit together, thus helping to carry out a thorough analysis of a situation. They are especially useful in identifying problems and sources of gaps in strategic forecasting performance, as shown in Figure 2.1.

C. *Conceptual modeling.* It is the practice of creating descriptive, high-level models of a business situation based on qualitative assumptions about its components and interrelationships. Conceptual models are created without implementation details, that is, without mathematical relationships, and their purpose is to help forecast stakeholders discuss the operating environment, industry, and market developments, and to determine real relationships between different elements of operations.

D. *Integrative forecasting.* It is forecasting based on integrative thinking, which is a critical type of thinking and skill that all SDF team members are required to master. Integrative forecasters are comfortable with complexity, accept that uncertainty is always present, are able to manage conflicting viewpoints, and can come up with solutions to strategic forecasting problems. They look for relevant factors that influence outcomes and not just the obvious factors. That is, they see forecasting problems in their entirety and how they fit with the decisions that need to be made and investigate multidirectional relationships between market drivers. Integrative forecasting in its basic form is exemplified in all SDF practices and forecast triangulations.

Integrative thinking extends beyond forecast solutions to factors clients consider when making decisions. The four steps involved in integrative forecasting are as follows:

1. Observing the operations of the business closely, understanding the decision needs, outlining the forecasting problem and its salient features, and determining what to focus on from a myriad of possible factors and what not to consider
2. Making sense of how and why things work the way they do by identifying relationships underlying the workings of the company's business and the markets, discovering what is causing what and in what ways, and gauging the strength of the causal relationships

3. Understanding and visualizing the different components of the decision and how everything fits together by considering alternative frameworks and creating a mental model based on the conclusions of the earlier steps
4. Searching among possible alternatives and synthesizing an eclectic solution, which includes the best of each when deciding on methodology, models, analyses and evaluations, and scenarios to use in developing the SDF solution

E. *Reframing matrix.* This is a technique that enables SDF team members to look at many forecasting hypotheses from different perspectives so that alternative solutions may be evaluated. It is used to enhance communications and facilitate cooperation among forecast stakeholders through the McKinsey 4P (product, planning, potential, and people) approach or the approach of asking for the organizational, functional, professional, or specialist perspectives and inputs. In the 4P approach, the strategic decision question or problem is posted in a box in the middle of a piece of paper with four spokes to the four Ps representing possible perspectives, such as the product, the planning, the potential, and the people perspective. In each alternative approach, different perspectives are sought such as (a) the strategic planning, the portfolio management, the R&D and product development, and the business development perspective; (b) the market researcher, the forecaster, the marketer, and the sales people perspective; and (c) the client, the senior management, the forecaster, and the customer perspective (Morgan 1993).

Strategic project stakeholders see decisions, issues, and solutions differently and the reframing matrix allows different perspectives to be used in creating strategic forecasts and helping to manage decision-making processes. The reframing matrix technique is particularly useful in the following situations:

1. Ensuring completeness of the information and intelligence set and balance of perspectives
2. Validating stakeholder understandings and assumptions underlying SDF solutions, the forecast methodology, and models used and eliminating flaws
3. Providing benchmarks and sanity checks and developing more reliable inputs to different models and techniques
4. Developing and evaluating different future scenarios and assessing forecast risks, sources, and causes
5. Selecting the best SDF solution among competing options and identifying extenuating circumstances impacting the forecast solution
6. Developing the SDF organization's balanced scorecard, the project performance dashboard, and the early warning system (EWS)

F. *Six thinking hats.* This is a parallel thinking technique to give SDF teams a way to reflect together with clients and improve communications and the quality of assessments that impact the development and selection of scenarios. It is a technique to check the validity of assumptions, scenarios or stories about future states, assessments of risk, and the implications of SDF solutions. It uses the metaphors of hats to reduce conflict and facilitate collaborative thinking, produce innovative ways of looking at forecasting problems, and provide direction and focus for project teams (De Bono 1985).

This technique separates fact from opinion, helps SDF team members and clients espouse different perspectives on issues than they would normally do, and brings balance to assessments and decision making. Wearing different color hats, team members assume different roles and think as if they saw the world from those perspectives:

1. White hat represents an objective and neutral perspective, and focuses on facts and where to find what is needed to get to answers, solutions, and closure.
2. Red hat stands for thinking from the feelings' and emotions' viewpoint with no proofs or justifications, but keeping things short and to the point.

3. Black hat symbolizes the somber experience, critical, judgmental, logical, and cautious but negative standpoint.
4. Yellow hat thinking is the logical, positive, and even optimistic perspective looking for benefits and values.
5. Green hat thinking is about creative thinking, new ideas, hypotheses, scenarios, and possibilities.
6. Blue hat thinking embodies control of thinking and processes, drawing conclusions, and making plans.

G. *Strategic intent.* It is a clear statement of the direction the SDF team intends to take to create value in the decision-making process, and it is the starting point of the SDF plan. Its definition is expressed as a set of propositions concerning capabilities, linkages, and incentives. Strategic intent is

1. A compelling target for organizational commitment and effort and an ambition out of all proportion to resources and capabilities.
2. A constant long-term objective that defines winning and is worthy of an expanded attention span.
3. A means to force leveraging of resources to reach intent rather than fitting objectives to limited resources.
4. An approach to fold the future back to the present; that is, it answers the question: What must be done now to get closer to that future target.

The SDF team's vision is developed before strategic intent and its business definition, goals and objectives, and the strategies to achieve them follow it. Strategic intent helps clarify the SDF team's purpose and turn its intentions into appropriate and applicable activities and results. A key characteristic of strategic intent is that it produces a customized sharp focus on the end results and leaves the means to achieve them up to the discretion of SDF team members. Implementing the SDF strategic intent comes with a number of requirements, such as senior management and SDF team members understanding it well, SDF team members appreciating the challenges set by the strategic intent, and key project stakeholders realizing the impact of strategic intent on the scope of SDF team activities. In addition, the SDF team must stretch its resources and capabilities to meet expected and unexpected challenges and redirect focus and emphasis on effective diagnostic tools, new skills and innovation, elimination of process flaws, and productivity enhancements.

H. *Total quality management.* Total quality management (TQM) is a management approach popular since the early 1980s that requires quality in all aspects of operations, with work done right the first time and defects and waste eliminated from processes. It is a method by which senior management and SDF team members become involved in the continuous improvement of the support the SDF team provides. It is a combination of quality and management tools aiming at increasing performance and reducing inefficiencies of wasteful practices (Ishikawa 1987).

TQM describes a philosophy that makes forecast quality the driving force behind SDF team leadership, design, planning, and improvement initiatives. It also describes the culture, attitude, and governance of the SDF team required in order to provide clients with support services that satisfy their needs in the most effective manner. In the context of SDF, TQM requires several key elements: ethical and professional team member behavior, integrity and trust in all dealings, training, and knowledge building. It also needs support commitment by senior management, leadership by example, stakeholder teamwork, benchmarking, and continuous improvement. To make it work, high levels of communication, cooperation, coordination, and collaboration (4Cs) are essential to reduce SDF solution development cycles, achieve minimum time to decision, and meet client needs in the most effective manner with performance measurement, incentives, recognition, and rewards attached to it.

6.2 ANCHORING OF SDF ANALYSES AND EVALUATIONS

Anchoring in SDF refers to various reviews, analyses, assessments, and evaluations, which create a secure foundation for modeling and assessment, build sound solutions, and select the correct decision options. The following are specific techniques and tools deemed necessary to achieve desired results.

A. *Competitive analysis.* It is a crucial component in developing SDF solutions and refers to the process of gathering information, converting it into intelligence, and then helping utilize it in decision making. It is also called "the process of early signal analysis," and it is carried out to answer questions like who are the competitors, what products or services they sell and their market share, what market strategies they have been using, what are their strengths and weaknesses, and the potential threats they pose. Its value lies in insights produced in the identification and description of factors that cause differences in performance between the company and its competitors. Competitive analysis is different than benchmarking, which compares company processes and performance metrics to best-in-class organizations.

B. *GE–McKinsey nine-box matrix.* It is a systematic approach originally developed for companies to allocate resources among different divisions using two factors to determine future prospects: the attractiveness of the industry and a division's competitive strength in that industry measured by low, medium, and high levels. It has also been used successfully in succession planning, and it has a place in SDF as well. The analysis that goes into developing the nine-box matrix is extremely valuable to and is used extensively in the development of SDF solutions and the selection of best options for two reasons: It provides the results of in-depth analysis of the factors that contribute to market attractiveness and the company's competitive position in the industry, and it serves as a reference point for the forecast implications and a benchmark against which to judge value created by the decision supported by a certain SDF solution (www.quickmba.com/strategy/matrix/ge-mckinsey).

C. *Megatrend assessment.* Megatrends are widespread forces of societal, demographic, economic, political, and technological developments that have major impacts on the future over the next one to two decades. They are broad shifts in thinking paradigm, assessment, and approach that affect companies, industries, and nations, and are made up of subtrends, which may also have large impacts. Megatrends are forces defining the present and future states of business, which the SDF team monitors and evaluates along with subtrends impacting the company and the industry. SDF team members monitor, evaluate, and interpret general megatrends, but also focus on the subset of megatrends that are relevant and have a major impact on the company's products or services and its future state. Megatrends are thought of as key drivers of probable futures, unlike events and black swans that enter the development of future scenarios because they can change the impacts of megatrends. Hence, megatrend evaluation enters the SDF methodology and scenario development process along with industry and market trends.

Some examples of developments in general megatrends currently affecting all industries and societies and expected to do so over the next several years are the following:

1. Societal and cultural shifts such as government involvement in social policy and redefinition of the family and its role
2. Demographic shifts such as aging population and differential growth rates among different races
3. Economic order developments, such as health insurance coverage and globalization of trade
4. Scientific and technological developments such as alternative energy sources and bioengineering
5. Political condition shifts such as spreading of democracy and privatization of resources and service provisioning

D. *PEST analysis.* It is the analysis of political, economic, social, and technological factors (PEST) with its variants that include legal, educational, and demographic factors. PEST analysis helps in understanding the external, uncontrollable forces of change that impact the company's operations. It is a situation analysis in which these factors are investigated to enable SDF team members to create a more informed view of the future environment expected to affect significantly company operations.

PESTLED analyses used in strategic projects require a common project participant understanding and a clear purpose of the SDF solution. This means that SDF team needs to ensure that relevant factors that impact only the company's business are included in the assessment. Once these factors are determined, the information pertinent to them is identified, a critical evaluation is performed, and conclusions are drawn about how each factor impacts the business and the SDF solution. While PESTLED analysis is useful to evaluate the impact of environmental factors on the company and the forecast solution, it is important that it is used in conjunction with the evaluation of megatrends impacting the company. Obviously, appropriate company responses to those factors should be planned to ensure forecast realization. Therefore, expert PESTLED analysis and megatrend evaluation are crucial elements in developing forecast assumptions, gauging forecast scenarios, planning responses to changes, and taking advantage of oncoming changes. And, with proper response planning, PESTLED analysis and megatrend evaluation help avoid selecting forecasts from erroneous scenarios and making wrong decisions.

E. *Industry analysis.* This tool is also known as Porter's five forces and facilitates the evaluation of a company's position relative to competitors that produce similar products or services. Understanding the forces at work through industry analysis is crucial since the findings constitute an important input to effective strategic forecasting and are a key component of strategic planning and decision making. Industry analysis is ordinarily performed after the megatrend and environmental analyses, and its goal is to establish the attractiveness of the industry by its profitability. It is a thorough analysis of the five factors that determine it: the likelihood of new market entrants, the power wielded by suppliers, the power of the buyers, the threat of substitutes of the company's product or service, and the degree of rivalry in the industry (Porter 2008).

F. *SWOT analysis.* Traditional strengths, weaknesses, opportunities, and threats (SWOT) analysis is a tool to evaluate internal strengths and weaknesses and external opportunities and threats. In the strategic forecasting context, it is a tool for evaluating internal strengths and weaknesses, currently not considered alternative opportunities and the threats from maintaining the status quo or from executing current plans. That is, it is a method to identify factors that are favorable or unfavorable to implementing a project and achieving the forecast and decision objectives. The main purpose of the SWOT analysis performed by SDF team members is to evaluate the prospects of sound project execution and the surrounding uncertainty as well as the achievability of future-state forecasts. In doing so, it makes the company aware of the internal factor deficiencies and the changes needed to face its challenges and helps clients develop ways to turn them to their advantage and create contingency and response plans (David 1993).

The hierarchy of SDF solutions development is as follows: assessment of megatrends and subtrends, PESTLED analysis, industry analysis, and SWOT analysis. The inputs used in SWOT analysis come from many different sources such as internal staff experts, outside consultants, industry analysts, competitive analysis team, project stakeholder brainstorming, and industry benchmarking. The SDF process used to conduct SWOT analysis is outlined by the following steps:

1. The client and the forecaster agree and communicate the project or decision objectives that drive the SDF team's objectives.
2. Company, management, employee, and project team strengths and weaknesses are assessed with respect to the ability to implement successfully a strategic decision.
3. Contemplated potential alternative opportunities and options outside the set currently being considered are examined.

4. The risks of avoidable threats due to limiting company actions, projects, and decisions to those that maintain the status quo are determined.
5. Viable ways to capitalize on company strengths and lessen weaknesses are explored and evaluated.
6. Feasible alternative options are determined and the implications for maintaining the status quo are assessed.
7. Findings and assessments are organized, verified and validated, and summarized. Then, usable intelligence and insights are distilled to be used in conducting the forecast due diligence and sanity checks.
8. The extent to which project or decision objectives and forecast realization are achievable is determined based on the SWOT identified.
9. Areas of improvement, changes needed, and additional resource support needed to achieve stated objectives are identified. Also, project or decision objectives and client and senior management expectations are reset.
10. Critical success factors and updated SDF, operational, resource, and project implementation plans are created.

SWOT analysis begins with finding out where to collect data and understand how to do it effectively. Data collection and SWOT analysis have to be specific to the project or decision objective and to accomplish that the SDF team follows certain guidelines. They include ensuring a common understanding of issues and solutions among forecast stakeholders, being specific and analyzing SWOT components relative to key competitors, and being honest and objective about the company's strengths and weaknesses. Additional guidelines include applying SWOT analysis at the level dictated by the project or the decision, distinguishing today's state versus the expected or desired state in the future, and reducing subjective judgments as much as possible. Also, avoiding the extremes of oversimplification and overanalysis and complexity, keeping the SWOT analysis and report short, direct, and simple to explain, and using SWOT analysis as a guide to assess forecast achievability and not as a solution are crucial in doing it well.

The company strengths and weaknesses commonly assessed for the SWOT analysis to be used as a reference point in reality and sanity checks commonly involve the following components:

1. Company size and ownership, organizational structure, and reporting relationships
2. Corporate culture, risk tolerance, reputation as well as customer relations, service, and care
3. Management team and employee attitude, knowledge, experience, skills, and training
4. Product line lifecycle positions, attributes, quality, pricing, and market acceptance
5. Condition of production facilities, equipment, and other assets as well as production and operations capacity, flexibility, distribution channels, supplier arrangements, and their efficiency
6. Information systems and human resource, financial and operational organization computer systems, and database capabilities
7. Patents, know-how, and other unique intellectual property and the state of technology platform employed in the company and future plans
8. Internal financial resources and access to capital markets and external funding
9. Company location and access to qualified labor and production inputs
10. Corporate-wide planning functions, key processes, and communication channels

6.3 SDF FORECASTING DUE DILIGENCE

SDF due diligence refers to the process of in-depth investigation and evaluation undertaken by SDF team members into the details of the company's operations and material facts. It is done exercising reasonable care to avoid leading clients and senior management down to wrong decisions. It is an

are not apparent and to reach consensus in the presence of massive amounts of information. The ways they work in strategic forecasting are the following (http://asq.org/learn-about-quality/idea-creation-tools/overview/affinity.html):

1. Describing the forecasting problem in as much detail as possible
2. Conducting an SDF team brainstorming session to generate ideas about the nature of the forecasting problem, its sources, and possible solutions
3. Sorting the ideas into major categories or related themes
4. Developing SDF team consensus and limiting the number of common themes
5. Creating header theme cards with description for the relationship among them
6. Continuing grouping themes to reach broad categories and have a better picture of the likely relationships

B. *Attribute listing.* It is a tool to ensure that the various aspects of a strategic forecasting problem have been examined, which lead to better forecasting solutions. This creative activity is useful when dealing with new product launches and works as follows: list the attributes of the product in question vis-à-vis competing products, consider the value of each attribute, and alter attributes to enhance its value. This technique is a precursor of value engineering, and its significance is in taking an existing product, service, system, or process; breaking it down to various sizes of parts; finding different ways of creating each part; and then reassembling them to come up with new forms.

C. *Causal loop diagrams.* They are a type of systems thinking tool that helps the SDF team understand how complex business operations systems work and visualize how interrelated variables affect one another. They consist of a set of nodes representing the variables in a system and of arrows connecting them and showing how one variable affects another in a reinforcing or balancing manner. They are also known as system diagrams and are powerful tools to model and show how a change in one factor can have impacts elsewhere in the system and to flesh out the long-term impacts of changes (Sterman 2000).

D. *Five whys method.* It is a systematic problem-solving method for investigating market and business relationships and identifying sources and causes. In strategic forecasting, this method begins with the desired result or expected scenario, considers possible causes, and questions the answer given five times in a row. The steps involved in this simple but effective technique are as follows (Serrat 2009):

1. Developing the decision problem jointly with the client and senior management
2. Asking forecast stakeholders the first why they are facing this problem or why they are pursuing a project and taking down and confirming the answers
3. Asking why four more times for each answer, taking down the answers, and confirming the plausible answers and explanations
4. Looking for systematic causes, sharing with project stakeholders the logic of the process, and agreeing on the most likely causes
5. Obtaining buy-in on the likely causes and the logic of the analysis

E. *Market research.* It is a subset of marketing research, which is a methodical, objective, and focused collection, tabulation, validation, and analysis of primary and secondary market research data related to marketing a product or service in a target market. The purpose of market research is to enhance the company's understanding and generate insights about its markets, define opportunities, identify problems, and generate marketing actions. The results of qualitative and quantitative market research methods are used in SDF because of their value in identifying markets and estimating their sizes. They are useful in discovering their characteristic needs for certain products or services and identifying desired attributes for each market segment and the uses and effectiveness of distribution channels. Market research also reveals the nature of tastes, preferences, motivations, and buying habits of targeted customers; the customers' willingness to pay for different product or service attributes and price elasticities; and the company's promotional spending requirements in light of competitor promotional strategies.

Market research results drive marketing strategy, which, in turn, impacts SDF solutions as well as project implementation and forecast realization. Therefore, distilled market research results provide a reference or benchmark for the SDF forecast solution, help in gauging and influencing the selection of support levels for a particular SDF solution, and are used in developing competitive response models and plans. Notice, however, that SDF solutions are also used as sanity checks and reference forecasts for projections coming from market research studies.

F. *Reality and sanity checks.* In the field of SDF, reality checks refer to the procedure of reconciling the SDF solution with the current reality of business operations. It is the process of examining closely inputs to and outputs of SDF solutions and eliminating unrealistic elements. For that, SDF team members clarify information and validate data, processes, and scenarios to ensure conformity with reality and correct misconceptions. Sanity checks, on the other hand, require validating the logic and rationale of assumptions and the positions of project team members. They also require checking the implications for the forecast solution on the company's operations and resources and determining if they are feasible and consistent with the company's current capability situation.

Sanity and reality checks in strategic forecasting are another layer of the verification and validation tests, are used to justify the forecasting solution, and ensure that there are no careless errors in the forecast solution components. They confirm that the underlying assumptions are reasonable and acceptable, the computer models produce reasonable outcomes, and the forecast results behave according to general expectations. Sanity checks also ensure that the generated SDF solution falls within ranges the project team considers achievable, it is fairly stable, and the scenarios and models used do not display extreme sensitivity to their driver values. Sanity and reality checks also ensure that the impacts of the forecast on human, financial, and physical resources are sensible and that the forecast risks, sources, and causes are well understood and within tolerance levels, and the company has the ability to manage them.

6.4 SDF PLAN AND PROCESSES

The SDF plan is the blueprint of the forecasting project management plan put together by the team members responsible for the project; SDF processes refer to the sequence of activities required to execute that plan. Following are some techniques and tools used to create effective SDF plans and processes, while others are discussed in Chapter 7 or deferred to specific application chapters.

A. *Appreciative enquiry.* It is a preventive technique whose basic premise is to develop SDF processes and solutions around what works, rather than trying to fix what does not. Appreciative enquiry works by focusing the attention of the SDF team on its most positive core values, which are crucial for its success in applying the team members' collective knowledge, strengths, capabilities, and potential. Appreciative inquiry is about seeing what others may not see and the idea is to identify and describe the forecasting problem to be solved using five steps (Brittain 1998):

1. Defining the forecasting problem so as to be able to analyze the situation
2. Discovering what happened in past strategic projects and what was done right
3. Visualizing future-state scenarios by thinking about and building strengths on the positives and eliminating the faults found in the SWOT analysis
4. Designing and building models, systems, and processes that support the envisioned approach to an SDF solution
5. Delivering a proposed solution with sufficient detail to the project team to make efficient decisions with positive results

B. *Flowcharts*. They are visual, schematic representations used to outline the structure of a forecasting problem, the sequence of steps and activities, and the paths to be followed in a process or plan to create an SDF solution. They are easy-to-understand diagrams and useful in defining the nature of a problem, delineating its boundaries, and designing plans and procedures to get to the solution. They are helpful in identifying and describing the elements of SDF process and understanding the end result and the nature of the deliverables. Flowcharts communicate effectively to clients and other project stakeholders how the SDF process puzzle pieces fit together and help develop a common stakeholder understanding of the SDF plan, its rationale, and how it is implemented. They also help to determine required informational inputs and their sources and identify the most qualified team members for the project.

In addition to helping determine the course of project activities, flowcharts are used to show how responsibilities are assigned to appropriate stakeholders, identify dependencies and potential bottlenecks, uncover redundant steps, and discover areas of needed improvement. By documenting in easy-to-understand terms complex SDF methods and techniques, they assist in managing the process effectively. How is it done? It is done by seeing the entire picture and focusing on the flowchart steps without being crashed by the complexity of the totality of strategic project decision processes. There are several types of flowcharts with various degrees of complexity, and strategic forecasters are familiar with their properties and uses in different situations.

C. *McKinsey 7S model*. Applied to SDF, the McKinsey 7S (strategy, structure, systems, staff, skills, style, and shared values) model is used to analyze and ensure that SDF processes work as designed and the group is well positioned to achieve its goals and objectives. It is used to determine how to implement SDF strategies, align its functions with other corporate planning functions, evaluate how to enhance the performance of SDF team members, and assess how revising processes and resources would affect its overall performance (www.vectorstudy.com/management_theories/7S_framework.htm).

The McKinsey 7S model is used to identify gaps in capabilities, competencies, and resources, and determine misalignments of goals and objectives, inconsistencies of processes, and discrepancies in the evaluation of strengths and weaknesses. It also makes possible a better assessment of the current state versus the desired state and how the interrelatedness of the seven factors may be used to align them so as to increase the chances of achieving the desired state. In addition, the 7S model is used to guide a comprehensive competitive analysis evaluation along the seven elements. Notice, however, that the McKinsey 7S model's elements are interdependent and reinforce each other's impacts.

D. *Scenario development, analysis, and planning*. Scenarios in the SDF paradigm are schemes, concepts, sketches, outlines or plans of the sequence of events, their timing, and what happens when decisions are made and begin to get implemented. They are models of assumed or expected sequence of decisions, inputs, actions, reactions, and events constructed for the purpose of capturing their effect on a target variable. They show how a hypothesized chain of events leads to future states in a structured way of seeing past the current state, creating descriptions of future states, and describing how they unfold. They are also used to explore wild card possibilities and black swans and quantify their impacts. In defining possible futures, scenarios help SDF team members to understand the time-ordered events and causation from current to the end states and to create strategies and options to deal with uncertainties. They build "flight simulators" used to create learning and sound project implementation strategies by articulating clearly the events and processes generating the future states, simulating them, and answering critical questions.

Good scenarios are stories of plausible, divergent but deterministic futures, and they capture project team biases and different points of view. They help to see vividly what drives the business and what is required to achieve the objectives of a project, stimulate discussion, question assumptions and the model of business operations, and increase learning. Scenario building is a structured approach to predict the future by assuming a series of alternative possibilities instead of forecasting the future on the basis of extrapolated historical or analog data alone. Scenario planning, also called scenario thinking or scenario analysis, is a corporate planning method employed in making strategic decisions and long-term plans. It is an adaptation of the methods used by military intelligence and strategic planning that relies on model simulation tools and controllable factors to manage the future.

Key takeaway: *The most valuable application of scenario development, analysis, and planning is that they can be used as a flight simulator which*

1. *Enables clients and senior management to visualize how to get to the future state and what is required to achieve it.*
2. *Helps create the conditions, commitment, and support levels that have to be in place to achieve the projected future state.*
3. *Allows the creation of the desired future state by altering drivers and inputs in model form today before decisions are made and are implemented.*

When other methods of forecasting are not appropriate, scenario development and planning is a major tool used to solve strategic decision problems and create future-state forecasts. It is a method to learn about the future by understanding the impact of the most uncertain and important forces driving the business. Scenario development is based on the belief that strategic decision forecasters are not at the mercy of fate; instead, they use this method of envisioning future states to incorporate them into scenario models to be simulated. Common steps involved in scenario development include the following activities:

1. Starting with an accurate description of the current state of the business and defining the internal environment, the context of the decision to be made, and its major goals and objectives
2. Defining the scope of each scenario and brainstorming on the megatrends and the macro external environment driving forces surrounding the project
3. Developing major assumptions about timing, causality, and strength of relationships, gathering information, and evaluating industry and market trends and structural changes
4. Engaging an independent facilitator to screen and provide suggestions on the driving forces and events and ensure objectivity and reasonableness of model building
5. Determining the extent to which scenario driving forces can be predetermined, projected, and fixed and how steady is their influence on the projected state
6. Creating distinct and convincing stories based on the effect of driving forces and critical uncertainties and eliminating similar narratives
7. Weaving hypotheses and plots to the stories to fit the identified events and forces and creating three or four plausible scenarios
8. Performing sanity checks on the assumptions, actions, reactions, events and their timing, and on the processes that generate the future states and their logic
9. Developing decision trees and influence diagrams based on modeling, simulations, and the Delphi technique and performing trend analysis
10. Focusing on historical analysis of discontinuities of trends, cross-impact analysis, and analog experiences to identify unexpected driving factors and uncontrollable events
11. Assessing the financial, human resource, competitive, operational, and strategic implications for each scenario and evaluating differences with project or decision expectations
12. Comparing the scenario-generated future states against the status quo projection and estimating the contribution of each driving factor to the additional value created by the project

13. Creating a system of early warning indicators to monitor each scenario's performance as they unfold through time and adapting the scenario to fit circumstances and approximate reality
14. Using the decision selection matrix to identify, evaluate, and rate the economic value of each future-state scenario and the likelihood of achieving them

E. *Swim lane charts.* They are a type of process flow diagrams where processes and decisions are grouped and shown visually in horizontal, parallel lanes. The lanes are assigned to the participating groups or functional roles. Drawing swim lane charts creates a visual story of the SDF processes showing the key steps involved in developing solutions. Process flows that change lanes signify handoffs, which are points of possible communication and coordination problems. This makes swim lane charts crucial in creating, auditing, communicating, and managing the forecasting function end to end.

Useful applications of swim lane charts include creating processes that are consistent with the corporate culture and the SDF team's mindset and mapping process flows showing the activities and steps involved in them. Swim lane charts are valuable in

1. Describing the roles and responsibilities that go with different activities and processes so they can be seen clearly by all stakeholders.
2. Organizing, designing, and optimizing complex processes across functional silos so as to increase communication, collaboration, and coordination.
3. Identifying and eliminating current process flaws, non-value-adding activities, steps, dependencies, and handoffs.
4. Simplifying workflows, reducing time required to develop SDF solutions, and enhancing the team's productivity.
5. Determining the appropriate methods and tools to use to increase efficiency and expand the SDF team's capabilities.
6. Reengineering SDF processes to consolidate functions, reassign roles and responsibilities, and reconsider how value is created.
7. Obtaining project team agreement, client buy-in, and senior management support for the SDF team plan and its implementation.

F. *Benchmarking.* It is the process of assessing the SDF team's internal processes and then identifying, understanding, and adapting outstanding practices from other appropriately selected organizations considered to be best in class. It is a systematic way to compare the SDF team's processes and performance in order to create target metrics for improving processes to levels comparable to industry bests or best practices in other industries. Benchmarking enables the SDF team manager to determine how well the SDF team members are doing their jobs relative to others considered best in this area and how to enhance their communications, collaboration, skills and competencies, and SDF processes overall.

The major types of SDF benchmarking are the functional and generic approaches. The former type benchmarks strategic forecasting processes and practices within the industry, while the latter compares performance with similar groups in other industries. For benchmarking to produce significant positive results in strategic forecasting, deep senior management commitment is needed. Such support is needed because benchmarking is costly and should be done well a few times going through the following activities:

1. Defining the scope of the effort and choosing an outside expert to get access to companies to benchmark, identify the specific areas to benchmark, and determine the measurement methods and variables
2. Selecting the data collection approach, getting the needed data, and verifying the collected data by comparing them to the SDF team's data, analyzing differences, and developing insights

3. Communicating the findings and implications of the reporting structure with recommendations of areas to pursue improvements and show expected benefits
4. Setting some initial goals and targets to reach at different time intervals and checking their feasibility vis-à-vis cost and time constraints
5. Developing and implementing new SDF approaches, processes, and techniques and tools in areas in need of improvement
6. Monitoring progress against the initial set of performance targets and plan for ongoing performance improvements once benchmarks are adopted

6.5 SDF METHODS, SYSTEMS, AND IMPLEMENTS

A number of suitable methods, tools, and systems are used by SDF team members to come up with sound forecast solutions, which form the basis of two key decision support elements: the project feasibility study and the project business case, which include some common implements.

A. *Feasibility studies.* These are an effective tool to achieve early on several objectives, such as clarifying client and senior management needs, goals, and objectives and ensuring alignment with corporate-wide strategic goals and objectives. They help in developing standards to use in evaluating the findings and identify criteria for go/no-go decisions; screening ideas, proposals, and alternatives; and eliminating those that do not hold value creation promise. Feasibility studies precede business cases and serve as an introduction to a major decision, proposal, or investment opportunity. Their purpose is to present an interim assessment to determine whether a project is doable and makes economic sense, given the corporate strategy and resource constraints. As with business cases, the contribution of the SDF team is in assessing the operating environment, determining fit, developing preliminary projections of key variables, and assessing potential risks and their impact on the project.

B. *Business cases.* They are part of a decision or project's mandate, are produced once a project is well under way, and address the business need the project seeks to meet. A business case evaluates the project rationale, the methodology used to project the future state, the expected benefits and costs of the project, the alternatives considered, and the expected project risks and how they are to be managed. The findings of the business cases are based on the forecast solution developed for a specific decision or project needs. As such, they include valuable information, benchmarks, previous studies, and intelligence and insights created earlier, which are documented in a clear and concise manner in the business cases. Business case documents are used to communicate the value of projects to decision makers and obtain corporate approvals to proceed with project implementation. A detailed discussion of the role the SDF team plays in the development of business cases is found in Chapter 19.

C. *Mapping and modeling.* It is a practical schematic system dynamics, process-oriented technique to help visualize the whole industry, see the big picture, and model company operations using the following:

1. Stock and flow diagrams that provide insight into key business processes and show causal loops and illustrate feedback effects
2. A schedule of trigger events showing the timing, the source and nature of the event, and its potential impact on the projected future states
3. Logical relationships, mathematical equations, statistical descriptions, and models of business processes
4. Scenarios of the processes represented in the models developed, hypotheses testing, and simulations
5. Sensitivity analysis to determine key influences, leverage points, and optimum conditions

D. *Early warning systems*. They are systems of steps, procedures, and interactive project components created to identify emerging threats or problems, provide timely and reliable information, and warn off impending risks being realized. EWSs enable clients and forecast owners to take action to avoid or minimize business risk by implementing appropriate forecast realization plans. In strategic forecasting, EWSs are an integral part of risk identification and management. They use performance measures or key indicators related to preparedness to reduce risk and reinforce other activities such as identification of sources and causes of risks, scenario simulations, and coordination of responses. The approach to EWS development and use is determined by the corporate culture, strategic priorities, company resources and capabilities, as well as client or senior management needs. Conceptually, they work as follows: The precursors to uncontrollable events are monitored continuously, and the data and information are analyzed and distilled into intelligence, which is used to generate a forecast of risk factors. In case a forecast for a significant event occurrence is produced, a warning or alert is issued.

The key benefits of integrating EWS in strategic forecasting are obvious, which are as follows:

1. Creating actionable intelligence in matrix form consisting of a column of events or developments and corresponding indicator thresholds and actions recommended to offset them
2. Helping clients and senior management to discern what to look for in the streams of data and information flowing in continuously
3. Having preset triggers of actions to manage risks as they appear on the horizon and achieve forecast realization
4. Reducing time to response while at the same time helping the company deal with strategic surprises and prepare more effective competitive responses

E. *Forecast conditioners*. These are factors that influence the development of forecasts and reference points, which are used as gauges to adjust forecasts in different stages of the project process. They are based on analyses, past experiences, seasoned judgments, and tools proven useful, and include items such as the following:

1. The order of entry matrix used to gauge market share relative to other entrants in a new product introduction
2. Industry standards, averages, and benchmarks, which serve as guides in judging the reasonableness of a forecast solution and as sanity checks
3. Share of voice information, that is, spending levels of advertising dollars on different types of media with different impacts
4. The product attribute matrix, which shows properties of the company's product versus those of competitors
5. Best practices from different industries and geographies that are adapted in the development and used in the selection of scenarios
6. Relevant earlier feasibility studies and business cases that could provide reliable guidance in scenario selection and risk assessment
7. Competitor experiences, which can provide objective reference points in judging the reasonableness of a forecast when differences with them are accounted for
8. Forecast triangulation, which is the art of integrating forecasts from different sources, methodologies, and models; assigning proper weights to each; and obtaining a more reliable forecast
9. Expert opinions, which are independent evaluations bearing on a forecast solution used to fine-tune a long-term forecast for a strategic project
10. Current resource constraints on human resources, financial muscle, and management capabilities considered in reallocating support for forecast realization

11. The levels of senior management support for the project and resources allocated to the evaluation and implementation of the decision to the end
12. The preparedness, competencies, and capabilities of the project implementation team and, when the project creates a new entity, of the new entity management team

F. *Qualitative and quantitative forecasting methods.* Qualitative forecasting techniques are used in the absence of historical data and are based on opinion and judgment of experts. When historical data and statistical forecasts are available, they are used to incorporate valuable insights into the future and triangulate those forecasts. Qualitative methods used in developing SDF solutions include the following:

1. Surveys of customer or user expectations and intentions
2. Executive opinions and senior management judgment
3. The Delphi consensus development forecast method
4. Development of distinct and plausible future-state scenarios
5. Sales force composites, which are based on polling individuals close to customers
6. Subjective experiences, feelings, and ideas of key strategic decision stakeholders
7. Product lifecycle analogy used in cases of weak similarity to other products
8. The cross-impact analysis method based on actual or potential relationships between events and situations

Quantitative forecasting methods are based on statistical models and used when sufficient and accurate historical and cross-sectional data are available. In the case of causal models, due to the length of the forecast horizon, projections of the driving variables do not exist that far out and statistical forecasts cannot be developed to make strategic decisions. They may, however, be used to project the status quo of business operations to get a sense of the difference with alternative future states that may be created by a project. The spectrum of quantitative techniques used by SDF team members includes the following methods:

1. Variants of autoregressive or moving average models
2. Exponential smoothing method weighing recent experiences more heavily
3. Time-series analysis, also known as autoregressive integrated moving average (ARIMA) or Box–Jenkins methods
4. Diffusion models and trend extrapolations of the linear and nonlinear types
5. Variants of curve-fitting methods based on the degree of similarity to other products or services
6. Operations research models that are abstractions of real business processes used to make future projections
7. Quantitative research models and techniques used to identify, assess, and predict buyer attitudes and behaviors
8. Causal demand forecasting models based on regression analysis
9. Management system dynamics methods used in simulating operational processes and forecast based on feedback loops and information systems
10. Input–output matrix, an accounting framework of the industry that describes intra- and intermarket flows and demands

Notice that regardless of the kind of quantitative models used, forecasts for outer years are always supplemented with qualitative model forecasts for outer years.

G. *Six Sigma forecasting.* Six Sigma is a structured, disciplined approach adapted in the SDF project management approach that gives SDF team members techniques and tools to reduce deficiencies, improve processes, and create new processes as needed. It also provides the practices to set challenging SDF team goals and objectives, evaluate processes and practices, and analyze results in order to reduce faults in them. SDF teams apply Six Sigma principles in managing the forecasting function and as a capability measure approach to improve the efficiency of SDF processes.

Adaption of Six Sigma principles helps align SDF processes with corporate strategy, client needs, and objective assessment requirements by methodically eliminating frozen mindsets and reducing problems embedded in current processes and practices. Six Sigma rules are put into effect on the belief that if one controls the inputs, methods, tools, and processes, one is able to control the quality of SDF solutions. Because of that, they require that issues and deficiencies are addressed at the root cause level so that restructuring of SDF processes and rework of forecast solutions are minimized or avoided. The requirements to ensure forecasting excellence require the following of SDF team participation:

1. Clarity of purpose and strategic intent before anything else
2. Sense of commitment to excellence that permeates the SDF team and other project team members
3. Tenacity in the rigorous use of appropriate tools to assess and interpret data and information
4. Dedication to continuous innovation and improvement at all SDF team's organizational aspects and processes
5. Senior management support and participation of key stakeholders in developing SDF solutions and implementing SDF team solutions and recommendations
6. Tolerance for questioning long-held beliefs and their rationale by key project stakeholders

Six Sigma forecast quality elements, principles, and performance factors are further discussed in Chapter 20 dealing with best-in-class SDF organizations.

H. *Stepladder technique*. This technique is used by SDF teams to help clients ensure that all forecast stakeholders participate in the process and their perspectives are heard before they are influenced by others. It is similar to the Delphi technique, but it is used with small groups and it is quicker. The steps used in the stepladder technique to prevent groupthink are as follows (Rogelberg et al., 1992):

1. Sharing the project or decision issue or problem with project stakeholders before assembling as a team
2. Allowing time for people to form their own opinions and bringing the core group together—consisting of the forecaster and the client—to discuss the decision, project issues, and problems
3. Bringing in a project team member who presents his/her ideas prior to hearing what the core team has discussed and then all three talk about their ideas and opinions
4. Repeating the process until all project team members are included and setting aside enough time for discussion of the ideas of new participants
5. Evaluating all ideas and opinions presented and deciding which ones are the most valuable in the creation of the SDF solution

I. *System dynamics*. It is a method to study a business situation in order to better understand and manage the complex feedback systems of a strategic project and the decisions that need to be made. It is a method to think systematically and see the big picture in its totality and how the main parts are linked together. SDF teams use the system dynamics method to model and describe the behavior of complex business operations and specify explicitly the major influencing factors and the main aspects of the system's behavior. As importantly, it is a way to manage the development of SDF solutions that can handle complexity arising from interdependent components of large models, which involve hypotheses and assumptions, data and soft information, opinions and management beliefs, nonlinear relationships, and multiple feedback loops (Forrester 1991).

SDF team members use the system dynamics method to hypothesize and assess market relationships and feedback mechanisms and to understand the structure of the model describing the company operations and the behavior it can generate. In doing so, they can diagnose causes of problems, learn from causal loop diagrams, see more clearly causes and effects, validate ideas, perform sanity checks, and understand how earlier decisions and changes in input variables work themselves out. The uses of system dynamics in SDF include applications of the following practices:

1. Conceptualizing and modeling strategic projects or decisions and managing the development of SDF solutions
2. Using system dynamics models as "practice fields" by showing the effects of different inputs, decisions, and changes in specified relationships and the end results
3. Building models that can simulate company operations, the environment of a decision, or the components of a project, produce insights, create scenarios, and conduct forecast sanity checks
4. Simulating to understand, depict, and analyze market forces; identifying causes, interventions, and support levels; capturing their impacts; and testing different options, scenarios, and trade-offs
5. Detecting industry structure changes, assessing key sensitivities, and helping set up defenses and contingencies in the future-state realization plan
6. Creating "what-if" scenarios to determine the impacts of key drivers, events, and leverage factors to improve forecast achievability or to create the future state today
7. Mitigating forecast risks and project uncertainty through process changes, initiatives, and other controllable factors included in the system
8. Using system dynamics model simulations as a forecasting tool in the event other methods are not appropriate

System dynamics modeling starts by identifying the problem or end objective, engaging the SDF support subteams, developing hypotheses that explain how the system works, and building computer simulation models. The models are then tested to ascertain behavior consistent with reality and assess the impact of different inputs and relationship parameters. The formal steps of the system dynamics process are as follows:

1. *Conceptualization*. This is the step where the problem and purpose of the model are defined, key variables are identified, and feedback loops describing behavior are diagramed.
2. *Formulation*. Here, feedback diagrams are converted to rates and levels of variables and equations, and relationship parameter values are specified.
3. *Testing*. This is where simulation and testing of assumptions, hypotheses, model behavior, and sensitivity to perturbations takes place.
4. *Implementation*. At this stage, model responses to changes and different scenarios are tested, usable insights are distilled, and future-state projections are made.

J. *Technology and technological forecasting*. Technology forecasting is about predicting the attributes of products, characteristics of machines, and applications of processes and uses of services and techniques. Technological forecasting on the other hand is forecasting an industry's future technology trends and focuses on predicting future technological capabilities, attributes, and parameters. SDF solutions often require forecasts of a technical nature and utilize both forecasting types.

Technology forecasting involves an assessment of the company's technological position and that of competitors in order to generate forecasts of new technologies on the horizon and their effects on customer demand. The main elements of the technology forecasting process include the following:

1. Evaluating current and planned company technologies and strategy
2. Gathering, evaluating, distilling, and disseminating relevant technology information to the project team

3. Following megatrends and subtrends and monitoring market needs, customer tastes, and preferences
4. Determining the company's current R&D directions and technological trends
5. Assessing various constraints on company and competitor technology evolution
6. Identifying risks associated with competitor new technologies in the forecast horizon and potential breakthroughs
7. Researching expert opinions and conclusions and generating usable insights for the project team
8. Adapting technology forecasting models, testing, and simulating scenarios, which fit the project specifics
9. Selecting a technology forecast consistent with the results of other analyses and evaluations and triangulating forecasts

Technology forecasting models and their uses are discussed in Section 12.3 of this book, which deals with technology licensing projects.

7 SDF Assessment and Implementation Techniques

The assessment and implementation techniques and tools used in strategic decision forecasting (SDF) are systematic, effective, and practical methods. They can be structures to evaluate a situation and systematic processes to investigate issues. They can also be tools to help visualize the operating environment and company operations, or to provide a structured way to organize ideas and information. In some instances, they are procedures used to enhance the SDF team members' understanding of the nature and the causes of problems or they can be efficient ways to assess a problem and come up with alternative solutions. Often, these SDF techniques and tools are aids used to make sense of huge volumes of data and information, critical thinking methods, and sound analytical methods to evaluate information and create intelligence, insights, and solutions.

In this chapter, we are concerned with the various techniques and tools used by SDF teams in five major areas: forecasting function project management; selection of solutions, options, and recommendations; performance measures and incentives; future-state realization (FSR) planning; and institutionalizing learning and knowledge. The discussion of this chapter centers around techniques and tools used in assessing and implementing SDF solutions, with particular emphasis on the following key takeaway points:

1. Manage the forecasting function competently because it is crucial to project success
2. Focus on creating sound SDF solutions; they are necessary. But because they are not sufficient to influence decision makers, concentrate on effective presentation of forecasts, options, and recommendations
3. Use performance measures wisely to judge forecasts, the effectiveness of the SDF team, and the contribution of the project to corporate value creation
4. Ensure that forecasts indeed materialize by making the creation of the FSR plan an integral part of the SDF process
5. Increase the success rate of strategic projects through augmenting learning and knowledge and disseminating to those in need to know

7.1 FORECASTING FUNCTION PROJECT MANAGEMENT

Project management is the discipline of planning, organizing, and managing activities, timelines, and resources to achieve effectively the project goals and objectives. The two challenges of project management are to achieve all of the project goals and objectives under scope, time, and budget project constraints and optimize the allocation and integration of resources necessary to meet predefined objectives.

A. *Strategic forecasting project management*. It is the art, discipline, and process of managing the scope of an SDF project, the expectations of clients and senior management, the human resources assigned, the costs involved, and the schedule of developing reliable forecast solutions and recommendations. The means used to manage forecasting projects is the SDF project plan, which is made up of the following elements:

1. The definition of the project or the decision to be made, their purpose, and major goals and objectives to be achieved
2. The scope of the SDF team support and its participation goals and objectives

3. The specification of deliverables, their format, and timelines associated with them
4. An outline of the processes and procedures to be used in performing the work in order to provide a common understanding for the project team
5. Agreement on the assignment of clearly defined roles and responsibilities of project team participants
6. Specification of major milestones, estimated costs, and resources to achieve them
7. Key success criteria and performance measures for each major phase of the SDF process
8. Briefings, updates, and reports to clients and senior management on progress to date, accomplishments, and unresolved issues that need their intervention

Swim lane charts and the tools mentioned in this section are an integral part of the forecasting project management function and are extensively used because of the clarity they provide on the sequence of process steps, which party is responsible for what activity, and the timelines involved.

B. *Forecast RACI model.* RACI stands for responsible, accountable, consulted, and informed, and the RACI model is a useful forecast project management tool. It is also called a responsibility assignment matrix that helps to define, communicate, discuss roles and responsibilities, and obtain agreement among forecast stakeholders. In strategic forecasting, the forecaster is responsible for developing the forecast solution and the client is accountable for the completion of the work and signing off on that solution. Senior management and outside experts are consulted and their opinions and inputs taken into account. Other stakeholders are informed about progress and the forecast solution, but do not have direct inputs into the process (Brennan 2009).

The way the RACI model works is based on a sequence of steps SDF team members follow in assigning ownership. These steps are as follows:

1. Defining the SDF process and identifying the activities to be performed on the first column and all stakeholders across the top of the matrix
2. Determining the role of each stakeholder in each activity and assigning appropriate designations when filling out the RACI matrix
3. Ensuring each activity has one R (for responsible) and resolving overlaps of responsibility when there is more than one R
4. Reviewing the responsibility assignment matrix to ensure that there are no gaps in R and identifying stakeholders to take on the responsibility for that activity
5. Sharing the matrix with project stakeholders, explaining the logic behind it, and obtaining agreement prior to implementing the SDF plan

C. *PERT and Gantt charts.* Project evaluation and review technique (PERT) and Gantt charts are visualization tools to define a forecasting project management plan and to show the activities and steps involved in developing a solution. The PERT chart, also known as a precedence diagram or network chart, shows all the tasks and dependencies between them and enables SDF team members to organize, schedule, and coordinate tasks and project manage large, complex forecasting projects. A PERT chart is similar to the critical path method and is visual illustration of the total strategic forecasting process that consists of numbered circles that indicate events or milestones, directional lines showing sequential and dependent tasks, and diverging directional lines showing tasks that take place simultaneously (Project Management Institute 2003).

The Gantt chart is another useful SDF project management tool that works particularly well when dealing with straightforward forecasting projects. This chart lists sequentially the activities required to develop the SDF solution and the time needed to complete them, which are represented as horizontal bars on the *X–Y* chart. The Gantt chart emphasizes the time needed to complete activities and tasks by the length of the horizontal bars corresponding to each activity or task. In addition to helping manage SDF projects, PERT and Gantt charts help SDF team members to communicate to clients, senior managers, and other stakeholders the activities they are involved in and the time required to complete each one. They also help to assess the status of SDF projects quickly and pinpoint possible bottlenecks.

D. *Stage–gate methodology.* It is a rigorous opportunity evaluation and screening process particularly well suited to determine if a new product, venture, or project has the potential to become a profitable undertaking. It is a well-structured method that includes some Six Sigma principles and provides a checklist to ensure that best practices and essential activities are integrated in SDF processes. Stages are the parts of a project where the various external environment assessments, R&D, market research, competitive analysis, industry analysis, information gathering and evaluations, and risk and alternatives investigations take place depending on the stage in the process. Gates are key decision points where a set of criteria is used to judge the degree to which they have been fulfilled and decide if it is advisable for the project to move to the next stage (Edgett and Kleinschmidt 2002).

SDF solutions are an integral part of the stage–gate methodology where different SDF team activities are performed at different stage gates using the latest information, evaluations, and results of analyses. The four major versions of SDF solutions developed under the stage–gate methodology are the following:

1. An initial forecast solution based on early product concepts, market research, and preliminary environmental analysis and a preliminary strategic project definition
2. A better grounded forecast solution based on more complete product and project definitions; an independent, critical, and objective assessment of the business; and the results of the feasibility study
3. A baseline solution built on firm assumptions, forecasting due diligence, sanity checks, tested scenarios, and risk assessment used to make a decision
4. A postimplementation updated forecast based on monitoring actual forecast performance, analysis of current environmental conditions, and the results of competitive response plans put into effect

Notice that SDF solutions are one of the inputs for the decision to proceed to the next stage, others being technical feasibility, production capacity, and timing of project execution. The particulars of the SDF solution for application in the stage–gate methodology are discussed in detail in Chapter 9, which deals with new product development and introduction.

7.2 SDF SOLUTIONS, OPTIONS, AND RECOMMENDATIONS

In many projects, the primary method of identifying and defining future states of the business is scenario development using system dynamics model simulations and testing of hypotheses. These methods were discussed in Chapter 6 and the following are SDF techniques and tools used in selecting forecasts and providing options and recommendations to decision makers.

A. *Force field analysis.* It is a tool used in SDF to identify and investigate forces in support of and against a scenario or forecast solution, which enables the company to strengthen the supporting forces and to offset or reduce the impact of the opposing forces. By listing the factors for and against the future-state forecast, a thorough evaluation of each force and their interaction determines the sources of risks and potential impacts as well as the likelihood of the forecast being realized. In doing so, this technique focuses the project team on dealing with the crucial forces to ensure decision and project success (Lewin 1951).

B. *Decision tree analysis.* It is the evaluation of decision trees, which are a tool for creating and choosing between alternative initiatives, actions, plans, scenarios, and options through time. Decision trees provide a structure to the SDF team to evaluate those scenarios and options and the underlying drivers at the level of detail needed. They also help to investigate the possible outcomes and create a good perspective of the risks and rewards associated with each alternative. As such, they are a useful tool for evaluating the probabilities of occurrence and the values of uncertain outcomes; that is, the value created and the risks associated with each alternative scenario considered (www.mindtools.com/dectree.html).

C. *Forecast story telling.* It is a presentation tool for delivering SDF solutions in words and vivid images with emphasis on the key points. It is a useful communication tool to create interest in the SDF solution that results in a more efficient acceptance and trust of the forecast state. Instead of walking clients and senior management through various mundane forecast processes and details, SDF team members use this approach to communicate in visual and vivid terms the solution, what went into developing it, and how it is to be used. Namely, they explain to clients and senior management through the use of stories the methodologies, the high-level processes, the scenarios employed, the sanity checks performed, and the forecast risk evaluations conducted. Similarly, they use stories to describe the forecast solution and explain the logic behind the selection of forecasts, scenarios, their implications, and the set of recommendations flowing out of the SDF solution.

D. *Grid analysis.* It is a useful tool to reach agreement and select the best among several SDF solutions where there is no clearly superior option and where each forecast has different strengths and weaknesses. It is also known as decision matrix analysis and works as follows (www.mindtools.com/dectree.html):

1. In the first row of the matrix are listed the forecast attributes that are considered important.
2. In the second row, weights from 1 to 10 are assigned to each attribute.
3. In column 1, starting with row 3, the different forecast options are entered.
4. For each forecast option, each of the attributes is scored on a scale of 1–10.
5. Across each forecast option, the score of its attributes is multiplied by the weights assigned and adding them across.
6. The forecast option with the highest weighted score is then selected.

The grid analysis tool appears easy to use, but experience shows that picking the right forecast attributes and their scoring is a complex process. This is because there are several forecast attributes that may be relevant in each case such as reliability of the forecast, usable information generated, timeliness of the forecast, and forecast accuracy. Other forecast attributes considered are the quality of model and scenario formulation, stability of the forecast to small changes in its drivers, and reasonableness of the projections. These are forecast qualities related to passing of sanity checks, ability to predict extreme outcomes, and forecast values outside past experiences. An equally important set of attributes consists of the importance of the forecast to the client and users; triangulation of forecasts from different approaches, sources, and models; easy-to-understand forecast logic and processes; and instructions on how to use forecasts. Additionally, the consistency of forecasts with industry benchmarks, averages, analogs, and experiences, the cost and effort of producing the forecast, and whether the forecast incorporates independent external expert review and feedback are important determinants.

E. *T-charts.* T-charts are a tool used by SDF team members to compare and contrast information, opinions, and results of alternative SDF solutions by placing them in two separate columns. The purpose of the two columns is to enable forecast stakeholders to see clearly forecast pluses and minuses, compare different ideas, and draw better conclusions. T-charts are also useful in displaying key features of distinct scenarios and their pros and cons, assessing potential value and risk, and selecting the scenario that makes more sense given the results of external environment, industry, and strengths, weaknesses, opportunities, and threats (SWOT) analyses.

F. *Risk assessment and management.* In strategic projects, risk refers to uncertainty of outcome and the threat that some event or action will affect adversely the company's ability to achieve its decision objectives. Risk elements are sets of interconnected conditions, events, circumstances, and actions resulting in adverse consequences on the decision or project objectives. Hence, risk management is an ongoing process of identification of forecast risk elements; assessment of the sources, causes, and impacts; and the prioritization of risks. It is a well-coordinated and communicated effort to minimize, monitor, and control the likelihood of occurrence and impact of these elements. Forecast risk management is an integral part of the SDF process that provides valuable inputs, but it is not as broad as corporate risk management, which includes managing the acceptance, sharing, shifting, or mitigating strategic, operational, and financial risks.

Forecast risk determines the value created by a strategic project decision and the reward from it. Hence, the purpose of risk management is to create a framework to reduce business risk and attain the decision or project objectives by implementing an effective forecast risk control process. However, successful risk management is achieved when there is a balance between risk and control. That being the case, the objective of forecast risk management is to create an environment where clients and senior managers feel comfortable making decisions based on SDF solutions quickly and efficiently.

Due to its crucial nature in SDF, uncertainty and risk management are discussed at length in Chapter 14, which deals with the application of the SDF paradigm to strategic project uncertainty and risk. The SDF organizational structure and qualification requirements to successfully manage forecast risks are discussed in Chapter 20.

7.3 SDF PERFORMANCE MEASURES

In this section, we review techniques and tools used in developing and monitoring SDF solution and team member performances as well as incentives and rewards for work done well. A more detailed discussion of performance measures can be found in Section 20.5.

A. *Performance metrics.* These are indicators of how well the forecast, the SDF team, and the project are performing; they show trends and allow comparison with industry benchmarks and best practices. To be effective, the development of performance indicators requires participation of the SDF team, clients, and project team members, each party representing their perspective and interests to ensure that indicators support their needs.

The process of creating strategic forecasting indicators covers the areas of forecast performance and quality of recommendations, client satisfaction, turnaround time and meeting of commitments, cost of developing forecast solutions, and overall SDF team performance. The major steps involved in this process are the following:

1. Identifying client or senior management requirements for an SDF solution and developing the processes needed to meet those requirements
2. Describing the form of an SDF solution and its attributes in concert with client requirements
3. Creating performance indicators for SDF team performance in key phases and for the quality of results
4. Defining project goals, targets, and benchmarks for those indicators consistent with company competencies and capabilities
5. Measuring, monitoring, and reporting forecast performance indicator deviations from established targets and benchmarks
6. Incorporating selected performance indicators in the SDF team's balanced scorecard and the management dashboard

B. *Forecast performance measures*. These are indicators that measure the performance of SDF solutions. They are based on ranking performance over several dimensions, which include the following:

1. Length of forecast turnaround time and timeliness of delivery of SDF solutions
2. Accuracy of numerical projections versus reported actuals on a year-over-year basis
3. Validity of the assumptions driving the forecast that supported the strategic project or decision
4. Reasonableness and soundness of scenarios and events and estimated impacts used to generate SDF solutions
5. Number of forecast revisions and updates in handoffs over different project phases
6. Clarity of forecast processes and models, client learning and understanding of the forecast solution, and instructions for its uses
7. Reasonableness and practicality of SDF solutions, options, and recommendations
8. SDF process completeness, efficiency, and transition from stage to stage and understanding by all stakeholders

C. *SDF team performance measures*. These are indicators of SDF team member performance on several dimensions. Ordinarily, they include the following measures:

1. Accountability of SDF team members and professional conduct throughout the project
2. Client satisfaction with all aspects of the SDF solution and decision options created
3. Cost per associate participation in SDF projects
4. Communication, coordination, cooperation, and collaboration (4Cs), and support in all the phases of the process among project team members
5. Effectiveness of monitoring forecast performance, variance analysis, and forecast realization plan implementation
6. Thoroughness and efficiency of SDF team project postmortem review

D. *Project performance dashboard*. This is a visual representation tool of project performance information linked to financial and project reporting systems, which provide real-time feeds of data and events. It is used to communicate a consistent view of performance to project stakeholders and, in the absence of an early warning system (EWS), to indicate when actual performance deviates significantly from expected levels, and to alert when actions and changes are needed. The common project dashboard elements include the following indicators:

1. Actual versus planned transition timelines from one project phase to the next
2. Cumulative cost of project implementation versus projections
3. Actual value created by the project versus SDF forecast-based value
4. Project goals and objectives accomplished against major milestones
5. Deviations of key forecast variables from corresponding actual data
6. Early warning information, alerts, and timing of changes

High-quality SDF team dashboards are characterized by properties such as keeping the number of measurements and indicators small, selecting indicators that give the needed information on factors affecting client satisfaction and forecast solution performance, and using metrics, which are simple to explain to others. Additional quality attributes include shunning metrics for which data cannot be collected or the data are incomplete, not using indicators that would complicate operations and create excessive overhead, and avoiding indicators that cause stakeholders to act not in the best interest of the business.

E. *SDF team balanced scorecard*. It is a strategic planning and management system adopted by SDF teams to align activities to its vision and strategies, improve internal and external communications, and monitor performance against key goals. The SDF performance dashboard differs from its balanced scorecard, which displays performance related to the SDF organization's goals and objectives and progress made toward them over time. Balanced scorecards are snapshots in time linked to team performance, whereas dashboards provide actionable project information. The SDF balanced scorecard is similar to the Hoshin planning system, and its value is in providing the foundation for organizational, process, and measurement feedback (Kaplan and Norton 1996).

Definition of Hoshin planning system: *It is a corporate planning system using a systematic and disciplined process to align, communicate, and implement strategy by focusing on few critical objectives and deploying and monitoring few strategic initiatives that give the company a competitive advantage* (Akao 1991).

In many cases, the SDF team's balanced scorecard consists of an octagon on which each performance indicator is plotted on a scale from 1 to 10. This gives SDF team members and senior managers an instant picture of the SDF team's performance and also shows how it has evolved through time when performance measurements at different points in time are plotted. The SDF team balanced scorecard is based on indicators of performance such as those commonly shown along the axes of the octagon along with their ratings, which are as follows:

1. Number of client requests for SDF team participation in strategic projects
2. Functional links and personal relationships developed internally and externally
3. Levels of SDF 4Cs with project stakeholders
4. SDF process completeness, effectiveness, and understanding by project team members
5. SDF team member professional training and capability and competency development
6. Successful applications of postmortem lessons learned in earlier projects in realizing SDF plan goals and objectives
7. Ease of doing business with and client satisfaction of SDF team support in all phases of a project
8. SDF team productivity gains from quality improvement initiatives

F. *Results-based management*. It is a useful tool adopted by SDF teams to ensure that its processes and forecast solutions contribute to accomplishing clearly articulated decision results. Used along with postmortem analysis and the SDF team balanced scorecard, it enhances accountability, learning, and implementing changes to improve performance. The foundations of SDF results-based management are as follows:

1. Articulation of SDF team goals and objectives in a project and the strategies that bring focus to effective implementation of the SDF plan
2. Specification of the form and nature of expected deliverables, which contribute to the SDF goals and align functions, processes, and project stakeholders behind them
3. Enhanced SDF team member and client accountability based on ongoing two-way feedback to improve performance
4. Continuous monitoring and evaluation of performance and integrating lessons learned in new projects

Results-based management helps to improve communication, cooperation, and coordination of effort and incorporates elements shared with SDF clients. It is affected through an SDF team governance structure that permits (a) authority and responsibility to be aligned with resources and results; (b) definition of relevant, measurable, and monitorable results; (c) adequate processes

and incentives and rewards are part of the solution development and option selection package. They are based on performance and the awards are judged against four major criteria of SDF team performance:

1. High levels of 4Cs and support to project stakeholders
2. Effectiveness of projected scenarios in generating understanding of what it takes to make them achievable and helping to manage risks associated with them
3. Timeliness of solution, that is, minimizing time to decision and effective project management
4. Client, senior management, and other key project stakeholder satisfaction

For the SDF function and the incentives and rewards to be effective, improvement should be demonstrated through balanced scorecard measurements and forecasting function management dashboard indicators vis-à-vis expected targets and goals. Further, they require setting and resetting performance goals on a regular basis. The incentives and rewards to SDF teams which are present in well-functioning groups include the likes of the following:

1. Secondments of SDF team members to client organizations or other planning functions
2. Pay increases and annual bonuses according to performance ratings
3. Involvement in new project experiences and expanded responsibilities
4. Training in different management areas and in particular SDF functions
5. Access to external contacts through industry conferences and active participation in analyst meetings
6. Introductions to senior managers and business unit heads and exposure to upper management thinking
7. Invitations by clients and senior management to participate in strategic project decision discussions

7.5 INSTITUTIONALIZING LEARNING AND KNOWLEDGE

Institutionalizing learning and knowledge refers to information, intelligence, insights, and experiences gotten from prior project participation and current involvement of SDF team members in strategic projects, which are saved, cataloged, and made available to all who have a need to know. The techniques and tools used to achieve broad building and sharing knowledge and learning include the following:

A. *Review of empirical and case studies.* The reviews of earlier company business cases, empirical studies, and industry-sponsored research and analyses are valuable sources of data, intelligence, and insights. They are an important tool used in the SDF anchoring and forecasting due diligence areas and in building organizational learning, knowledge, and insights that are created from them. Reviews of selected case study materials are a secondary form of benchmarking, which can be well researched, are inexpensive, and yield important benefits that include the following:

1. Ability to compare descriptions of needs, decisions to be made, the context, and specific findings and experiences
2. Identification of the main issues, questions, challenges and problems, and how they were dealt with
3. Explanations of the methods, techniques, and tools used in resolving various issues and problems
4. Descriptions of the analyses, evaluations, and assessments done in different projects and the underlying rationale

5. Important findings and conclusions applicable to the current or future projects
6. Elaboration on the failures and successes and the reasons for them, the lessons learned, and how to leverage that learning

Uploading data and qualitative information on the company Intranet SDF database is the first step in affecting the sharing of knowledge obtained from reviews of earlier company business cases, empirical studies, industry-sponsored research, and analyses by the SDF team. The second step is cataloguing the intelligence and insights so that they can be searchable, and the third step is the ability to diffuse knowledge effectively throughout the organization.

B. *Intranet SDF database.* Enterprise management systems such as SAP and Hyperion are large systems that integrate different modules of company operations from sales to demand planning and logistics to financial performance reporting. They are useful sources of company sales, quantity, price, advertising, and other marketing and financial data. However, they are not designed to handle the concepts, distilled insights, and tools required to create sound SDF solutions and manage project uncertainty and risk such as assumption descriptions, methods used, and forecast realization plans. On the other hand, Intranet-based SDF databases are designed to handle external environment factors; competitive data and intelligence; reports and insights; and industry reports, data, evaluations, findings, and special studies. Also, they may record political, economic, social, technological, legal, educational, and demographic (PESTLED) developments and industry and market events impacting the business and record relevant megatrends and subtrends for the forecast horizon and technology evolution assessments and futurist reports.

In addition to external environment information, Intranet SDF databases include benchmarks and reports of performance indicators, most recent business intelligence, structural market changes, and consumer feedback; EWS indicators used, and alerts of earlier projects. They may include forecast update information, the nature of decisions made, and the actual use of forecasts by project. Included are also corporate strategy, project goals and objectives, and client and senior management expectations along with assumption sets used in different forecast stages, people who provided input, and the rationale behind assumptions and hypotheses. These are useful pieces of information to track as are identified known and unknown, knowable and unknowable, and controllable and uncontrollable factors as well as forecast changes between stages along with their timing and reasons for changes are stored.

The methodology, models, techniques, scenarios, tools, and data used to develop different forecast solutions are stored in the SDF database by project along with stakeholder forecast inputs and feedback at different forecast stages. The selected baseline, any combined forecasts, and detailed description of models and tools used, and rationale for the selection along with sanity checks performed are all included. Also, competing forecast solutions rejected and reasons for along with objections to rejected forecast solutions are shown. Brief descriptions of selected scenarios and their impacts and rationale, and evaluation of risks, their sources, and causes are given. Lastly, decision maker, client, or senior management adjustments to forecasts, timing of changes, and their rationale are recorded as are the consensus or consolidated forecasts that ended up being used in the decision.

Other useful Intranet SDF database material for strategic project team review are forecast presentation materials, senior management reactions and reasons for them, and meeting notes. Inquiries from internal sources concerning forecasts, assumptions, models and scenarios used, benchmarks, sanity checks, and risk analysis may also be included. Uses of forecasts and anything related to recommendations derived from forecast solutions are often included. Next to monitoring forecast performance measures and indicators are found variance analysis, performance of driving and enabling factors versus expectations, and progress reports. Lastly, the individual forecast project postmortem analysis, results and findings, and lessons learned are included as are forecast updates and revisions in the postdecision implementation period. It is important to include the reasons for forecast changes, possible adoptive remedial models created, and actions actually taken to ensure forecast realization.

Advice on structuring the SDF database: *Develop a well-thought-out, structured, and searchable SDF database with definition and remark elements because it is one of the most useful tools in effective forecasting function management, knowledge management and sharing, and improving the quality of SDF solutions. It is also valued because it is*

- *A project management tool that can trace all the elements used in the development of long-term forecasts.*
- *Capable of easy storing and retrieving data, information, insights, opinions, client expectations, and senior management directives.*
- *A complete record of processes, activities, assessments, evaluations, results, findings, and conclusions.*
- *Able to query and sort out database elements by specific criteria that may include project type, SDF team composition, timing, key assumptions, approach and methods used, solution developed, implementation success, and so on.*
- *Capable of identifying sources of oversights and project team errors as well as capturing SDF team contributions.*
- *A tool used in conjunction with the project postmortem analysis to identify areas of improvement.*
- *Growing the collective knowledge, skills, and competencies as well as productivity and effectiveness of the company in evaluating and implementing strategic projects.*

C. *Postmortem analysis.* The postmortem analysis is an in-depth review performed jointly by the project team leader and the manager of the SDF team soon after a decision is made and its implementation is on track. The purpose of this analysis is to find out and report on what took place from the time a need for SDF team participation was identified to the point when SDF solutions were created, and the decision was implemented. The postmortem analysis begins with a review of what was done and which entails in-person interviews with SDF team members, the stakeholder team, clients, and senior management to determine the following:

1. What was done, why, how, when, by whom, and how well
2. What methods, techniques, and processes were used and the rationale
3. Which approaches, tools, and processes worked well, which did not, and why
4. What were the important lessons that were learned by the stakeholder team and can be institutionalized and used in future projects

There are several benefits from an SDF team postmortem analysis when done properly and communicated well throughout the company. They include accumulation of practical and usable insights, knowledge, and experiences to be used in SDF projects and sharing the learning with other organizations, which can apply it to their work and save resources that would otherwise be expanded to get that knowledge. Informal and qualitative, yet effective information gathering on SDF team member performance, the project team, processes, and methods is carried out to identify areas of improvement and ways to go about achieving that.

Part III

Strategic Decision Forecasting Applications

8 Corporate Policy and Strategy Evaluation

Corporate policies are written and unwritten principles, rules, and guidelines created, adopted, and enforced by management to define the scope of work, activities, and behavior in the company's daily operations. Generally, policies support what senior managers believe to be best practices and focus primarily on procedural issues, with emphasis placed on efficiency of operations and obtaining desired results. Corporate policies evolve over time, and some examples are human resource policies, code of conduct guidelines, corporate social responsibility policies, customer relationship management policies, and corporate security rules.

Strategies refer to the methods or approaches, the plans of action, and the combination of initiatives used to achieve long-term company goals and objectives. The focus of strategies is on both internal capabilities and the external environment, and the emphasis is placed on creating capabilities to adapt to changing environments. For companies that are not diversified beyond their core business, corporate and business strategies are inseparable and the four key aspects of corporate strategy are as follows:

1. Management of the current businesses in the company's portfolio and the allocation of resources among them
2. Management of internal development and direct linkages between businesses and realization of synergies across businesses
3. Diversification of business, whether through acquisition or creation of shareholder value through strategic initiatives

In this chapter, we discuss the application of strategic forecasting to the first two aspects of corporate strategy and defer discussion of applications to diversification projects to Chapters 9 through 13.

Corporate policies and strategies interact, feed and reinforce each other, and share a number of common properties, and their impacts are often inseparable. They are indispensable controllable instruments to drive internal operations and external performance. At the same time, policies are affected by and, sometimes, determined by internal developments, while strategies are affected by changes in the external environment and market trends. Changes in one often result in changes in the other in ways not easily decipherable and quantifiable. But for purposes of brevity, we focus on strategy evaluations and abstract from inseparability complications and the fact that in the minds of some people, policies and strategies are considered one and the same.

Strategic planning takes on many different forms depending on the company, the organizational structure, and the decision needs. Strategic planning decisions range anywhere from contemplating an acquisition of a company, to licensing a product, to major restructurings, down to forming a new subsidiary to enter a new line of business. Each decision has particular issues and challenges associated with it; consequently, strategic decision forecasting (SDF) solutions must be customized to fit the situation. For that reason, there cannot be one way of doing things, but, rather, the basic SDF framework is modified according to the needs of the situation accompanied by the use of appropriate techniques and tools.

8.1 THE CURRENT STATE AND CAUSES OF FAILURES

Corporate policies are the result of interactions among historical reasons, accumulated experiences, trial-and-error experimentation, or, simply, management wishes and directives. They are developed, enforced, and modified by the organizations responsible for them. Policy change evaluations consist of general statements of impact, and the forecasting function is rarely involved in generating forecasts of policy change impacts on the business. Corporate strategies on the other hand are initiatives to address specific performance issues or responses to competitive or environmental threats. They are developed by Strategic Planning organizations (or senior managers) with or without forecasting support, executed by the owners of initiative implementations, and monitored and modified by both. And, from the interaction of policies and strategies emerge the operational practices that determine the performance of the company.

Objective observers have concluded that in the current state of strategy development and evaluation, there is little, if any, correlation between the amount of planning and results, but that the quality of the plan development process and the learning that takes place result in good strategic plans. Some strategic project sponsors have remarked that many companies do not really have a strategy; instead, they have broad statements of intent or general thrust. But, in our experience, strategic planning in large corporations makes a difference primarily because of the effort directed toward aligning objectives of diverse stakeholders and coordinating the activities and resources required to execute the strategy effectively.

The current state of strategy development and evaluation has many variants and takes on different forms, but the general approach used consists of steps beginning with performing a situation analysis, which entails a competitor evaluation and a cursory review of the external environment and industry conditions, and prioritizing and selecting strengths, weaknesses, opportunities, and threats (SWOT) areas in need of addressing. Then the activities of defining the goals and objectives to be reached, creating a set of assumptions based on those goals and objectives, and quantifying their financial impact take place. Based on these activities, the plan of actions making up the strategy is developed, which includes timelines, resource requirements, performance measures, and risks associated with the proposed strategy which are evaluated. The decisions required by the strategic plan are implemented and the performance of the strategy is evaluated by the degree to which expectations are met in terms of revenue, cost, profit, and market share targets.

Case 1: Inadequate uncertainty assessment. One of the major limitations of the current-state approach is that inadequate assessment of strategy change uncertainty in changing environments results in strategic planning building false expectations about future events. Because of lack of thorough megatrend and subtrend evaluations, early warnings, and inability to solve problems and issues as they occur, the value of strategic plans is questionable. Erroneously, some decision makers think that sequential short-term reactions to a problem constitute the company's strategy of addressing SWOT elements. Besides these issues, troubles with strategy are caused by several shortcomings in processes and practices such as the following:

1. Superficial understanding of the company's business, the external environment, and the markets it operates in
2. Long-term forecasts based on erroneous assumptions and biased scenarios, and without risk management plans
3. Strategic plans that cover only the first steps of their execution and rarely incorporate actions, reactions, and feedbacks beyond the first or second order
4. Divergent strategy stakeholder positions and unmanaged expectations, control issues, and processes along with limited management support
5. Excessive senior management influence on strategic initiatives

Strategy changes introducing new uncertainty costs and benefits are not well understood, quantified, and tied to market, customer, and investor reaction measurements. Also, forecasting is a separate, not well-understood function viewed as a necessary evil and brought in the process late in order to create long-term projections to value the strategy. Often, strategy clients use single-point, single-scenario forecasts because many decision systems cannot handle dealing with multiple scenarios and forecast ranges. In some instances, blind faith is placed on market research results, reams of untested data, and subject matter expert opinion, and the result is failure to see the forest for the trees. More importantly, however, there is inadequate scrutiny of the strategic plan's goals and objectives, the underlying uncertainty and assumptions, postulated paths of causality and feedbacks, long-term forecasts, and reality checks. That is, uncertainty is overlooked and risks are treated superficially or even ignored. In other cases, there is unwillingness to face the reality of the situation and engage in independent, critical, and objective strategy assessments of the firm's SWOT due to short-term priorities.

Case 2: Sticking to bureaucratic requirements. In large stable companies with low risk tolerance, strategies deviate little from the status quo as a rule; otherwise, they may be considered as too risky and difficult to undertake. As a result, strategic forecasting and planning is, in reality, done to satisfy bureaucratic requirements and not to establish sound foundations for large impact decisions. There is a general lack of strategic forecasting expertise and processes to create sound long-term forecasts associated with new strategies and which include external environment influences. Instead, long-term forecasts are simple extrapolations from short-term forecasts. Furthermore, mismatch of skills and management talent required to develop forecasts and strategies and evaluate those in place is prevalent. Also, narrow mindsets preoccupied with formal static strategic planning models coupled with confusion about processes, roles and responsibilities, and best practices pervade decisions.

In the absence of SDF team participation to perform well several important functions, they are currently either performed on an ad hoc basis or missing all together such as in the following cases.

Case 3: Lack of expertise. This is a major cause of policy and strategy development and assessment failure in a situation where there is no real expertise and support provided by the forecasting team. This gap included the following functions:

1. Getting a handle on uncertainty and risks surrounding a contemplated change, project, or decision and helping develop the company's strategic posturing
2. Identifying, prioritizing, and selecting large impact issues to address through modeling and assessing their impacts
3. Helping to identify major initiatives, levers, and controllable factors to ensure achievability of strategy and create a strategy realization plan with appropriate responses to internal and external changes
4. Helping the project team to think through possible good and bad outcomes, including black swans
5. Incorporating the quantified effects of external and internal changes via models and plausible scenarios to the projected status quo state to arrive at alternative future states

Case 4: Poor execution. Poorly executed functions are often the cause of failures in strategy and policy development and evaluation areas. They include meager execution of the following process activities that lead to failures:

1. Creating and testing a realistic set of assumptions, modeling current operations, and describing the future status quo state of the company, that is, creating and monitoring assumptions, generating future-state forecasts, and revising them as warranted
2. Describing, evaluating, quantifying, and projecting the effects of megatrends and subtrends, political, economic, social, technological, legal, educational, and demographic (PESTLED) environment changes, industry developments, and company SWOT vis-à-vis key competitor capabilities and advantages for the duration of the strategy in effect

3. Verifying the validity and accuracy of data, information, and assumptions; redoing reality and sanity checks on assumptions and factors impacting forecasted states; and creating actionable intelligence and insights
4. Sorting out and evaluating the impacts of selected scenarios on company performance through hypothesis testing, performing sensitivity analyses, simulating key driver ranges, and assessing their implications
5. Describing and quantifying the risks inherent in the projected future states due to strategy changes and identifying ways, means, and support needed to deal with them effectively
6. Assisting in the development of actions and timelines in the execution of strategy and developing the strategy realization plan, which includes measurements of performance, a strategy management dashboard, an early warning system (EWS), a competitive or crisis response plan, and strategy realization instruments to use in different circumstances

The gaps in the current state of strategic forecasting and strategic planning present a unique opportunity to apply the capabilities of the new SDF paradigm to improve the evaluations, foundations, and quality of the strategic plans themselves. For this to take place and make a difference, it is crucial for the SDF team to have an active role from the beginning of strategy creation and evaluation at the functional, business unit, and corporate-wide strategy levels.

8.2 PARTICULAR NEEDS AND CHALLENGES

Sound strategy development and evaluation is a broad and labor-intensive undertaking, and the top pressing need frequently articulated is the ability to verify existence of the right mindset, skills, and resources to assess the situation. It is also important to create a sound approach and plans to define policy and strategy states that lend themselves to practical quantitative analysis and execute strategy changes successfully. Other obvious needs include developing consistent approaches and comprehensive models of company operations to evaluate the effectiveness of current policy and strategy, processes and practices, and appropriate techniques to identify and prioritize policies and strategies in need of change. Most of these needs to effective strategy and policy development and assessment go unmet in the current state.

Reliable ways to determine the root causes of inadequate policies and strategies are needed badly as are effective processes to shape, change, and enhance policies and strategies. And, identifying the right approach to perceive, model, and test plausible scenarios; developing future-state projections; and creating strategic decision options are a huge undertaking. Also, establishing the means to obtain the needed level of senior management support to affect change and make operational improvements is another commonly expressed need. This requires development of indicators to measure true costs and benefits, that is, to measure how effective the proposed changes to policies or strategies are. Additionally, identifying controllable levers to use to manage risk, ensuring strategy achievability, and meeting expectations are universally unmet needs.

Issues commonly encountered in strategy and policy development and evaluation that present major challenges include a long list of shortcomings. First, there are interactions and feedback loops between policy and strategy changes and, often, the line between the two is blurred; that is, it is impossible to evaluate the impact of strategy changes in the context of policy changes that affect immediate operations. Then, certain corporate policies change in support of strategy changes simultaneously, and there is no time lag in which to assess the impact of strategy alone. This is complicated by the inability to distinguish the impacts of strategy changes from those of environmental changes or the effects of megatrends. A further complicating factor is that strategy changes impact not only operational performance, but also expectations of future company value and stock prices more that deliberate corporate decisions. Also, decision inertia in dealing with external threats and the discrepancies between corporate risk appetite and risk involved in developing and executing needed strategies are additional challenge. Furthermore, conflicting stakeholder objectives and undue senior management influence on key assumptions, creation of scenarios, assessment implications, and development of strategies are issues that come up often and are difficult to manage.

Solid grounding of strategy- and policy-related analyses, assessments, evaluations, and proper modeling and the creation of plausible scenarios in the midst of continuous internal and external changes are infrequent. And, assumptions about high levels of 4Cs required are difficult to establish, verify accurately, and ensure that they will hold true in the future. Additional complicating factors are strategy change expectations and announcements leading to swift competitor reactions, while there are longer lags between strategy changes and realizing changes in business performance. The complexity of strategy change effectiveness is increased by black swans which are rare, but must be considered, their occurrence included in strategy models, and responses to deal with them prepared in advance to avoid catastrophic eventualities.

Other challenges to effective policy and strategy formulation and evaluation are analyses with inadequate validations, incomplete forecasting due diligence, reality checks, and risk assessments due to lack of the right information and forecasts created under severe time constraints. Also, lack of strategy implementation planning and execution capabilities, inadequate support and unexpected resource constraints, poor or absent response planning capabilities, and lack of strategy realization plans render strategies ineffective. Furthermore, absence of sophisticated EWSs to detect risks and communicate the need for action to decision stakeholders in a timely fashion is a major concern. In addition, strategy reversals are costly, cause confusion and organizational upheavals, and are difficult to evaluate how quickly they can be made and the magnitude of consequences. These are real challenges to making strategy adjustments when needed.

Appraisal: *Are the current-state challenges and issues as ominous and threatening as they sound? Yes, even though in many cases some factors are offset by other factors often in an unpredictable fashion. But, if the objective is to make the strategy and policy development or evaluation more effective, addressing the current-state gaps, deficiencies, and problems increases the chances of achieving the intended strategy or policy objectives.*

8.3 THE ROLE OF THE SDF TEAM

The SDF team plays an important role in providing proactive and objective research, intelligence, and insights into issues aimed at improving the effectiveness and overall efficiency of strategic planning. The Strategic Planning group's need for SDF analyses, forecasts, expertise, and advice is in several areas including the following:

1. Helping to refine the objectives of corporate policy in view of current intelligence about the external environment and competitor practices and activities
2. Assessing the current and proposed strategy uncertainty levels
3. Defining corporate risk appetite, developing strategic posturing, and identifying risks associated with contemplated decisions
4. Providing assumption inputs and long-term forecasts to strategy development
5. Conducting sanity checks regarding implementation impacts and evaluation of forecast implications
6. Creating basic templates of competitor response and strategy realization when certain triggers are activated

Other important roles played by SDF teams are investigating and sharing competitor processes, methods, and benchmarks and performing policy evaluations related to the company's business definition. Determining the viability of company investments proposed by strategy and the feasibility

According to McKinsey Consulting, the profitability performance of a company is 20–40% determined by firm-specific actions and industry factors determine 10–20%. The remainder 20–40% is determined by external factors and their interaction with internal factors. Hence, the areas in which the SDF team plays a major role and adds substantial value to the assessment of existing and the development of new strategies include the following:

1. Determining senior management biases and expectations, preference for organic versus acquired growth, and risk tolerance and subsequent conditioning of their expectations
2. Grounding evaluations on thorough understanding of mega- and subtrends and their future implications for company operations, in-depth external environment assessments and objective and critical evaluation of SWOT, comprehensive forecasting due diligence, well-tested models of the company's operations, reasonable plausible scenarios, and well-balanced judgments
3. Identifying what needs to be known and what it takes to assess properly current policies, strategies, and changes in them
4. Determining what information, data, analyses, and evaluations are known, knowable, and doable and creating intelligence and insights in order to do them right
5. Identifying major challenges in developing company policies and strategies and assigning, through the project sponsor, appropriate tasks and responsibilities to project team members
6. Developing effective approaches, processes, and plans to perform the evaluation of current and future strategies and changes in them
7. Creating or selecting the right models, scenarios, and techniques and tools to perform the evaluations to fit the company's situation as closely as possible
8. Creating more efficient and effective customized solutions and future-state descriptions based on realistic assumptions and good understanding of business performance drivers
9. Assessing the resource implications, risks surrounding the future states, and creation of future-state realization plans, and determining support levels needed to succeed
10. Developing alternative decision options to address different issues and practical implementation recommendations
11. Creating a strategy management dashboard, an EWS, and a set of indicators to measure the extent that the strategy has realized its intended goals and objectives and initiate actions to ensure that they are achieved

Reality check: *Are there qualified strategic decision forecasters out there for the SDF team to play the roles discussed here? Currently, there are only a few, but an experienced client can compensate for that gap to some degree: first, by helping the SDF team to perform the prescribed functions in conjunction with Strategic Planning and other organizations; second, by assembling a good team of individuals currently dealing with strategic forecasting, external affairs, competitive analysis, tactical forecasting, and market analysis and research; and third, by having the project guided jointly by strategic planners, forecasters, and project clients.*

8.4 SDF PROCESSES AND TECHNIQUES

8.4.1 Corporate Strategy Evaluation

The generic SDF process for strategy evaluation parallels that of policy evaluation, but entails a more in-depth analysis and critical evaluations. The process starts with SDF team members getting a good understanding of the corporate mission, vision, core value statements, and corporate culture and their evolution through time. This is followed by developing an appreciation of the senior management vision of the company's future and their needs, goals and objectives, preferences, and risk tolerance. Then an objective and critical situation analysis and an evaluation of uncertainty and risks relative to proposed strategy are performed to help in the development of strategic posturing and strategic intent. By reviewing the effectiveness of current company strategies and comparing them with key competitors' strategies and experiences, SDF team members are able to assist strategic planning clients in creating a good strategy problem definition and evaluating proposed strategic goals and objectives.

Benchmarking best practices in modeling, projecting, and evaluating the effects of strategy changes on value creation and adopting the most effective ways of doing them are part of the SDF team's responsibility. This enables SDF team members to create their plan's goals and objectives and processes and support clients define both broad and specific performance indicators of strategy effectiveness. Adopting best practices also helps to project manage the research, analyses, and internal and external evaluations, and develop SDF forecasts and options effectively. SDF team members clarify the required support estimate and help in monitoring progress on the execution of strategic plans by creating a strategy performance dashboard and an EWS. And if warranted, they are involved in the monitoring and isolating strategy change impacts from those due to environmental changes and internal transformations and in developing adaptive strategy realization, competitive, or environmental response plans.

The specific SDF process steps involved in developing and evaluating strategy vary widely from case to case, but ordinarily include the following activities:

1. Starting with the company mission statement and senior management vision of the company's future and developing a thorough understanding of current strategies and how they evolved through time to ensure on noninterference or conflict with major current or proposed corporate policies
2. Conducting in-depth evaluations of megatrends, subtrends, and the PESTLED environment impacting the company, performing a thorough industry analysis, and carrying out independent, critical, and objective company and competitor SWOT evaluations
3. Determining the uncertainty surrounding expected or projected future states, reducing it to residual levels through analyses and evaluations, identifying the decisions required to succeed, and helping senior management and clients develop a corporate strategic posture
4. Reviewing current major product, price, place, and promotion and value chain practices vis-à-vis competitors to determine shortcomings and identify conflicts with current and proposed strategies
5. Identifying the major business performance drivers, instruments of influence, controllable factors, and resource constraints in the current and proposed strategies
6. Modeling the current state or status quo of operations and describing and modeling possible future states under the current strategy
7. Using the SDF strategic management approach shown in Figure 8.1 to link strategy to investor expectations, modeling strategy development, and evaluating the impacts of changes
8. Defining desirable future states, creating distinct plausible scenarios leading to those states, and identifying support requirements and constraints in the current and proposed strategies
9. Conducting multiple reality and sanity checks and assessing the implications of proposed strategy changes for corporate policies and resources
10. Evaluating risks involved in the future-state scenarios, their sources and implications, and potential impacts and planning approaches to avoid, shift, or mitigate them

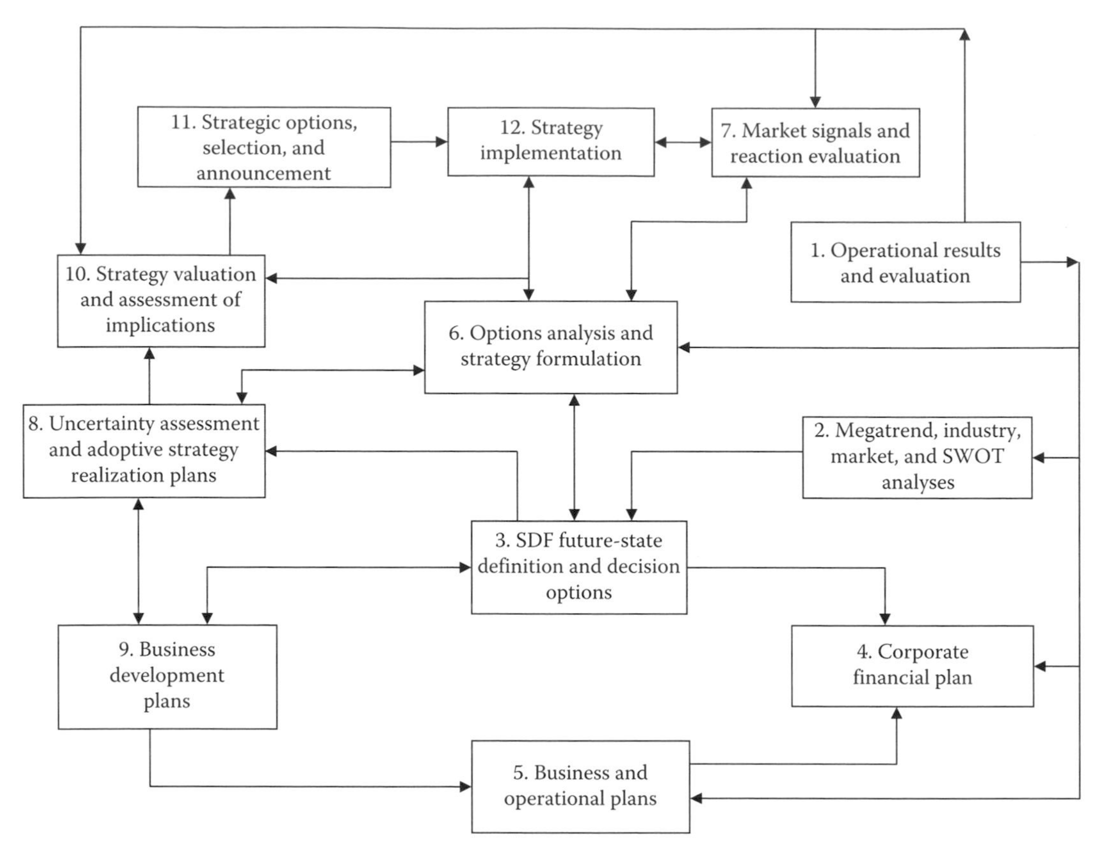

FIGURE 8.1 SDF management framework.

11. Presenting the plausible scenarios and identifying the support required to achieve in order to obtain feedback and agreement from the project team and, once more, assessing the competitive, financial, and human resource implications
12. Picking three achievable strategy scenarios, outlining actionable options, and developing strategy implementation and realization plan recommendations
13. Creating a performance management dashboard and defining multiperspective performance measures to judge strategy effectiveness and SDF team member's value-added contributions
14. Creating an EWS capable of detecting approaching risks and communicating their impending threats effectively to those responsible for initiating offsetting actions
15. Uploading data, intelligence, analyses, evaluations, findings, and insights on the SDF Intranet database and making them available to strategy stakeholders

The type of decision, the organizational positioning, and the relationships of the SDF team with clients determine continuing SDF team involvement in the strategy monitoring and response implementation process. At that point, if follow-up responsibilities are handed over to the client group, the SDF team assumes an advisory role. If it is determined that the SDF team will remain involved in the process, its activities consist of the following functions:

1. Monitoring company performance due to strategy changes and determining, to the extent possible, to isolate the effect of environmental changes from those of strategy changes on performance
2. Reporting progress in the implementation of a new strategy through the strategy management dashboard and the company balanced scorecard

3. Lending help to clients in the uncertainty and risk management area through analyses to reduce new uncertainties to their residual component, evaluating new options, bringing up-to-date strategy levers and implementation instruments, and creating competitive response plans
4. Conducting a strategy project postmortem analysis from an SDF team perspective and institutionalizing findings, knowledge, and learning from the project

SDF deluxe or nothing? *Since strategic projects are unique and have their own analytical needs, and because the version of SDF team support presented here is not always possible, what are the key analyses and evaluations one must do under time, experience, and resource constraints?*

Our advice: Involve the best people with some knowledge to do the few analyses the project leader considers important, assess project uncertainty and risks, pay attention to assumptions, understand implications, and conduct sanity checks.

8.4.2 Corporate Policy Evaluation

SDF processes in support of strategic planning must be parallel to and consistent with corporate policy evaluation and plans. Thus, the objective of the SDF team here is to ensure that (a) there are no conflicts of policies and strategies in the current or in the projected states and (b) company policies are at least as effective as those of competitors. For major policy changes, the objectives are accomplished with a progression of steps, which start with a review of key company policies and major competitor policies in large impact areas of performance. What is needed then is to use best policies and practices to develop an understanding of the evolution of the current and proposed strategies and the underlying rationale. Also, in cases where policies and strategies impact each other, the magnitude of potential impacts is identified and followed by a description of the influence of corporate culture on major business processes and an assessment of how they are affected by corporate policies and strategies. An appraisal of functional and business unit strategic initiatives to uncover possible current or future conflicts is conducted and the senior management team's policy predispositions, biases, preferences, stated positions, and past directives are validated as well.

The specific activities SDF team members are involved in developing and evaluating major corporate policies ordinarily commence with understanding of the corporate culture, the process for policy and strategy development and changes, and a review of key company and industry policies and assessment of their effectiveness. Then the rationale, evolution, consistency, and key drivers and influencers of corporate policies are determined and corporate policies are compared with those of competitors to identify key differences and their effects. But, prior to helping clients identify key corporate policies to be reviewed for possible update, change, shift of responsibilities, and organizational changes, SDF team members determine senior management beliefs, biases, directives, expectations, and assumptions. They also verify client needs and decisions to be made and help them define goals and objectives consistent with market realities, structure an executable project, and take on project managing the development of SDF forecasts, options, and recommendations. How do SDF teams do all that? They do these by modeling the policy implementation process and creating reasonable assumptions to quantify the impacts of policy changes and assessing the implications of results.

SDF team members assess the corporate policy development and change processes, practices, and past experiences and help clients responsible for policy formulation or change to assess the situation, define goals and objectives, and structure a project. They project manage the internal and external analyses and evaluations and the evaluation of current and future impacts, help create options to fine-tune policy and provide practical recommendations for a policy change decision, and assist in the implementation changes guided by the evaluations of impacts. Furthermore, as SDF team members identify policy impacts, they help to reconcile possible conflicts, minimize interference, and enable the client to set up corporate policy performance measures and a monitoring system to report progress. Then, they assist clients to assess the result of policy changes vis-à-vis the goals and objectives and determine the source of deviations and they provide support in the area of course correction and fine-tuning of the corporate policy to achieve stated objectives.

In other cases, a review of the current company policy development and change practices is conducted to help benchmark and adopt best policies and practices consistent with current and proposed strategies. In the process, SDF team members identify and develop a thorough understanding of the company and key competitor human resources and financial, competitive, and technology management policies. They also use the techniques and tools discussed earlier to develop a system dynamics approach to forecast the status quo and the new state under the revised policy regime. They model the role of each major policy in company performance separately and assess potential feedback effects to and from strategy drivers and strategic plans. Thus, well-considered scenarios of changes in corporate policies are created, the most likely are selected, their impacts are quantified, and their resource implications are assessed.

SDF team members communicate findings of their analyses and evaluations to the strategy stakeholder teams, obtain feedback and input, and then develop options on changing corporate policies as well as quantifying what will take to do it effectively. They identify possible risks in making large, discrete changes of policies, especially in new areas of operations, and present the SDF evaluations to clients with actionable recommendations. They also help create a system to monitor the implementation of new corporate policies and assess the effects of policy changes. Lastly, they help clients to fine-tune the new corporate policy by quantifying the impact due to the change from the status quo via scenarios created earlier.

Strategy or policy first? *The recommendation of strategy experts is that strategy evaluation should precede policy evaluation. A strategy evaluation takes place to see if it was implemented right, and if it was not, one investigates what was responsible for the ineffective strategy execution by reevaluating the policies that cascade the strategy down to the business functions. However, strategies and policies must be consistent with each other at all times.*

8.5 SDF TECHNIQUES AND TOOLS

The techniques and tools used by SDF team members in the development and evaluation of strategy are determined by the context, the specifics of the strategy or policy problems, and corporate needs. Additional techniques and tools to those mentioned in earlier chapters include the following:

A. *Situational analysis.* This is a deductive problem identification and definition tool used to identify and prioritize issues and problem situations affecting company operations that need to be addressed. It is a focused assessment of the key drivers, constraints, and issues that surround the problem, and it is the first step in answering basic questions, such as the following (Porter 2008):

1. What is the nature of the problem to be resolved, do competitors have this problem, and what are the root causes of the problem?
2. How much does it affect the company, can dealing with the problem be delayed or, if it can be ignored, what happens then?
3. What business units, functional areas, markets, and financial performance components are affected most by the consequences of the problem?
4. What has been done and what needs to be done to resolve the problem?

B. *Strategy and strategic posture under uncertainty.* This is a framework to develop strategy under uncertainty by first assessing it and then developing an appropriate strategic posture to enhance the chances of strategy success. The four levels of uncertainty and way to project the future as described by Courtney et al. (1999) are as follows:

1. *Clear future.* A single outcome future is, basically, a continuation of the status quo. It can be projected using statistical models and then future event adjustments can be added.
2. *Alternative futures.* A few possible future-state outcomes require tools such as decision analysis, option valuation, and game theory models to project them.
3. *A range of futures.* An array of futures is possible but no dominant propensities. Tools used to predict future states are consumer demand research, technology forecasting, and scenario planning.
4. *True ambiguity.* There are no data or previous experience and a basis for forecasting future states. In this case, analogies and pattern recognition are two possible techniques along with other appropriate qualitative methods.

C. *Strategic management model.* This is a model that links strategy with key evaluations, other corporate planning functions, future-state forecasts, and operational results, which include key financial measures and stock price performance. It is a systematic approach that allows identification of driving factors, causality paths, and interactions and feedback effects using the tools of analysis grounding. These tools include megatrend, subtrend, PESTLED, industry, and SWOT analyses and evaluations of impacts to and from a given strategy. The strategic management framework is shown in Figure 8.1 and its main advantages are as follows:

1. It lends itself to modeling interrelationships using different methodologies, although it is best suited for system dynamics modeling and scenario planning.
2. It portrays the interrelationships and enables the SDF team to trace impacts from operational performance through the different decision phases to strategy formulation, evaluation, and implementation, down to operational performance.
3. Its primary use is to formalize and manage the entire SDF process in support of strategic planning so as to create sound forecasts, add significant value, and enhance the effectiveness of the strategic plan and decision-making processes.

D. *Measuring the moat.* It refers to the analysis performed by SDF team members to establish how a company creates shareholder value and determine the extent of its ability to sustain value creation. Moat, according to Mauboussin and Bartholdson (2002), is the set of a company's unique attributes that enable it to prevent competitors from achieving comparable ability to create shareholder value. Moat-producing attributes investigated are factors such as the ability to innovate through process and technology, reinvesting profits to support a growth strategy successfully, and the capacity of a company to restructure and reinvent itself. Other moat-producing elements evaluated are expertise and talent in M&A and consolidation activities, creation of various types of barriers to entry, and superior new product and global market development.

E. *Strategy realization plan.* It refers to the specific steps to be taken to ensure that the strategy is executed properly and achieves the goals and objectives it set out to accomplish. It is developed as part of SDF solutions using applicable models and tools, it is one of the crucial determinants of strategy success, and it incorporates the following elements in order to achieve results:

1. Evaluating and extrapolating short-term company performance further out to assess continuity or a need for change in strategy; hence, the SDF team analyses must look at the near term and far into the future conditions and select an appropriate strategy management dashboard to focus on what is important.
2. Uncertainty and risk identification, quantification and management, contingency plans, and communication to all strategy stakeholders

3. Assessing variations in the company's moat and changes in corporate policies that impact performance and determining levers that may be affected, which, in turn, can influence the realization of the strategic plan
4. An effective EWS with internal monitoring, external scan, trend detection and evaluation, and reporting capabilities
5. A strategy communication plan delivered to internal stakeholders and the investment community that outlines key assumptions, future-state projections, risks associated with the proposed strategy, and how they will be managed
6. Strategy conditioners that are second-order actions the company can take to reduce or amplify the effect of certain events on strategy, which are controllable factors that are ordinarily not included in the scenario planning models, such as increased management support for the strategy and additional industry association intervention
7. Effective management of 4Cs that are a crucial part of the strategy realization plan because the main challenge of strategy is how to create value through configuration and alignment of its multifunctional activities
8. Competitive response preparation plans, actions, and moves to be put into effect as soon as competitor actions and reactions start taking place: competitive response levers include both strategic and tactical factors, that is, factors ranging from putting into effect product, place, price, and promotion changes to initiating a planned acquisition.
9. External environment interventions used to bypass strategy constraints, offset adverse impacts, and prevent undesirable futures from occurring: an example of this is the use of company or industry lobbying to change regulations to its advantage.
10. Managing expectations; modifying, refining, or changing strategic plan performance objectives; timing of implementation; and putting into effect alternative initiatives previously evaluated

Note: *Strategy realization has many elements in common with forecast realization, which contains several useful suggestions that enhance the chances of strategies achieving their objectives and which is discussed at length in Chapter 18.*

F. *Strategy wheel*. This is a visual representation and a quick tool to review the company's strengths. A strategy wheel presents the company's competences on the axes and the scores of the competences on those axes. The wheel is derived from a matrix with two rows showing the firm and its resources versus the external situation and two columns showing today's or current operating environment, and the environment five to ten years out. Each cell of the matrix is split into two segments containing key strategic questions that need to be addressed at each stage and the idea is that by turning the wheel once the first pass in developing strategy is completed. Additional spins of the wheel strengthen the foundations and refine the strategy developed (Taylor 1999).

G. *Quality of strategy evaluation*. This is a technique used to answer questions about the quality of strategy at the broad level and also serves as a strategy sanity check. The key questions it seeks to answer are the following:

1. How well are the corporate vision and strategic goals and objectives articulated in the strategic plan and communicated to decision makers?
2. Is the corporate strategy internally consistent; that is, are Business Unit, R&D, NPD, Business Development, Portfolio Management, Marketing, Finance, and other functional strategies consistent with it?
3. Is the corporate strategy in accord with policies, and might there be conflicts that would interfere with its successful execution now or in the future?

4. What criteria were used to judge the proposed strategy as the preferred option, and what alternative strategies were considered and why were they not selected?
5. What is the level of risk in the proposed strategy, and is it consistent with corporate and senior management risk appetite or is it too risky?
6. How do the resource costs of executing the strategy stack against potential benefits to be obtained by it?
7. How well is the strategy grounded in reality, is it feasible and what makes it feasible, and how much senior management attention and support does it have?
8. Does the strategy resolve the problems it was intended to, can it produce a corporate advantage, and what competitor reactions are likely?
9. When and how would we know that the strategy has been successfully executed, and what metrics would be used to measure success?
10. What happens in the event the strategy proves ineffective in the near term, and is there a best alternative strategy to replace it?

Key points: *The three tools to use when applying the SDF approach in abbreviated versions of strategy and policy evaluation are*

1. *Measuring project uncertainty and defining strategic posture.*
2. *Applying the SDF strategic management model shown in Figure 8.1.*
3. *Insisting on the creation and implementation of a strategy realization plan.*

8.6 BENEFICIAL PRACTICES AND SDF TEAM CONTRIBUTIONS

8.6.1 Examples of Useful Practices and Factors

From an SDF perspective, corporate policy and strategy evaluation success factors are those key practices that lead to the realization of future-state projections and the strategic goals and objectives in the most effective manner. Because each policy or strategy change needs and situation are unique, experts in different industries state varying useful practices and success factors, which include the following:

1. Knowing the industry and company business, understanding the current strategy and how it evolved, and sensing customer and employee satisfaction levels
2. Understanding the corporate culture, current policies and practices, and their historical context and evolution as well as those of competitors
3. Effective communication, coordination of effort, and senior management support determine strategy and policy realization.
4. Ability to interpret the effects of megatrends and subtrends on the industry and the company; visualize the industry's evolution; and perform independent, critical, and objective assessment of internal versus competitor SWOT
5. Skill to synthesize multiple evaluations, results, and future-state projections into decision options and practical recommendations
6. Selecting the right measures of performance to evaluate forecasted states that result from policy and strategy changes
7. Expertise in interpreting market signals and incorporating them in models and scenarios, and the capacity to separate the long-term from the short-term impact factors
8. Skillfulness in identifying and communicating the sources and causes of strategy failures, not the symptoms

9. Determining and managing effectively the controllable factors impacted by strategy, identifying black swans, and assessing competence in creating adaptive company responses
10. Identifying key performance drivers and levers to use to impact them and developing actions to compensate for risks or for enhancing performance
11. Understanding customer needs and the company's capabilities, and performing sanity checks throughout the entire process
12. Creating a flexible and adoptable approach to execution of strategy and policy changes
13. Ability to isolate policy or strategy change effects from those of the external environment and to manage conflicting stakeholder interests, goals and objectives, and senior management expectations
14. Help from the SDF team to assess and create viable strategies consistent with the likelihood of achievability, sustainability, effect on the industry structure, and competitor actions
15. Well-considered policy and strategy and SDF team communication, implementation, monitoring plans, and variance analysis
16. Competitor choices against the firm's competencies need to be evaluated because industry analysis alone is inadequate foundation for strategic forecasting and strategy.
17. The positioning, functional links, the SDF team members' mindset, and their strategic intent orientation
18. Understanding the customer's value chain and how the company can create additional value for them
19. A superior EWS feeding a management dashboard and giving ample warning of impending threat realization
20. Effective process, models, tools, and scenarios to capture the essence of strategy and describe future states and effective forecast and strategy presentations internally and to the investment community
21. Ability to identify and measure industry reaction and prepare efficient strategy realization plans, effective risk management implementation, and balance of costs and benefits

Word of caution: *Since the SDF discipline and applications to strategic planning and policy development or change are not fully developed and the number of experienced strategic project participants is small, the shared beneficial practices and success factors are not all considered critical success factors. Hence, they should be viewed as factors to mull over, and each strategic forecaster and project team leader must determine project-specific success factors.*

8.6.2 SDF Team Contributions

The unique SDF team contributions in the corporate policy and strategy development and evaluation areas according to participants in our discussions include the following:

1. The approach, process, and techniques used, which enable decision makers to trace the effects of changes in SDF models on company operations
2. The SDF team's internal and external links, its personal relationships, and access to perspectives that create intelligence and insights
3. Independent, critical, and objective assessments of internal and external factors and the creation and validation of assumptions, data, methods, and forecast implications
4. The clarity of description of future states and the communication of how policy and strategy changes affect them as well as risks associated with changes in them
5. Reducing uncertainty to residual uncertainty and helping develop appropriate corporate posture through research, analyses, and assessments

6. The ability given to clients to shape the future with scenario model simulations before implementing strategies through hypothesis testing and scenario planning
7. Thorough analyses, evaluations, and sanity checks performed in a transparent and collaborative manner that result in better recommendations and quicker decisions
8. The direction and help provided in strategy realization planning and linking it with the strategy management dashboard and the EWS
9. Knowledge and learning shared through the central repository of the SDF team on the company's Intranet
10. The active contributions of SDF team members in the postmortem evaluation and their recommendations for future process improvements

9 New Product Development

New product development (NPD) is the set of functions and processes involved in bringing a product or service from the idea stage to the market. On average, new products account for more than 30% of sales and profits, and this explains the 2.5–3.0% R&D share of GDP spent in developed countries. Companies are constantly fighting to enhance or maintain their market share position, and new products are viewed as competitive weapons and tools to achieve differentiation and gain competitive advantage. And, since NPD and product innovation are recognized as crucial elements of growth and business strategy, senior management places emphasis on them and they are given a high priority.

The NPD process is a long series of commonly taken steps, which includes all the activities that take place from the initial idea or invention to the time a product, service, or technology are launched. It combines different disciplines in order to meet the goal of commercializing an idea, and the functions involved in NPD include strategic planning, R&D, design and engineering, portfolio management, forecasting, finance, manufacturing, marketing and sales, and supply chain management. It is a discipline used by multifunctional teams in large companies; but in small business environments, NPD is many other things, the process is less structured but more creative, and all functions are performed by a small project team.

New product or service development involves similar processes and they are treated in the same manner, but when there are no upfront or significant future capital requirements, they are considered extensions or simple innovations. The overall process of developing new technology is similar to the NPD process, but forecasting for it has its own peculiar needs and challenges discussed in Section 9.3.2 and further examined in Chapter 12.

In addition to the current state of NPD forecasting and the causes of failures, in the sections that follow we address the approach, processes, and techniques used in the new strategic decision forecasting (SDF) paradigm for NPD project support. Then, we discuss the SDF analyses and evaluations performed in the different stages and corresponding gates. Lastly, we present case examples of sound practices and factors and share project participants' accounts of SDF team contributions.

9.1 CURRENT STATE AND CAUSES OF FAILURES

The NPD effort is commonly led by the R&D or the Product Portfolio Management (PPM) groups working closely with the Strategic Planning and Sales and Marketing organizations. One of the striking observations is that despite the importance of forecasts in NPD decisions, forecasting organizations play relatively minor roles; and in some cases, all forecasting is either done by the lead organization or is outsourced. Another common observation is that while each function is managed by the owning organization, the overall NPD forecasting process does not have a project manager, but that function resides in a cross-functional team.

When it comes to current state and failure causes attributed primarily to forecasting, there are three major occurrences found with remarkable regularity:

Occurrence 1: Absence or late involvement. Since SDF team participation is absent at the start of most NPD projects, there is unclear forecast ownership from stage to stage. R&D and engineering groups usually drive the forecasting process and as a result, forecasting processes are fragmented, incomplete, broken down, and wrong. Yet, there are unreasonable accuracy expectations of forecasts produced by forecasting groups that did not participate in earlier project stages, instead of focusing on understanding what will shape the future of the business. Instead of trying to incorporate key drivers in the forecast and be directionally correct at the early stages, project leaders focus on accuracy of prediction not understanding that even if initial assumptions are correct, the assessment of the operating environment in the forecast horizon will be changed in subsequent stages and forecasts need to change as well.

Occurrence 2: Inadequate assessments. The evaluation of the external environment and the effects of megatrends and subtrends are given insufficient attention, and the strengths, weaknesses, opportunities, and threats (SWOT) analyses are consistently biased and incomplete. Most of the times, forecasting groups have little or no involvement early on in the process, and there is no systematic monitoring of all the factors that influence the outcome of long-term NPD forecasts. Additionally, there are fragmented and ineffective forecast realization management efforts: too little, too late. And, many experts believe that inadequate NPD forecasts are, to a good measure, the result of insufficient attention paid to how product features and attributes meet customer needs and what really drives company performance.

Occurrence 3: Organizational issues. NPD forecasts drive all capacity planning, logistics, marketing and sales force plans, and resource allocation to facilitate the project as well as corporate financial expectations. Yet, there is late engagement of forecasting teams; and in the rare instances where forecasting is present early on, there are problems with organizational reporting structures and what the forecasting team's role should be. More importantly, there are weak linkages between the forecasting and the other internal planning groups and nonexistent personal relationships with project stakeholders. Also, because NPD project stakeholder interests may diverge out in the future, it is very difficult to incent high-level communication, cooperation, collaboration and coordination (4C) behavior, which is crucial to creating sound forecasts.

The current state of NPD forecasting and failures that result from it are illustrated in the examples that follow, which are practices observed in actual NPD projects under groupings of prevailing project team actions.

Case 1. Unclear roles and responsibilities. The stage–gate methodology is widely used in NPD, and the idea originator usually drives the process and the forecasts. However, in most instances there are unclear assignments of roles and responsibilities to the stakeholders in the stage–gate process, and there is diffusion of responsibilities across the many organizations involved. Furthermore, there are multiple handoffs of responsibilities, assumptions revised between project phases, and priorities and forecasts changed from stage to stage as new players come into the project with incomplete understanding of the forecasting process in place and the forecast solutions developed. This does not help to ensure continuity of purpose and consistency of assumptions and forecasts.

Case 2. Expectations determine assumptions. Assumptions used in NPD projects reflect the lead organization's goals and objectives and are usually founded on expectations or other experiences and tend to be less than objective. More often than not, assumptions are not tested and monitored, changes are not well justified and recorded, updates are done on an ad hoc manner, and changes are not registered. Also, undue emphasis on industry analogs and prior experiences is observed across the board, and the current overall NPD process does not incorporate all strategic forecasting process elements in its requirements.

Case 3. Poor communications. Customarily, only cursory reviews of prior new product launches are performed; and in the current state, information and learning from past company NPD experiences do not flow out to all stakeholders equally well. As a result, NPD forecasts in different stages are heavily influenced by the prevailing silo mentality in large companies. Often, originating organization unreasonable expectations and politics, and vacillating positions on priorities and risk tolerance leave overseeing alignment across functions diffused. In such an environment, ensuring 4Cs is left to senior management; it is done at very high levels, and its effect is not cascading down to work levels.

Case 4. Marketing organization optimism. Financial and marketing considerations are behind most NPD projects, and one of the first challenges project team leaders face is the well-known fact that marketing organizations will never predict failure. That is, starting out all new product, service, or technology introductions are considered a success; and in most companies, strategy is already embedded in forecasts and forecasts into strategy except that forecasts are too optimistic and strategies are inconsistent with reality. What happens then is that optimistic new product long-term forecasts are an integral part of the strategic decision without a critical external environment assessment, appropriate launch support, and forecast realization plans. As a result, 6 out of 10 new product ideas make it to development, and 1 out of 4 products in development succeed commercially.

Case 5. Exercises in futility. The most common NPD experiences are of the mediocre type, result in many "me-too" product launches, and are attributed to a number of causes such as the following:

1. Lack of complete and effective NPD forecasting processes, planning and support, and coordination between various stakeholder activities
2. Unclear future vision, untested and poor NPD strategy, and frequent, hastily conceived strategy changes from idea to product launch
3. Failure to identify, assess, and manage uncertainty and project risks correctly early on in the process
4. Lack of senior management commitment and support stemming from internal politics and diverging interests of project stakeholders
5. Unclear project participant roles and responsibilities, lack of understanding of forecasting basics, and unreasonable expectations of the forecasting organization
6. Short-term focus in each stage driven by the priorities of the organization in charge of that stage, due to absence of a forecasting project management agent
7. Inadequate internal and external assessments, especially those related to project execution and implementation capabilities that result in poor product quality, attributes, and pricing
8. Inadequate uncertainty and risk assessments, absence of early warning systems (EWSs), and no competitive response and forecast realization plans

Case 6. Project team turnover. Assembling a good cross-functional team and ensuring the required degree of 4Cs being practiced is a major challenge. The many questions, issues, and areas of impact vary by stage of NPD, but because of personnel changes in the different phases, there is an ongoing need to redefine the roles and responsibilities of team members across all stages and obtaining agreements is difficult. The situation is further complicated by the fact that in such environment there can be no meaningful single definition of the support role of the forecasting group. Forecasting support has to change in order to meet the needs of the particular NPD project stage and the new project participants.

Case 7. Focus on speed of execution. The duration of a project is significantly affected by several external factors; yet in all NPD projects, the focus is on speed of execution and internal coordination. At the same time, multiple revisions of assumptions and recasting of long-term forecasts occur as several handoffs take place, which make it difficult to identify team member expectations, biases, and contributions. Thus, getting to consensus on assumptions, key drivers, and likely future scenarios is extremely difficult and further complicated company politics. The end result is less reliable forecast solutions, options, and recommendations because keeping up with continuous external and internal changes requires continuous updates of assumptions, scenarios, SDF solutions, and options. This is a truly challenging task due to severe time constraints.

Case 8. Differences by industry. We observe that some industries do a better job in NPD and launching than others due to several reasons but four practices stand out:

1. Application of organizational learning from extensive past experiences in product innovations, introductions, and market management
2. Engagement of the SDF team supported by the external environment assessment, competitive analysis, market research, and tactical forecasting teams right at the start of the process
3. Organizational infrastructures that allow for functional alignment, high levels of 4Cs, and intervention to course correct efficiently throughout the NPD process
4. Management of assumptions and of the forecasting function by the SDF team across all NPD project stages and gates

However, even in those cases, the functional links and processes are inadequate and incomplete, which explains to a large degree the high failure rates and associated high cost of new product introductions.

9.2 SDF APPROACH, PROCESSES, AND TECHNIQUES

9.2.1 Forecasting for New Product and Service Development

Professional SDF teams are well qualified to lead the NPD forecast process, and the purpose of their involvement early on is to help answer three basic questions: What factors will shape the future of the business? Does the company sell what it develops or develops what it sells successfully? How can the company capitalize on those factors to shape the future of the business? To help answer these questions, SDF team members use an approach, which is based on the Product Development Institute's stage–gate process with a number of modifications as seen in Figure 9.1. Notice that in the new paradigm, SDF team member involvement and forecasting project management are present in each decision, and the activities of the SDF team members are shown under each stage and gate of the NPD process.

The choice of the SDF method in NPD projects is influenced by a number of factors such as the nature of the product, the level of capital expenditures required, the amount of uncertainty and importance of the forecast in the NPD decision, and the degree of accuracy required due to risks associated with a decision. Also, client decision-specific needs and timeline requirements and the number of controllable factors that could make a difference are important considerations. Other factors weighing on the SDF approach in each project phase include the cost of developing forecasts vis-à-vis expected benefits, availability and quality of data, and reliability of intelligence sources, the level of forecaster skills and qualifications, and client sophistication.

One of the first SDF team objectives in NPD projects is to understand what decisions need to be made and how its forecast solutions will support them at each stage. To achieve that objective, the SDF team structures the NPD forecasting problem solution as shown in Figure 9.2, and it follows a flexible, tailor-made process to meet the current project needs. The SDF approach is overlaid on generic stage–gate process and creates a method to handle new product launches more effectively.

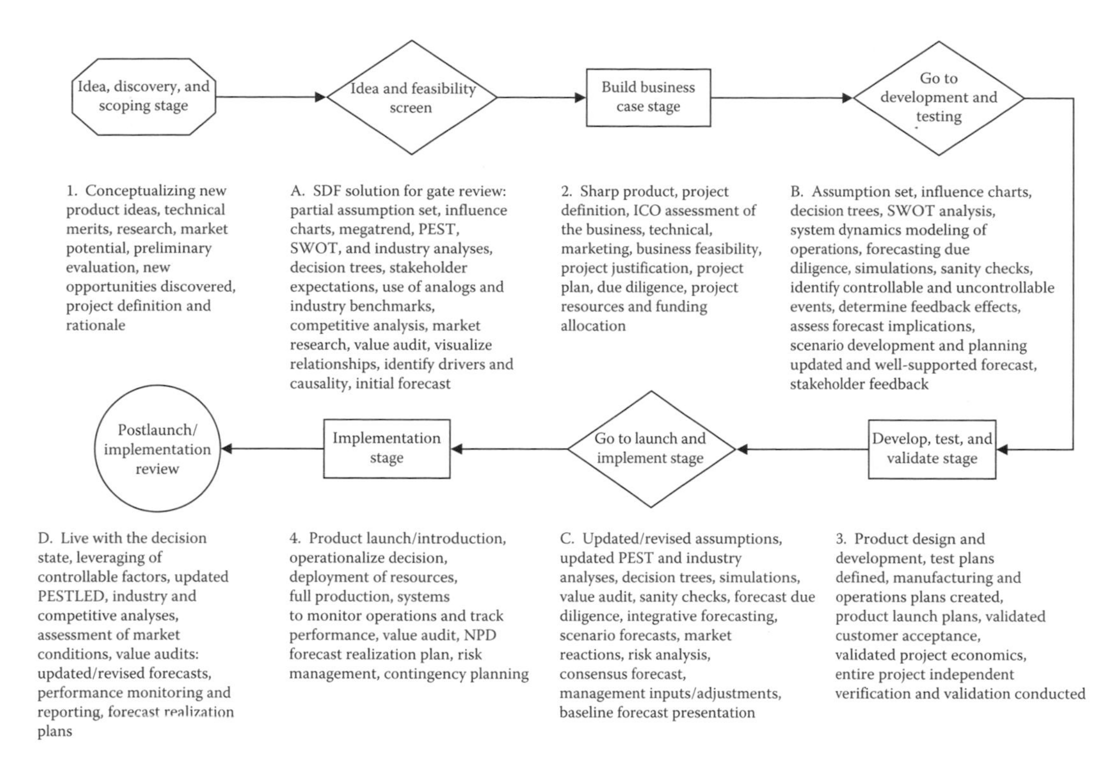

FIGURE 9.1 SDF approach for NPD. (Adapted from Cooper, R. G. and Edgett, S. J., Your roadmap for new product development, New Product Development Institute, www.prod-dev.com/stage-gate.)

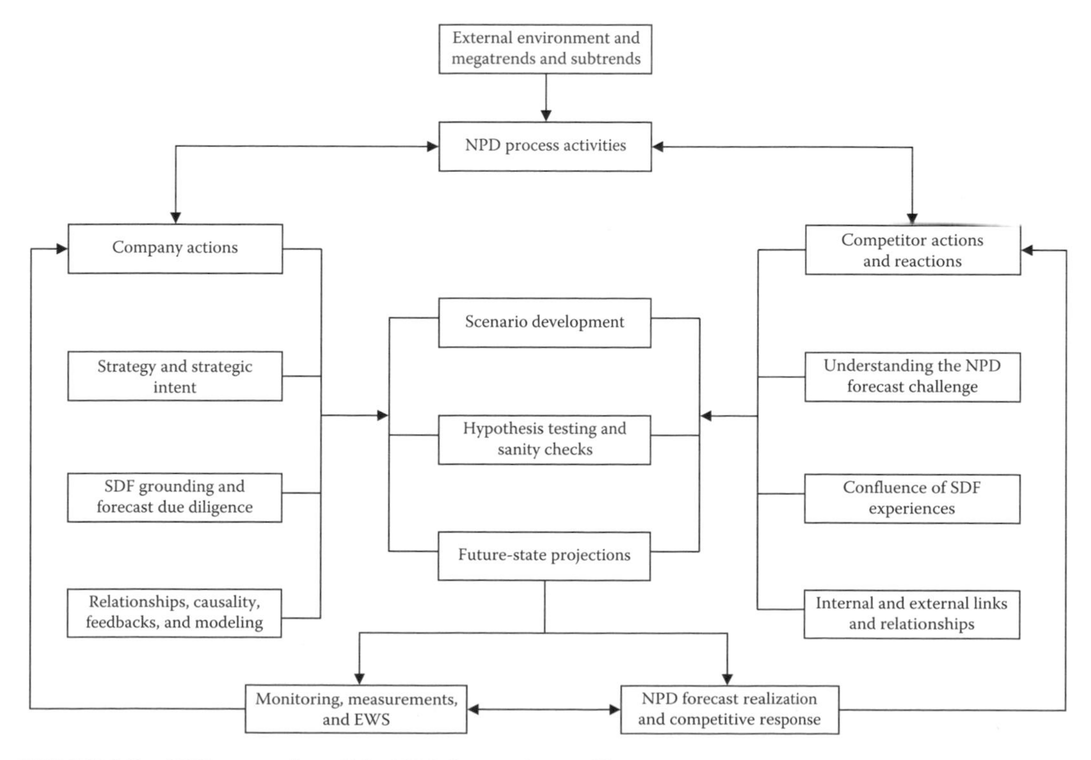

FIGURE 9.2 SDF structuring of the NPD forecasting problem.

Following the outline of the SDF process, the most notable steps and activities, whose order may be changed as needed, are

1. Assigning experienced and dedicated SDF team members to the NPD project right at the start of the new product assessment.
2. Developing close business links and personal relationships with all functions represented on the NPD project team and with external sources of intelligence and expertise.
3. Establishing an end-to-end SDF process customized to fit each stage of an NPD project from the idea to the product launch phase.
4. Identifying the long pole in the tent; that is, identifying what needs to be known, determining the activities of highest importance and longest duration, and prioritizing them.
5. Doing a thorough assessment of the current state of the company business, identifying the degree of corporate risk tolerance, and creating an Intranet database to house data, intelligence, and insights created in the project.
6. Being actively involved in project team meetings and activities and a valued business partner providing intelligence, insights, and expert opinion.
7. Validating the rationale, consistency, and feasibility of the stated NPD project objectives vis-à-vis market realities and internal competencies and capabilities.
8. Performing megatrend and subtrend, industry, political, economic, social, technological, legal, educational, and demographic (PESTLED), and SWOT analyses, and reviews of corporate and portfolio management strategies.
9. Identifying current corporate expectations of NPD and reconditioning them by prior company experiences, appropriate industry analogs, and external and internal SWOT influences.
10. Assessing what the new product launch will change, what will take to do it well, and how it will impact the company business. That is, evaluating key drivers: external factors, corporate strategy, and product attributes and characteristics; internal priorities, policies, capabilities, and funding; people assigned to the NPD team; product pricing, promotion, and advertising; market size; and product acceptance and uptake.
11. Creating, testing, obtaining project team agreement on, documenting, communicating, updating, and maintaining the project's forecasting assumption set as it moves across stages.
12. Ensuring that assumptions are not in conflict with external trends and internal policies and strategies and that at each stage all NPD stakeholders understand and support the updated assumptions.
13. Developing a consensus view on the operating environment and how it is likely to evolve over time in light of findings from earlier steps.
14. Ensuring that the customer value analysis and conjoint marketing research are performed well, and the results are reasonable in order to understand how customer choices are made, the importance of pricing, and other key factors.
15. Developing the rationale for and describing distinct and feasible scenarios of the future business caused by the NPD project and helping to quantify associated financial costs and benefits. And, just as important is estimating the opportunity cost of not doing the NPD project.
16. Assessing the implications of forecasts, conducting sanity checks, and highlighting the risks involved in each scenario, potential impact magnitudes, and their chances of occurring.
17. Focusing on the key project drivers and aggregate long-term projections and leaving the new product forecast breakdowns and details to the Marketing and Tactical Forecasting teams.
18. Evaluating the new product positioning, launch plans, public relations campaign, and other launch support needed for a successful product introduction.
19. Reevaluating the opportunities and risks incorporated in the SDF future-state projections and communicating them to decision makers.

20. Defining key measures to track in the EWS, such as strategy changes, channel distribution developments, reported or pending events or changes in the operating environment, press releases, pricing and promotional activities, team member changes, and controllable and uncontrollable events.
21. Ensuring effective SDF solution transitioning from stage to stage, flawless handoffs of forecasts and related documentation, and defining SDF team deliverables for the next decision point.
22. Conducting independent verification and validation (IVV) to ensure that there is no overcommitment of company capabilities and resources and no timing and magnitude benefit overcommitment to customers and industry analysts.
23. Performing an independent assessment of functional alignment and 4C practices across the different project phases.
24. Building an EWS that feeds the NPD management dashboard and creating a competitive response and forecast realization plan with clear team member roles and responsibilities, triggers, and possible interventions.
25. Assisting the NPD project team in conducting a thorough postmortem review and developing SDF team recommendations for future projects.

Sometimes, in new product or service development, forecasts from different short-term qualitative and/or quantitative forecasting methods are used and adjusted to account for product or service maturity, competitor reactions, market trends, and lifecycle management considerations over the long run. In all cases, however, forecasts from more than one method are combined to generate more informed, balanced, and more accurate new product forecasts. Popular techniques used in the creation of NPD forecasts include the following:

1. Various types of diffusion models used to estimate the rate or speed of adoption over time, such as the Bass, Fourt and Woodlock, and deterministic models
2. The Delphi technique used to obtain consensus forecasts through comparing forecasts, their rationale, and justification iterations
3. Senior management opinion based on their own experiences, assessment, input from advisors or experts, and gut feeling
4. Appropriately adjusted market research-based forecasts from focus groups, market testing and conjoint analysis, and commercially available services such as ACNielsen BASES
5. Analogous or look-like forecasting, which draws from the experiences of similar products in similar markets and time series that are thought to have similar patterns
6. System dynamics modeling, which is a method to conceptualize, describe, and model the workings of the NPD process, make transparent the assumption set, identify key drivers, develop and test hypotheses, and simulate alternative scenarios that yield desired states

A number of useful tools are used by SDF teams to search for and identify NPD analogs, establish where a new product stands relative to competitor products, and determine the validity and reasonableness of a new product forecast. The most popular tools are as follows:

1. *New product positioning matrix.* This tool helps forecasters to visualize a product's position in its market by showing (a) market newness across the matrix rows for existing market, strengthen a market, or new market and (b) the extent of new product innovation along columns for no product innovation, improved product, and new product. When populated, this matrix then shows the new product's classification as enhanced or product line extension, replacement, and diversification, each of which has different growth characteristics.

2. *Product-specific need–attribute matrix.* This tool is a matrix of customer needs listed across rows and product attributes along columns. It is used to check the extent to which the new product

attributes satisfy customer needs. This is probably the most important and improperly used tool in evaluating product demand.

3. *Product-specific satisfaction map.* It is a variant of the need–attribute matrix where rows show the customer's product requirements and columns display satisfaction ranges. For each requirement, customer satisfaction levels are plotted for the new product using market research information versus those of competing products.

4. *Satisfaction ranking matrix.* This is a technique that first rates and then plots satisfaction levels (high, medium, low) across different user requirements of the new product as well as the satisfaction levels of competing products.

5. *Product attribute ranking.* This is a chart that plots the rating of the company product and that of competitors and enables the project team to visualize how they stack against each other. Attributes may include product design, quality, functionality, price, safety, and other factors.

6. *Product competitive position–market share matrix.* This tool displays market share ranges of high, medium, low across rows and product competitive ranges of high, medium, low along columns. When the company's new product expected performance and the competitors' products are placed in their respective elements, the forecaster obtains a sense of the new product's likelihood of success.

7. *Market growth–market share expectations matrix.* This tool is a variant of the product competitive position–market share matrix, except that market growth replaces competitive position along columns. By placing the company's own product and that of competitors on this matrix using actual and market research data, the forecaster has a better idea as to whether the product's history may be a good analog to use in the final forecast view.

8. *Order of entry effects matrix.* This tool is a matrix, developed by McKinsey Consulting, which displays average experiences of product market shares for the number of products in the market across rows and the order of entry in the market. It has been validated by empirical studies and is used across many industries as an input to new product forecasts and it is a valuable sanity check instrument.

9. *Product promotional frontier.* This is a plot of expected and actual market shares obtained from different levels of share of voice (advertising expenditure percentages) for several competing products over time. This is used in the selection of appropriate product analogs with similar share of voice paths to the ones used in the NPD forecasts.

10. *Overall position summary matrix.* It is a technique to rate the company's and key competitors' standing listed along two columns, in key functional areas listed across rows such as brand equity, R&D, engineering and technology, marketing and sales, NPD capabilities and competencies, and product launch experience. This is also used in the capability assessment part of the SWOT analysis and adjusting analogs.

11. *Consistency of NPD objectives versus due diligence conclusions.* This is a list that shows for each NPD objective, such as customer satisfaction, market share, pricing, and return on investment (ROI), the forecasting due diligence findings. It is an instrument used to determine to what extent objectives can be met under the current project plans and identify additional resources and efforts needed to achieve the stated objectives.

12. *New product barrier analysis matrix.* This tool helps the NPD team to identify problem areas, evaluate their impact, and focus on their resolution. The objectives of the NPD project team are listed across

rows and the different barriers along columns; and for each objective, barriers are identified and their severity rated. The results are used to determine actions to eliminate barriers and limit their impact.

13. *Share of voice matrix.* It is an expanded version of the promotional frontier used to determine new product success by plotting share of voice data against market shares of competing products in the market over their lifecycle. This matrix gives an indication of promotional spending levels required to obtain different market share levels based on actual experiences in the early, mature, and declining stage of product lifecycles.

14. *Product-specific SWOT analysis*. This tool is an extension and in-depth analysis of the company SWOT by focusing on the particulars of the new product in each category. In the product strength category, the evaluation deals with items such as marketing and sales experience, the product team experience, and the extent of verified 4Cs; attributes of the product, design, quality, and pricing; resources allocated to the NPD project; and the level of senior management commitment and support.

In the weakness area, the assessment identifies factors such as lack of product management experience, product development costs and pricing, production and logistics capabilities, brand equity, and product launch support. Under opportunities, the assessment concentrates on factors such as new markets for the product under development, new distribution channels, new product applications and uses, product licensing options, and product acquisition alternatives over development. Finally, the focus of evaluation in the threats category is directed toward issues such as new competitor products; substitute product enhancements; new, low cost market entrants; product quality and performance issues; and negative customer and press reactions.

15. *Seven Ps check matrix.* A tool developed by McKinsey Consulting and used by SDF teams to assess NPD strategies across seven business areas, namely the product, promotion, price, people (project team), process, practices, and physical evidence.

16. *Cross-impact matrix.* This is an extension of the Delphi technique and a tool to identify how the occurrence of a future event can affect the likelihood of certain events and assign probabilities in the presence and absence of other events. It also forces the project team to look at the consequences of forecast changes between stages.

17. *Forecasting due diligence.* The ultimate objective of the NPD forecasting due diligence is to validate the depth of understanding of the NPD strategy and the project management and execution capabilities across all functions involved in implementing a new product introduction. Moving from the big picture assessment of the early stages to the practical details level, it is crucial to determine what needs to be done to reach the desired levels of NPD performance. In the forecasting due diligence effort, SDF team members look for unreasonable stakeholder expectations, mental barriers, resource constraints, inertia, fear of failure and risk aversion, organizational politics interference, and unwillingness to hear the customer voice. On the other hand, they verify the company's NPD commitment in terms of persistence and follow-through of corporate strategy, actions that are hard to reverse, and specialized NPD investments. As such, it is probably one of the most important contributions of the SDF team to project success.

The forecasting due diligence also extends across the 4Ps of NPD, which are portfolio, people, project, and process areas across the four stages of NPD: product strategy, development, production, and launching. To establish to what extent the company is likely to have a successful new product launch, SDF teams engage in conjunction with the project team in assessments that cover key project areas, such as the following:

a. Verifying the company's competencies; that is, market access, customer perception, 4Cs, and ease of reproducing its NPD processes
b. Confirming the firm's capabilities, namely agility, process consistency, cross-functional capabilities, speed of execution, and customer focus

c. Verifying corporate-wide alignment, which consists of coordination, directed commitment, balanced project team perseverance, concentration of effort, and strategic intent
d. Reassessing project uncertainty and risk, validating assumptions, data and information, models, and scenarios used, and assessing the forecast implications and performing sanity checks

In support of new product pricing and financial evaluation, the forecasting due diligence also includes validation of product, place, and promotion plans, total market potential, target market, and product adoption rates. Tests of expected market share target feasibility and support requirements to achieve are also performed against committed pre- and postlaunch support levels, along with reality checks. Lastly, evaluations of the new product unit and revenue forecasts achievability, management of risks, and human and financial resource support commitments are performed.

18. *Product analogs.* It has been said that the decision to invade Cuba's Bay of Pigs relied heavily on analysis of analogous situations in trying to forecast possible outcomes to various strategies. Analogies are also used in creating SDF solutions in NPD projects, but are conditioned by findings from a number of sources, which include industry experiences, averages, benchmarks, and practices; the order of entry matrix, critical path analysis, grid analysis, the cross-impact matrix, paired comparison analysis; and the logistic function, the Gompertz curve, and the Gaussian curves. Furthermore, the quality and effectiveness of analogs' use in forecasts is improved by pooling industry and company experiences and those of complementary products. Additionally, forecast triangulation at each stage leads to faster consensus views and enhances the accuracy of projections.

9.2.2 New Technology Development Forecasting

The methods used in new product or service forecasting are often used in technology forecasting areas as well. However, technological forecasts involve more detailed descriptions of the introduction, the features, and impacts of technology. That is, they identify and investigate the different paths of going from the current to the future state and attempt to give a consensus view of the future science and technology to decision makers. The common qualitative technological forecasting methods used in conjunction with quantitative methods include the following:

1. *Monitoring.* In this method, the competitive analysis group collects, screens, and catalogs information appropriately from internal groups and systems responsible for reviewing and interpreting scientific and technical literature, patents, and from well-informed US government agencies about changing technologies. The most important element of this method is that it can generate reliable data, which can be used to forecast future trends using quantitative techniques.

2. *Substitution model.* This model treats technological innovations as substitutes of one way of satisfying a need for another and assumes that the rate of substitution is determined by the amount of the old to be substituted. It also assumes that once substitution begins, it will complete the process and will progress along an S-shaped curve.

3. *Limit analysis.* This is more of a sanity check than a forecasting method, but the idea here is that technological growth in one field is limited because there is a limit to progress in another forecast time range. Therefore, forecasts should reflect the fact that technological improvements should get close to that limit sooner or later within the forecast period.

4. *Trend correlation.* This method is used when a technology is a precursor to one or more technologies and when advances in the precursor can be adopted in the follower technologies reasonably well. In such cases, precursor technology improvements are used to predict, with appropriate lags, the path of the follower technologies.

5. *Network analysis.* It is an extension of monitoring and a complex analysis technique used in two ways: (a) as an exploratory forecasting tool to investigate future technologies and capabilities and systems resulting from new scientific research and (b) as descriptive forecasting tool to define desired future capabilities and determine what research results are needed to reach a desired system or level of capability.

6. *Scenario development.* This is a familiar method in NPD forecasting that seeks to describe a future technology, surrounding events, and the operating environment. Scenarios are based on and draw from company past experiences, subject matter expert opinions, and analogs. They cover long-time horizons and are viewed as management "flight simulators" whereby different impacts are simulated before actual decisions. Scenarios begin with a set of facts to which knowledge, reasonable assumptions, and conjecture are applied to develop extrapolations of future technologies.

7. *The TRIZ method.* This is the most common technological forecasting technique whose basic function is to answer the question: What changes should be made to a product or process to move them to the next position on a specific predetermined or desired line of evolution? This method uses technology roadmapping, which starts with defining the scope of investigation and moves along the following steps:

a. Functional abstraction
b. Applying evolutionary patterns
c. Generating adequately detailed technology roadmaps
d. Extracting and clarifying process, product, and service ideas

TRIZ (the Russian acronym for theory of inventive problem solving) is a method used to improve the efficiency of technical innovation and is especially helpful in identifying the logical order of steps and evolutions a product or technology may go through. It is also used to sort through and prioritize future research and alternatives. The four major elements of this method are analytics, knowledge building, development of appropriate analogs, and future vision. In addition to the technology roadmap, the TRIZ method generates a list of desired innovations, improvements, and ideas to integrate technology roadmapping into operational plans. These plans, in turn, define a good foundation for future technology forecasts.

8. *Morphological analysis.* This is a systematic, prescriptive technological forecasting tool that uses a matrix of technology stages or parameters across rows and alternative ways to achieve those stages or parameters along columns. Morphological analysis is a valuable creativity technique and forecasting tool because it seeks to identify ways to achieve certain capabilities and estimate the availability timelines of the required technology.

> ***Note:*** *Combining long-term NPD or technological forecasts is useful in cases of uncertainty about the future environment, when large errors translate to large losses, and when there is no clarity as to which method is best. Forecast accuracy is usually improved when combining forecasts from several different methods that use different sources of information. Averaging of long-term forecasts is considered least biased, and assigning weights to individual components based on previous performance and expert opinion can further improve accuracy.*

9.3 SDF TEAM ANALYSES AND EVALUATIONS

By virtue of its position in the company and expertise residing in it, the SDF team is best qualified to lead and project manage the development of NPD forecasts. Determining what needs to be known, what is knowable, and what is not knowable is the starting point of all its activities, analyses, and evaluations performed. The main considerations deciding the scope of SDF team involvement are

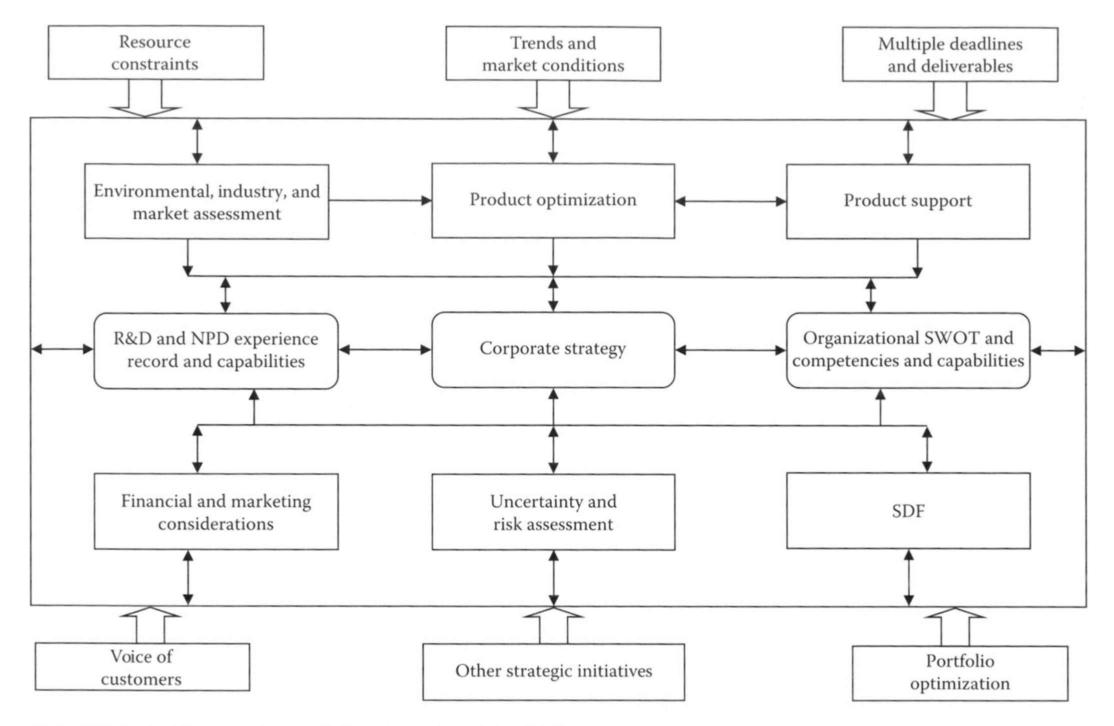

FIGURE 9.3 Factors determining the role of the SDF team.

the factors shown in Figure 9.3, but the context of an NPD project and the SDF team's functional linkages and personal relationships dominate other factors.

Typically, SDF project plans are driven by the team's goals and objectives. The SDF plans that support NPD projects include several activities. The activities performed in conjunction with the project team are highlighted:

1. Assessing project uncertainty and risks and providing inputs for the development of strategic posture
2. Validating client needs, senior management risk appetite, and product launch support requirements
3. *Assembling the support subteams, assigning forecast-related responsibilities to project team members, securing necessary resources, and assuming ownership of the NPD forecast process*
4. Establishing close links with stakeholder organizations and personal relationships with counterparts in those groups as well as external sources of expertise
5. Customizing the SDF processes and functions to be performed at each stage and gate decision point
6. Generating, maintaining, updating, obtaining buy-in, communicating the assumptions, and performing required analyses and evaluations
7. Evaluating the internal and external environment, product attributes, and the product introduction factors determining market acceptance
8. Utilizing the modified stage–gate model, shown in Figure 9.1, to explain to the project team the SDF activities, inputs, and outputs needed at each decision gate
9. Identifying what factors are controllable and the extent of control and what factors are not controllable and their likelihood and timing of occurrence
10. Reviewing of corporate and product strategies, the NPD project rationale, and goals and objectives; clarifying of issues; and validating assumptions and inputs in each stage of the process

11. Managing the information knowledge creation process and ensuring consistent dissemination across all participants through an Intranet-residing SDF database
12. Conceptualizing and modeling the NPD process, determining appropriate drivers and inputs to shape desired future states, verifying organizational capabilities, and identifying and helping remove barriers to successful project execution

The organizations supported by the SDF team in the NPD process are the key stakeholder groups, namely R&D, NPD, PPM, Strategic Planning, Finance, and Marketing and Sales. The activities of the SDF team and other project team members in the different stages and gates are listed in the sections that follow. Those SDF functions and activities that are commonly performed jointly with project team members are highlighted.

9.3.1 Idea, Discovery, and Scoping Stage and Idea and Feasibility Screen

The idea discovery and scoping stage is where new product ideas are introduced and their market potential or technological merits are evaluated. Alternatives are considered and a project definition takes place. The analyses and evaluations performed here are used to create the initial forecast, which is used to screen the new product, service, or technology idea.

A. *Major activities of other NPD project team members.* A lot of effort is expended in this phase, with participation and support from the SDF team, which includes the following:

1. Assessing corporate strategy and conducting business situational analysis
2. Creating a forum to bring in alternative product and project ideas
3. Determining target markets, geographies, and customer population
4. Identifying the degree of corporate risk tolerance relative to the project uncertainty
5. Evaluating the feasibility of bringing the new product to market profitably
6. Conducting own company and competitor PPM review
7. Defining new product features and characteristics and establishing how the new product is to be positioned
8. Estimating resources required and their costs to bring the project to completion
9. Determining internal capabilities and logistics feasibility and identifying possible risks
10. Performing an initial and preliminary financial evaluation of the project

B. *SDF team functions.* The role of the SDF team in this phase consists primarily of performing the following activities:

1. Getting a good understanding of the organizational dynamics in each of the particular NPD process phases
2. *Developing the initial assumption set and considering alternatives to NPD, such as acquiring a product line already in existence*
3. Developing internal functional links and personal relationships, external contacts, and industry sources of intelligence and insights
4. Determining what needs to be known to project manage the forecasting process effectively and identifying the known and unknown, knowable and unknowable, controllable and uncontrollable factors
5. Determining overall project uncertainty, reducing it to residual uncertainty, and identifying risks to help in the subsequent development of strategic posturing
6. Appraising the company experience in NPD projects over the past five to ten years vis-à-vis those of key competitors and ascertaining stakeholder expectations
7. Checking independently the consistency of NPD strategy with corporate, portfolio management, and product strategies

8. Initiating or reviewing market research to better understand new product attributes, quality, and performance or to adjust forecasts based on market research results
9. Obtaining or performing the first round of competitive analysis focused around new product attributes and issues

Note on strategic posture: *Once project uncertainty is reduced to its residual component, a strategic posture defines the project's strategic intent and its implementation with one of the three basic postures:*

1. *Shaping the future where the firm plays a leadership role in setting standards, creating new markets and innovative products, or introducing disruptive technologies*
2. *Adapting to the future that requires determining the speed and agility to take advantage of opportunities*
3. *Reserving the right to play in which the company invests enough to stay in the game, but avoids large or premature commitments*

10. Evaluating megatrends and subtrends impacting not only the industry but also specific company operations, plans, and products and undertaking thorough PESTLED environment, industry, and SWOT analyses
11. Identifying, visualizing, and modeling relationships and determining the key market drivers, causality, and feedback loops
12. Estimating total market, new product uptake, and adoption and acceptance, and assessing technology evolution and complementary products
13. Deciding what SDF method and techniques to use in a specific NPD project forecast development
14. Identifying industry analogs, norms, benchmarks, and best practices using the tools described earlier
15. *Validating the NPD project residual uncertainty and risks and revising the strategic developed earlier*
16. *Revisiting the assumption set, updating or revising it as needed, and determining how the new product is to be positioned*
17. Using influence charts and decision trees to show the project plan logic flow
18. Conducting data and intelligence collection and validation and the first round of the forecasting due diligence
19. Creating and describing the initial forecast scenario, generating the first NPD forecast, assessing and communicating forecast implications, and obtaining feedback and input from the project team
20. Conducting sanity checks on assumptions, data inputs, and implications of the initial forecast and incorporating project team feedback in the assessments
21. Revisiting the forecast risk factors and their sources as well as the timing of trigger events, and potential impacts
22. Developing a baseline consensus forecast and uploading to the Intranet database and making it available to the project team
23. Conceptualizing the elements of a response plan to competitor reaction

9.3.2 Business Case Stage and Go to Development and Testing Gate

In this stage, independent, critical, and objective SDF team assessments take place that lead to sharper project definition, an NPD project plan, and allocation of resources to the project. Forecasts are updated and used to decide whether to move on to the next stage.

A. *Major activities of NPD project team members.* In this stage, the focus of project team members now shifts to the following activities:

1. Evaluating and incorporating the findings of market research and developing a sharp product definition
2. Assessing the technical, marketing, logistics, and business feasibility of the NPD project
3. Defining in detail the NPD project structure and a realistic project plan and obtaining project team buy-in
4. Providing the project justification in terms of rationale, alternatives considered, and financial results for this stage
5. Identifying resource requirements for each stakeholder organization and product introduction
6. Securing funding for the resources required to implement the decision, launch the new product, and manage the future-stage realization plans

B. *SDF team functions.* The activities, analyses, and evaluations performed in this phase of the NPD project by SDF team members consist of the following:

1. Deepening the personal relationships with the NPD project team members, clients, and the external network contacts
2. Redefining the SDF team's project goals, objectives, and strategy and taking over all the NPD forecast project management responsibility; that is, fine-tuning the SDF project plan, assigning additional roles and responsibilities to participants, identifying milestones and timelines, defining inputs and outputs, and verifying resource requirements
3. Undertaking an independent, critical, and objective assessment of the business and NPD processes, and other strategic initiatives under way
4. Continuing the forecast analog, benchmarking, and due diligence efforts and updating the competitive analysis findings with new insights and conclusions
5. Updating the megatrend, subtrend, and industry analyses and PESTLED environment evaluations
6. Verifying and validating the market research results vis-à-vis updated findings of the own company and competitor SWOT analyses
7. *Creating an updated and complete assumption set and obtaining project team agreement and support*
8. *Examining the differences between the initial and the present set of assumptions and explaining the reasons for the differences*
9. Adopting a system dynamics approach to handle the NPD forecasting problem and begin the modeling of bringing the new product from this stage to market using influence charts and decision trees
10. Evaluating the roadmapping of markets and segments as well as technologies to verify the NPD project feasibility
11. *Undertaking an independent validation and verification of the cost modeling to provide insights into product pricing and price change effects*
12. *Reexamining and ensuring consistency of NPD strategy with PPM objectives and expectations*
13. Updating the market research input on customer and market needs and product attributes relative to competitors
14. Identifying black swans and possible future event interactions using the Delphi or the cross-impact matrix method
15. Creating and describing a few distinct, plausible future-state scenarios, developing future-state projections, and assessing their implications

16. Obtaining project team and senior management buy-in on the inputs used in the scenarios and conducting simulations, hypothesis testing, sensitivity and "what-if" analyses, and sanity checks
17. Generating scenario forecasts, obtaining NPD forecasts from other sources, and combining them according to confidence of realization in each
18. Identifying forecast risks, their sources, implications, and ways to handle them and initiating the future-state realization (FSR) planning effort
19. Examining the differences in future-state projections between the first and the current set of forecasts and validating the reasons for them
20. *Developing a consensus view of future-state projections and noting differences in expectations among different team members*
21. *Creating a management dashboard and an EWS for the NPD project*
22. Identifying SDF team inputs and resources needed for the next stage of the NPD process and ensuring smooth transitioning of forecasts to the next stage, continuity of purpose, and senior management involvement and support

9.3.3 Develop, Test, and Validate Stage and Go to Launch and Implement Gate

This is a crucial phase where detailed plans are laid out and tested and IVV takes place. Assumptions and inputs to scenarios are firmed up and a consensus forecast is created, which is used to estimate the value created by the NPD project.

A. *Major activities of NPD project team members.* The work of the project team members in this stage is in preparation for the launch and implementation gate and centers on the following activities:

1. Conducting a new situational analysis and reflecting on recent market developments
2. Assessing different new product launching, pricing, and promotion schemes
3. Deciding on the final product design and specifications and attributes based on the market research and conjoint analysis results
4. Creating the manufacturing, marketing, operations, and logistics plans
5. Validating the NPD project financials, identifying remaining uncertainty and risks, and developing plans to manage them
6. Conducting an objective new product value audit vis-à-vis future PPM adjustment plans
7. Planning for the new product launch and assessing possible market and competitor reactions to it
8. *Testing the project management dashboard and the EWS*

B. *SDF team functions.* The major activities of SDF team members in this NPD project phase include the following:

1. Cultivating further relationships with project team members and external sources of intelligence, insights, and advice
2. *Validating customer acceptance of the new product and customer feedback*
3. *Creating an updated and revised set of assumptions and documenting differences over the earlier set and the motives and reasons for the changes*
4. Updating, verifying, and validating the market research, competitive analysis, forecasting due diligence, and the megatrend, subtrend, PESTLED, and SWOT analyses
5. *Creating comprehensive measures of overall NPD project, SDF team, and forecast performance*
6. Simulating and testing the reliability of the EWS with inputs from real-time operational data and feeds to the management dashboard

7. Evaluating the roadmapping of markets and segments as well as technologies to ensure the NPD feasibility and viability
8. Revisiting the megatrend, subtrend, and market trend evaluations and the assessments of implications of the PESTLED, industry, and SWOT analyses
9. Applying lessons learned from company and competitor product introductions in the past five to ten years to the development of the current SDF solution
10. Modeling the new product launch by market and stage and updating identified key drivers, causal paths, and feedback effects
11. Obtaining market research inputs on customer and market needs and product attributes relative to competitors and reestimating total market, product uptake, and adoption and acceptance
12. Probing the answers given to the basic questions raised in conducting IVVs in the idea screening, concept development and testing, the business analysis, the beta testing and market testing, the technical implementation, and the launch phases
13. Finalizing the forecasting due diligence, reevaluating market research results, and providing adjustments based on more recent facts and assessments
14. Recreating or refining sound plausible scenarios of future states and describing the sequence of events leading to those states
15. Using an integrative forecasting approach and obtaining forecasts from different models, sources, and external expert opinions and creating composite forecasts
16. Assessing the implications of different forecasts and the risks involved in each one and conducting further sanity checks
17. Communicating the results from different forecast models or sources and obtaining stakeholder feedback on the combined forecast
18. Creating a consensus view of future-state projections, obtaining senior management inputs, and making appropriate adjustments
19. Identifying risks inherent in the latest consensus NPD forecast, assessing potential impacts, and determining the likelihood of occurrence
20. Undertaking a critical and objective evaluation of cost model results to determine appropriate pricing ranges and changes in planned promotions
21. *Reassessing alternatives to NPD, such as acquiring a product line already in existence or a company with an exceptional product line*
22. Uploading the baseline forecast and details associated with it to the company Intranet for access by the project team
23. *Monitoring market developments and creating a complete product forecast realization and competitive reaction plan with specific strategies, moves, and actions*

9.3.4 Post New Product Launch Review

Ordinarily, after the introduction of a new product, service, or technology, the project lead organization assumes that the responsibility of tracking performance and SDF team's involvement is reduced to an advisory role, if any. However, in some cases, its continued involvement is warranted because of risk management considerations and SDF value added in implementing competitive response and forecast realization plans and because product modification assessments may be required.

A. *Major activities of NPD project team members.* Their activities in this project phase focus on NPD performance evaluation and include the following:

1. Evaluating the effectiveness of marketing and advertising programs assumed in the analyses
2. Fine-tuning the new product features, pricing, and sales and marketing
3. Tracking new product actual monthly results against projections
4. Conducting variance analysis and identifying problem areas to be addressed

5. Ensuring that all logistical glitches are resolved successfully
6. Assessing whether changes are needed in any aspect of the new product launch support

B. *SDF team functions.* The supporting functions performed by SDF team members in this phase of the NPD project consist of the following:

1. Reassessing the validity of assumptions used in the forecast based on most current information, recording findings and causes of errors, and making adjustments
2. *Assessing the levels of market acceptance and customer responses against prelaunch expectations*
3. *Monitoring market and competitor reaction as product modifications and price and promotion changes take place*
4. Monitoring performance of the long-term forecasts used to justify the NPD project, but handing over to Tactical Forecasting the responsibility of creating market segment and product subcategory forecasts
5. Keeping scores of major experiences and lessons learned and industry NPD experiences from the NPD project
6. Monitoring the EWS and the NPD management dashboard, generating reports, and communicating pending or risks appearing on the horizon
7. Considering conditioners or leverages such as public relations and additional product support to use when needed and providing expertise and assistance in the execution of the FSR plan
8. *Conducting a project postmortem review and communicate findings to the NPD stakeholders with recommendations for future projects*
9. Obtaining project team, client, and senior management evaluations of SDF team performance and identifying areas of improvement
10. Uploading data, intelligence, insights, models, and results of evaluations to the corporate Intranet database

9.4 SOUND PRACTICES AND SDF TEAM CONTRIBUTIONS

9.4.1 Sound Practices and Success Factors

A recent McKinsey survey of the US and European companies found that the companies best at NPD do three things better than those less successful, namely, they

1. Define clearly and communicate well project requirements to stakeholders at the project's start.
2. Nurture a strong project culture company wide.
3. Keep in close contact with customers in the entire product development process.

Our discussions with NPD project participants concerned with issues around success factors produced a long list of factors considered essential to success of NPD projects. This is an indication that each industry and company has its own needs and ideas about what constitutes best practices and success factors. Following are groupings of factors considered critical to different sets of participants in our discussions and by industry experts.

Category 1: Comprehensive success factors. In this category, participants thought as essential to success the following aspects of NPD projects:

1. Identifying success factors, obtaining project team and client buy-in, evaluating their impact, and prioritizing the project team's focus and activities
2. Securing senior management involvement, commitment, support, and the necessary resources dedicated to the project

3. Having sound, clearly articulated corporate strategy and a well-defined new product strategy driving the project
4. Creating complete and tested NPD and strategic forecasting processes, ensuring quality of execution in all stages, and providing the required support for a successful product launch
5. Placing heavy management emphasis on upfront homework before the development stage
6. Managing with strong market orientation, hearing the voice of the customer, and creating sharp product definition before development
7. Adopting flexible system dynamics models and quality of scenarios to produce reliable forecasts, decisions, and implementation plans
8. Performing sound external assessments, forecasting due diligence, and ongoing vigilance
9. Possessing superior project, knowledge and forecasting function management capabilities
10. Ensuring high levels of 4Cs and company-wide participation, alignment, and support

Category 2: Healthcare industry factors. For discussion participants from the healthcare industry, strategic forecasting support was very important, but it did not make the short list of critical success factors. The success factors common in all responses were excellence in strategy, planning, and project management. However, there were different success factors among companies in that industry as shown by the following statements of participants with NPD experiences: a sense of urgency in getting a product to the market, outstanding public relations, and being first to market but flexible as well. Also, the degree of R&D productivity matters as do communications and teamwork up and down the chain of command, quality of business intelligence, and functional expertise of the project team.

Category 3: Implementation factors. This group of participants was most concerned with the execution aspect of the NPD process and indicated that for their needs, the following practices were required to achieve success:

1. Identifying the long pole in the tent and centering attention to highest importance and longest duration elements to avoid diffusion of effort
2. Creating a good SDF teamwork plan starting from decision gates and working backward
3. Investing twice as many upfront homework person days than those of average projects are required for successful NPDs
4. Providing reliable inputs to feasibility studies and the business case: data, assumptions, models, analyses, evaluations, scenarios, projections, and synthesis of the results
5. Defining performance measures, indicators, and the scope of monitoring, and conducting thorough variance analyses
6. Leveraging the capabilities of Enterprise Resource Planning Systems, such as SAP and Hyperion, to synchronize forecasting with marketing events at the product launch phase

Category 4: Project and forecasting management. A group of project sponsors and industry experts considered project and forecasting function management in NPD of the highest importance. In their view, effective forecast and overall project management should include the following elements to ensure NPD project success:

1. Project team excellence and reasonable planning, execution milestones, and deliverables
2. Monitoring the deliverables, timelines, costs, and quality of inputs to and outputs from the SDF and the NPD processes
3. Integrating SDF with the broader NPD processes and managing the SDF team's activities, workflow, and handoffs across the project consistent with NPD process swim lane charts
4. Creating, documenting, updating, communicating, and managing the assumptions and updating forecasts when conditions and new intelligence warrant it

5. Managing effectively issue resolution and process control functions at every stage by ensuring no strategy conflicts and the project team exhibiting high levels of 4Cs
6. Building and updating periodically the Intranet database to house and share knowledge created in the project

Category 5: Best practices. Based on benchmarking experience, this group of participants gave the following as best practices and success factors in their NPD and launching experiences:

1. Sound corporate strategy, resource commitment, constancy of purpose and strategic intent, agility, flexibility, alignment of functions, and effective project execution
2. Early and sharp product definition and product superiority in quality, differentiation, and unique benefits along with sound pricing and launch strategy
3. Market attractiveness vis-à-vis product attributes, internal capabilities, and adequate and dedicated resources allocated to the NPD project
4. Attention paid to the support required and enablers in the launch phase of the new product and to controllable levers to use in the postlaunch period
5. SDF team positioning and structure, expertise, capabilities, and close business and personal relationships with other NPD project stakeholders along with high levels of 4Cs practiced throughout the project across all functions
6. Independent, critical, and objective assessments of the firm and competitors; complex project modeling capabilities; and scenario planning to simulate the future environment
7. Alignment and coordination of plans and processes with broader NPD processes, SDF team forecast ownership, and actual senior management support consistent with planned levels
8. Ongoing sanity checks and forecasting due diligence in each NPD stage
9. Institutionalized learning and knowledge sharing through a well-structured Intranet data and intelligence warehouse
10. Ability to identify impediments and constraints, monitor progress within and between stages, and validate NPD customer and market orientation
11. Effective competitive response and FSR planning and management
12. NPD management dashboard with feeds from operations systems, the EWS, the project manager, market research studies, and financial reports

9.4.2 SDF Team Contributions

The value added and major contributions of SDF team members supporting NPD projects are summarized from participant statements and include the following:

1. Developing, testing, monitoring, updating, and reporting on assumptions used on a consistent basis throughout the company
2. Modeling the NPD process beginning with the current state or status quo and NPD goals and objectives and linking them to the next stage sequentially
3. Conducting independent, critical, and objective assessment of the business, the company's and product's standing vis-à-vis competitors, and the external environment
4. Developing and project managing the entire SDF process; performing reality checks on an ongoing basis; and describing, testing, and revising scenarios of the ultimate future states
5. Building simulation models to test hypotheses, inputs, and random event occurrences; identify black swans; and guide scenario planning exercises
6. Helping the project team to understand through scenario simulations what events, plans, support, and conditions are necessary to achieve the desired project outcomes

7. Identifying internal bottlenecks and constraints and helping in the identification, sharing, allocation, avoidance, and management of project risks
8. Creating an EWS to track and communicate risks and helping to create the FSR plans

The SDF team's contributions and value added in new product launches are further demonstrated in the case of product introductions in new markets and geographies, which are addressed in Chapter 10.

Key lessons learned in NPD projects:

- *Research the subject before engaging in forecasting for NPD projects and master the basics.*
- *Study the techniques and tools used in NPD projects; use only the ones you need.*
- *Strive for consensus on assumptions because they create forecasts and forecasts create conflict.*
- *Understand the NPD process well enough so that you can take shortcuts if you need to.*
- *Learn from the experience of others, but establish your own project success factors and create appropriate performance measures.*
- *Project manage the forecasting function that requires technical qualifications and experiences, interpersonal skills and emotional maturity, and strong functional links and close personal relationships with key project stakeholders.*

10 New Market or Business Entry

This chapter is an extension of the new product development (NPD) discussion and builds on and supplements those analyses and evaluations for new market or new business entry projects. Entry into a new market means starting operations in a new geographic area or market strata, with a reference point being the experience in the markets the company is currently operating. Entry into new business areas on the other hand entails sailing into unknown waters with no reliable reference points other than the decision maker's, the forecaster's, and the project team's investigations, and their prior experiences. The former case requires an SDF solution that is highly intensive in its environmental assessment, while the latter entails that plus more in-depth understanding of trends and objective evaluation of internal competencies and capabilities in the context of the new company strategy.

The focus of this chapter is on new market or business entry that is strategic in nature, which means they are long-term commitments that involve substantial financial and human resource investments. Under this criterion, entry into new parts of a country the company has a presence in, entry into a new market stratum, and market development are not part of this discussion. However, entry into a foreign market or new business is a major piece, and the underlying reasons for such projects include

1. Severe domestic market crowding and pressures, aging or outdated products, services, and technologies.
2. Preemptive competitive positioning to meet international customer needs, introducing new products, services, technologies, or ideas and testing their acceptance.

Entry into a new similar business is undertaken to establish long-term market presence, or due to cost and portfolio management considerations. Entry into an unrelated business area is motivated by factors such as sourcing low-cost production inputs, unique or untapped market opportunities, and market exploitation with high profit margins. Entry into a peripheral business takes places primarily for two reasons: horizontal integration growth or vertical integration control.

Both of these growth endeavors employ most of the analyses and evaluations performed in the NPD area, and it is worth noting that, for most practical reasons, the difference between the new market and the new business area is blurred. For example, entry of a US company into the Chinese market in the 1980s and entry of a telecommunications company into the credit card business in the United States involved customized approaches, but similar tools, modeling methods, analyses and evaluations, uncertainty and risk assessment, and scenario development and planning. All analyses and assessments, however, need to be conditioned to fit the different operating environments.

Sometimes new market entry projects may involve new ways of doing business or some kind of technological innovation and, infrequently, product development for the new market. For the purposes of our discussion, new market entry refers to international market entry and is an option for company growth often viewed easier to accomplish than domestic NPD, along with product licensing, technology innovations, acquisitions, and joint venture (JV) projects.

10.1 THE CURRENT STATE

Observations of the current state of forecasting for new market or business entry reveal that they are mostly driven by Marketing and Sales organizations and globalization efforts with short-term financial objectives. The second observation is that forecasts are based on less than clear long-term strategic and future growth objectives and use simple transfers of domestic market experiences and evaluations to international markets. Hence, the intense focus is on market research, market share analysis, distribution channels, and marketing plans, and the less emphasis is on understanding the peculiarities of the host country's business environment.

There are a number of practices common to most new market or business entry projects, such as those often being initiated by business unit interests and financial factors, and, in reality, long-term strategic issues are secondary considerations. These projects are characterized by inadequate environmental and incomplete industry assessments, and little attention is paid to entry mode assessment, entry strategy, and consistency with corporate strategy. The country analyses performed are quick and cursory, and no in-depth evaluation of risks takes place. Instead, backward looking, biased, and domestic market-oriented strengths, weaknesses, opportunities, and threats (SWOT) analyses of the company and competitors are common occurrences. And, since these practices do not address entry uncertainty and assessment of the risks, they result in wrong strategic posturing. Also, there is indiscriminate use of market research results, inflated market potential estimates, and undue reliance on known market and industry analogs, norms, and experiences.

There is a shortage of experienced personnel assigned to evaluate and implement these projects and, in the majority of foreign market or business entry cases, there is heavy reliance on external advisors. Costly new market entry affected through product upgrades, extensions, lifecycle management, and pricing strategies is a common practice in the current state as is the mostly "go-it-alone" approach in markets or business areas of no recognized company expertise and competencies. Assumptions are usually created by the originator of the idea without supporting evidence or the participation of the forecasting group. And, because of lack of strategic forecasting experience, long-term forecasts are simple extrapolations of short-term forecasts derived from domestic experiences with few sanity checks being conducted. Also, minimal testing of assumptions is performed, and project uncertainty and risks are not really investigated; they are simply ignored.

Forecasting organizations are brought into the project late in the process to rubber stamp client expectations and forecasts based on domestic industry analogs, which are routinely applied to new market or business entry cases. In most cases, there are no established processes in place for new market or business entry projects or even a forecasting management function. Forecasters have a limited involvement in the ongoing monitoring of market conditions and foreign operations performance, and ordinarily ad hoc approaches are used to assess competitor reaction and create risk management plans. Furthermore, there is little scenario development, hypothesis testing, and planning done to simulate the conditions, events, and project support necessary to achieve desired results.

The results of the current practices of developing long-term forecasts for new market or business entry projects are reflected in a number of observed consequences, such as the following:

1. Long-term company growth problems are treated as short-term opportunities to enhance financial performance through new market or business area entry projects.
2. Scenario planning, risk assessment, and competitive response planning are not part of the new market or business entry or the current forecasting processes.
3. Forecasting processes are absent, incomplete, and broken down; analyses are flawed; and poor evaluations lead to wrong decisions.
4. Lengthy and delayed project implementations and, sometimes, reversal of decisions with correspondingly high exit costs
5. There is costly experimentation with and unmanaged engagement of external consultants, foreign business agents, and subsequent intellectual property losses.
6. Overly optimistic market potential estimates drive the forecasts, while product quality, features, and customer acceptance take second place.
7. Long-term forecasts are deterministic and simple baseline, best-case, and worst-case projections with no credible scenario behind them or assumption testing involved.
8. Forecasts are frequently revised according to the project sponsor's needs and to justify certain decisions.

9. A less-than-expected performance materializes and, often, failure to capture sufficient market share is required to maintain profitable presence.
10. Evaluations of new business entry proposals are delegated to and conducted by Marketing and Finance organizations or by external advisors with conflicting interests.
11. Commonly occurring lack of operational control, quality standards, service excellence, sound performance measures, and costly interventions to remedy problems
12. Once entry is achieved into an international market, the project is treated as a domestic market initiative.
13. Frequent new market or business entry projects produce a false sense of experience in a company and inadequate problem understanding persists.
14. Exiting from new market or business areas is viewed as a part of market management; that is, they are considered a short-term problem not impacting corporate strategy, image, and brand.

10.2 SPECIAL NEEDS, ISSUES, AND CHALLENGES

The first major issue in new market or business entry projects is resolving differences of interests among project stakeholders and getting consensus on the approach to move from the current state to the future state of business, given the increased levels of uncertainty and business risks associated with such projects. The second major issue is the absence of established long-term new market or business entry strategy, policies, processes, and practices.

Other issues include determining which market or business to screen for possible entry and choosing the criteria to use to select among competing alternatives, countries, entry modes, strategies, and opportunities. Understanding the new market and customer motivations, tastes, preferences, and buying behavior; interpreting and making sense of international market research results; and understanding their implications are crucial to successful entry, but difficult to do properly. Hence, the importance of recognizing the intricacies of international marketing and advertising as well as learning from the experiences of other entrants.

Success in this type of projects requires fully appreciating the differences between domestic and foreign business practices and ways of getting things done, knowing how to detect unfair competition, lack of law enforcement and corruption, and how to deal with such situations. It also requires getting a correct assessment of project uncertainty and risks in the context of unknown environments, developing strategic posture in unclear risk tolerance situations, and identifying and assessing correctly structural and strategic barriers to entry. At the execution phase of these projects, aligning functions, processes, and plans to execute strategic initiatives and objectives is a challenge as is developing, selecting, and implementing new market or business entry strategy when a project is driven by short-term considerations.

In new market or business entry projects, there is an ongoing need for identifying and evaluating objectively competitive threats, predicting competitive reactions, and having response plans ready to deploy. Such plans are difficult to manage due to the inability to ensure availability of a qualified local labor pool, materials sourcing, and supply chain considerations, especially in developing countries. Other challenges in foreign markets include getting and maintaining control of operational costs, quality standards, and service levels as is finding qualified management talent for overseas assignments and trained local employees to fill important positions. Using unreliable international benchmarks or best practices in evaluating new market or business entry, generating long-term forecasts, and dealing with the multiplicity of risks involved in international new market or business entries with less than qualified resources are treacherous and present formidable challenges.

Accounting for the variability of product uptake by country, type of product, and market sophistication in long-term forecasts is a major challenge that needs to be addressed properly to reverse the absurd practice of validating project rationale after decisions are made and before all entry modes and strategies have been evaluated. Furthermore, aligning functions and processes to strategic objectives, defining clear roles and responsibilities for all project stakeholders, and ensuring high-level communication, coordination, cooperation, and collaboration (4C) practices are important issues that go unresolved.

These problems are amplified by not addressing the lack of the right forecaster experience and the absence of internal and external links, relationships, and partnerships. Also, getting project participants to be objective and admit company weaknesses and competitor strengths is often next to impossible in many of these projects. And, introducing discipline in the forecasting and decision-making process, given the insistence of marketing organizations to project manage all functions, makes these projects difficult to create objective future-state scenarios and to execute them successfully.

With these limitations, special issues, and challenges as the background, we discuss the strategic decision forecasting (SDF) support and solutions for international market and the new business entry areas. The treatment of these projects starts with validating the project feasibility and assessing the project uncertainty and risks involved. Attention is given to treating new market or business entry as a long-term, strategic decision with a corresponding orientation and use of the right analytical tools. Once alternatives are assessed, clarifying the project rationale and developing project goals and objectives take place. Then, the creation of end-to-end SDF processes and objectives and project management of the strategic forecasting process are discussed.

10.3 SDF APPROACH AND PROCESSES

We mentioned earlier that forecasting for new market or business entry involves many of the analyses and evaluations in new product, service, or technology introductions and this discussion is a continuation of the NPD discussion. The approach SDF teams use for these projects is a complementary approach where one analytical model leads and complements another. As in the case of NPD projects, a key point is that the processes and analyses must be tailored to the specifics of the situation and the project strategic objectives and the project team orientation must be such that long-term considerations dominate.

The generalized new market or business entry approach to developing SDF solutions is shown in Figure 10.1 and the forecasts resulting from that approach are based on the following five major factors:

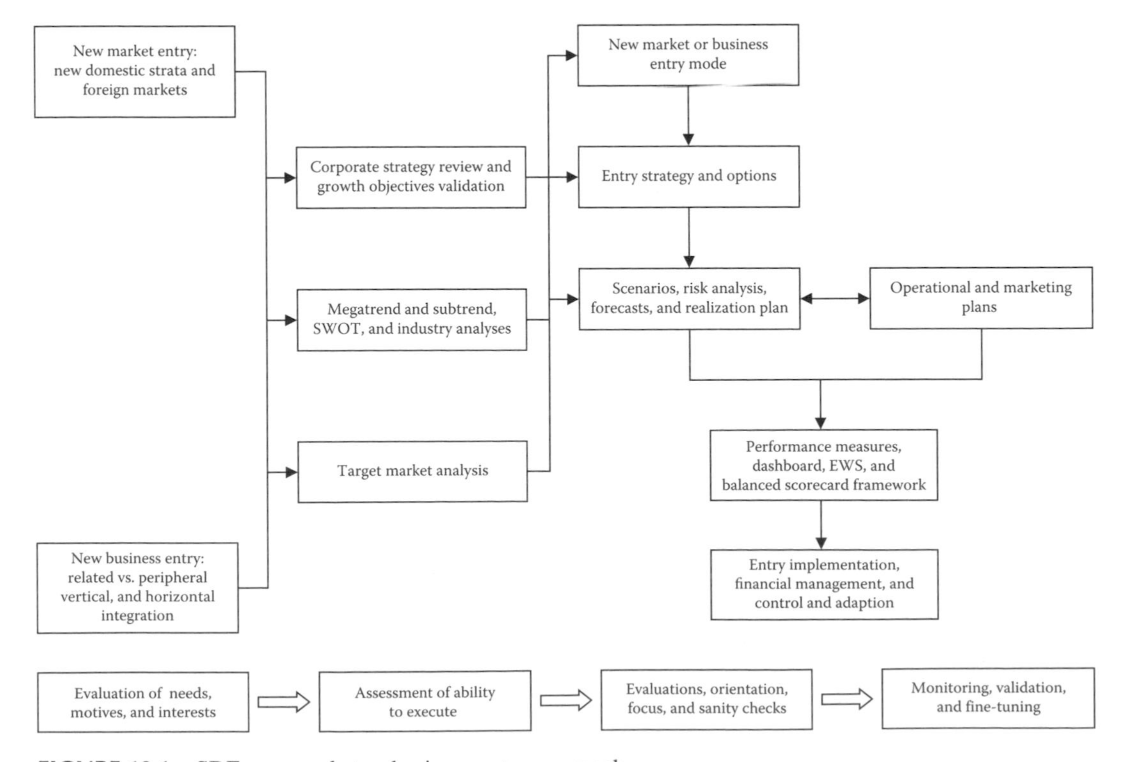

FIGURE 10.1 SDF new market or business entry approach.

1. Evaluation of customer, corporate, and stakeholder needs and interests in the context of corporate future growth strategy
2. Independent, critical, and objective assessment of the company's ability to implement the new market or business entry strategy
3. Proper project team and client orientation and focus leading to sound evaluations, thorough forecasting due diligence, and scrupulous sanity checks
4. Creation of reasonable assumptions and well-considered scenarios used to simulate the effects of different events and interventions
5. Close monitoring of the external environment, the industry structure, market developments, accomplishment of objectives, and overall project performance
6. Validating forecast realization and competitive responses, fine-tuning underlying models and forecasts, and revisiting entry strategies

Notice that domestic or international new market or business entry via acquisitions or JV projects requires a different approach, which will be discussed in Chapter 11.

The SDF new market or business entry approach shown in Figure 10.1 is the foundation used to develop the SDF process and includes the following activities listed by major groupings:

A. *Needs identification.* In this area, SDF team members first determine what needs to be known to make the right decisions, what is known or knowable, and what factors are controllable and uncontrollable. They review the company's current growth strategy and policies and goals and objectives early in the process and, later on, they provide support to project teams in terms of

1. Verifying customer requirements, needs, and preferences, and identifying market structure changes.
2. Determining the interests and priorities of the key stakeholder organizations and facilitating reaching a common perspective.
3. Establishing the levels of required inputs and support requirements to affect a successful new market or business entry.
4. Identifying sources of forecast risks, assessing probability of occurrence, and helping the project team develop a risk management plan.

B. *Ability to execute.* The ability to implement successfully new market or business entry projects using the approach shown in Figure 10.1 requires managing well the following SDF team activities:

1. Assigning the SDF team members to the new market or business entry project team at the inception of the project to validate project rationale, client needs, and objectives, and proper project structure
2. Helping define achievable project goals and objectives, establishing the SDF team's objectives, and conditioning sponsor and senior management expectations to new market or business area realities
3. Developing functional links and close personal relationships with counterparts in stakeholder organizations and with external partners, consultants, and experts
4. Creating and communicating a preliminary set of reasonable assumptions in conjunction with the lead organization and obtaining buy-in from all stakeholders
5. Performing an industry analysis with emphasis on barriers to entry, rules of competition, and business practices, and conducting megatrend, subtrend, and political, economic, social, technological, legal, educational, and demographic (PESTLED) environment assessments
6. Evaluating capabilities and competencies of new market rivals and potential entrants using objective analytical methods
7. Assessing the availability of company and new entity qualified managers to implement and manage the operations of the new entity

8. Providing intelligence, insights, and input and guidance to the project team evaluating different modes of entry, such as Greenfield operations, indirect or direct export, creating a marketing subsidiary, local partnership or JV, licensing, or acquisition
9. Helping in the selection of entry strategy and scope and developing strategic intent based on critical, objective, and honest internal and external environment evaluations

C. *Analytical support.* In this area, the SDF approach requires that team members perform a number of evaluations and checks that include the following:

1. Assessing overall project uncertainty, determining the degree of corporate risk tolerance, and helping develop a strategic posture
2. Scanning potential markets or business opportunities for the purpose of narrowing down the number of options
3. Evaluating the impacts of megatrends and subtrends on the products of the business created by the new market or business area projects
4. Conducting an in-depth country analysis to determine the impacts of possible country, systematic, systemic, credit, and liquidity risks
5. Undertaking a critical and objective internal and key competitor SWOT analysis with a forward-looking orientation
6. Engaging in a target market analysis with three objectives in mind: understanding customer needs, preferences, behavior, and uptake; estimating market potential; and appreciating how business is conducted in the new market or business area
7. Modeling the new operating environment and creating and testing of plausible, distinct scenarios describing the transition from the current state to the future state
8. Creating scenarios, assumptions, and events; determining their timing; identifying black swans; developing a set of forecasts; and testing their reasonableness
9. Performing hypothesis testing and sanity checks and assessing the financial and human resource implications for the different scenario forecasts

D. *Validation of activities.* In most cases, the involvement of the SDF team in monitoring, validating, and fine-tuning areas encompasses the following activities:

1. Confirming strategic, cultural, operational, and technological fits, and helping the project team define project performance goals and objectives
2. Identifying project performance measures and defining performance indicators and monitoring actual performance by the management dashboard indicators
3. Conducting variance analyses that seek to understand the root causes of forecast deviations and help in updating forecasts
4. Creating, along with the project team, a competitive response and forecast realization plan, a management dashboard, and an early warning system (EWS)
5. Creating the project's balanced scorecard objectives to be monitored over the forecast period

10.4 SDF ANALYSES, EVALUATIONS, AND TECHNIQUES

10.4.1 SDF Analyses and Evaluations

The first priority of the entire SDF team in new market or business entry projects is the development of relationships within the client, the project team, and other planning organizations and external experts. Prior to analyses and evaluations, SDF team members review the corporate long-term growth strategy, goals and objectives, and policies and practices. Then, they assess corporate risk tolerance and senior management comfort levels operating in environments very different from the present domestic operations. Following these activities, they perform an iterative determination of project uncertainty level and development of strategic posturing in conjunction with entry mode and entry strategy selection.

After SDF team members clarify customer needs and key stakeholder interests, goals and objectives, and expectations, they help develop consensus on project goals and objectives. This is followed by a review to determine the extent of prior new market or business entry project experience and skills required in the evaluation of industry analogs, benchmarks, standards, and norms at the business level. Development of the SDF team's goals and objectives is next, followed by dedicating forecasting, competitive analysis, and market research team members and assigning roles and responsibilities to participants in the forecasting function. The standard identification of what needs to be known and what is known, what is knowable and unknowable, and what factors are controllable and which are not happens next. Finally, SDF team members undertake the data collection, the verification and validation effort, and the development of intelligence and insights useful to the project.

The specific analyses and evaluations performed by SDF team members vary by type of project and across companies, but most common are the following activities:

A. *Megatrend and subtrend analysis.* Megatrends are significant and unmistakable forces in the global economic, political, cultural, and technological landscape. They are new patterns and directions that will define the business environment and affect the new geography or business area over the next seven to fifteen years. For new market or business entry projects, especially in foreign countries, the examination of megatrends and subtrends begins at the global level. Then, it moves on to the country level, to the industry group level, to the subindustry level, to the company, and finally to the product level. The systematic analysis of megatrends and subtrends involves the sequence of identifying the relevant trends, gaining understanding of their nature, assessing potential impacts on the company business and the project, and preparing to react to them. Reaction to external changes takes place through future-state realization (FSR) plans that include tactical or strategy changes, modification of plans, innovation, and improvement.

B. *Country risk profile.* The purpose of this analysis is to identify country risks, determine the company's ability to manage those risks, and determine national competitiveness. It entails an extensive review of a country's past and present economic environment, political stability, social trends, business practices, tax policies, import duties and export subsidies, and power of labor unions. It also requires an assessment of the legal and regulatory environment, safety and intellectual property protection laws, demographic profiles and trends, and availability of an educated, trained, and trainable labor force. Country risk analysis also calls for an evaluation of the country's overall state of technological development and sophistication, access to competitively priced resources, and presence of supporting industries.

C. *Industry structure analysis.* This important analysis is intended to make qualitative evaluations of a fi rm's strategic position in the new market, and the two primary functions are to assess the current intensity of competitive rivalry and to determine barriers to entry of the structural as well as of the strategic type. To determine barriers to entry entails assessment of economies of scale, product differentiation, capital investment requirements, customer acquisition/migration costs, access to cost-effective distribution channels, and government policies pertaining entry into its markets. As in the case of megatrends and country risk assessments, to be useful this analysis is performed at the line of business level and not at the industry group level.

D. *Market analytics.* The market analysis effort for these projects is equivalent to the grounding of the analysis and the forecasting due diligence discussed in Chapter 6 and mirrors the country risk and industry structure analyses; that is, it is done for markets in the country that is the target of entry, or the new business area, and the first step is to articulate the company's value proposition of entry. This involves a multiplicity of assessments, the most significant being the following:

1. Understanding the target country's culture and the forces shaping the market today and likely to do so in the future
2. Assessing the impacts of megatrends and subtrends on the new country, the new market, or the new business and those of the PESTLED environment changes
3. Assessing the new market or business tendencies and trends and understanding the implications of the differences with domestic practices
4. Using market research to identify customer needs, determine changes in tastes and preferences, develop customer profiles, validate results, and decide on the marketing infrastructure needed
5. Assessing the strengths of current competitors, potential future entrants, and how different local alliances may impact the project
6. Estimating the potential market size, target groups, and market segments in a critical and objective manner
7. Reviewing the company's and competitor's past experiences, and evaluating actions taken, timing of competitor entry, and scale
8. Validating industry entry benchmarks, norms, and analogs, and soliciting expert opinion concerning product uptake and market share estimates
9. Assessing market growth potential in the context of the industry and product life cycles and soliciting local market expert validation
10. Determining differences in marketing infrastructure and in costs for advertising and sales over current operations

E. *Understanding new market customers.* In all strategic projects, understanding the customer is a priority one and a critical success factor. To achieve sufficient level of customer understanding necessitates several assessments, the most important ones being appreciating the local customers' customs and habits, tastes and preferences, expectations, and buying motives. It also requires evaluation of the input of local market experts and opinion leaders and using it to adjust market research results. It is essential to address the differences and similarities between current market and new market customer behavior in selecting uptake analogs to produce long-term forecasts. It is vital to determine the right product selection for new market or business entry and whether product modifications may be in order. It is also important to ensure that entry is affected through a good product design, high quality, and features appropriate for the new market and understanding the local market or new business area dynamics in terms of customer price sensitivity and response to promotion and advertising.

F. *Assessing company competencies.* The assessment of company's competencies and capabilities begins with how the company creates value and identifies what the company could do better than competitors in a different environment. Then, it proceeds to evaluate the company's competencies and capabilities in a SWOT analysis format versus those of competitors. That is, in addition to assessing competencies and capabilities, the competitive analysis performed in this step evaluates the company's functional skills, the availability of financial resources, and the leadership capabilities needed for a successful new market or business entry. Thus, it establishes the company's competitive position through the following elements: (a) the product, service, technology, or cost differentiation; (b) the company's ability to reduce customer costs; and (c) the capacity to pursue system lock-in, lock-out, or proprietary standards.

G. *Cost estimate accuracy.* Reliable cost estimates are as important in determining new market or business entry viability as are demand and revenue forecasts. This type of cost analysis is done by Finance with support from the SDF team and begins with developing current market production costs and then determining additions and subtractions to get appropriate adjustments for the new market.

Understanding the upfront investment requirements and additional capital expenditures and the labor costs involved in the new market or business entry is crucial. Other cost components to be determined include product, service, or technology adaptation costs to the local market, import duties and export subsidies, tax liabilities, and labor contract obligations. This is especially needed in the case of new business entry. Once reliable cost estimates are developed, export prices can be compared to the foreign market production costs as inputs to forecasts supporting various decision options.

H. *Entry mode assessment.* At this stage of project evaluation, SDF team members examine different new market or business entry vehicles or options. The analysis entails strategic, operational, and financial considerations of the alternatives, which usually include indirect or direct export to the foreign country, establishing a marketing and distribution subsidiary, or licensing the product, service, or technology to a reliable local company. Other entry modes considered are creating a wholly owned subsidiary for local production and possible exporting, establishing a JV to carry out local production operations, or acquiring a local company capable of producing products, services, or technologies comparable to those produced domestically.

I. *Entry strategy selection.* This is a crucial element of the entry evaluation and it starts with a review of the current corporate growth strategy, goals, and objectives in order to ensure consistency with and to articulate the goals and objectives of the new market or business entry project. Then, an assessment is done of the new project uncertainty, which is needed to develop strategic posture and intent. A number of analyses are performed in this step and options are laid out concerning the following competitive strategy elements to fix entry strategy:

1. Market penetration and product, service, or technology development required
2. Market leadership, challenger, follower, or niche strategy player
3. Value discipline, which is product leadership, customer service, or cost efficiency
4. Waterfall, sprinkler, or wave line of entry strategy to a new market or business
5. Distinctive competence, which refers to resource deployment, scope of operation, and synergies
6. Cost leadership, product differentiation, or focus
7. Triangle strategic options, namely, custom solutions, best product, or system lock-in options

J. *Scenario development.* The intent of the SDF team in this effort is to create the equivalent to a flight simulator by modeling the implementation of the new market or business entry and determining different scenarios of event occurrences and interventions leading to the future state. The analyses and evaluations performed here begin with identifying the most important drivers of the new entry project, developing the assumption set and obtaining project team validation, and modeling and describing the transition from the current state to the future state. Then, the SDF team develops, in conjunction with the project team, a set of hypotheses and ideas to test, defines end goals, and identifies levels of inputs for key drivers needed to achieve expected results.

Creating distinct and plausible scenarios and performing simulations comes next along with extensive testing of the assumption set and the hypotheses and conducting sanity checks and sensitivity analyses. A set of future-state predictions for each scenario is generated and forecast risks are identified and quantified after black swans and random shocks to test the stability of the scenario models have been introduced. Assessing the implications of each scenario forecast for corporate resources is the next step followed by determining the risks inherent in the scenario forecasts, their timing, and probability of occurrence. The last step is selecting the three most likely scenarios and forecasts with input and guidance from key project stakeholders.

K. *Development of decision options.* By this juncture, almost all of the evaluations are complete and decision options need to be developed. To do that, SDF team members lend support to the project team in terms of ranking the selected scenarios by likelihood of forecast realization, identifying the key decisions to be made concerning support requirements, outlining possible competitive reactions to new market or business entry, and developing a set of implementable recommendations with appropriate caveats and qualifications.

L. *Monitoring competitive reaction.* This is an ongoing observing and assessing the effort of competitor reaction to the company's new market or business entry. Competitor moves and behavior changes are identified and changes in strategy, policies, and tactical operations are observed very closely. Based on these evaluations, SDF team members make judgments, predict future competitor moves, evaluate the occurrence of risks due to these moves, and assess their impact on the project. This information is then used in updating the initial forecast, in strategy and entry mode evaluation and revisions, and in creating FSR plans.

M. *Entry strategy and mode evaluation.* In the postimplementation phase, actual developments are assessed by SDF team members, in conjunction with the project team and key project stakeholders, against expectations and earlier project decisions. The evaluations that take place in this step are to assess if the selected entry mode was appropriate, establish whether the right strategy was used for the new market or business entry, and determine if strategic intent was aligned with new market or business entry implementation plans. Follow-up evaluations also include determining how well the entry strategy was supported and executed, establishing whether other options may have been more viable alternatives, and adjusting the mode and entry strategies based on the intelligence and insights that come out of these analyses and evaluations.

N. *Forecast realization plan implementation.* In this step, one of the SDF team priorities is to focus on monitoring forecast performance, conduct variance analysis, and implement the forecast realization plan. Of particular importance are establishing which of the three most likely scenarios selected describes and fits best the actual new state and determining what changes are needed in controllable factors to achieve the forecasted levels of performance. The objective of the forecast realization plan in the post project implementation period is to identify new risks, update the forecast and the competitor response plan, and revise the FSR plan. Also, reviewing performance with key project stakeholders and releasing a project performance report are crucial as is providing feeds to the project performance dashboard and the project balanced scorecard systems.

O. *Postmortem analysis.* This is the last element of SDF team involvement in the analyses and evaluations of new market or business entry projects, the results of which include the following components: establishing what was done right and what was done wrong and why; assessing gaps in processes, competencies and capabilities, and lack of team member qualifications; and identifying lessons learned in the project and what can be done better next time. Two best practices in postmortem analysis are conducting informational meetings with senior management and project stakeholders, and posting the findings and recommendations of the project postmortem analysis on the company Intranet database.

10.4.2 SDF Techniques and Tools

The techniques used to evaluate a market or business entry and develop appropriate strategies touch on many aspects of competitive behavior and the idea is that by the end of the analysis, competitor behavior can be predicted and reasonable long-term forecasts can be generated. These tools are not used in isolation, but in conjunction with the analyses and evaluations discussed earlier, and they complement each other in order to perform better assessments. Notice, however, that industry analysis has only a present-state orientation, the other analytical models have a present to short-term state orientation, and only scenario analysis has the long-term orientation needed in all SDF. Again, these techniques should be adapted to the particulars of the new market or business entry project needs and the most significant in evaluations of these types of projects include the following:

A. *The GE matrix*. It is also known as the GE/McKinsey matrix and it is an extension of the Boston matrix and a systematic way to gather intelligence, display it on a three-by-three matrix, and use it to make decisions. Market attractiveness (low, medium, high) is shown across rows and competitive position of a company, business unit, or division (low, medium, high) along columns. Market attractiveness measures considered are the size of the market and market growth; the nature of competition barriers to entry; the role of local governments; the customer acceptance and adoption of new products, services, and technology; and market maturity and profit margins. The competitive position factors used in the GE matrix analysis include the following elements: quality of the company's products, services, or technology; distribution channel efficiency; management capabilities and competencies; branding and promotion expertise in foreign markets; and market share and trends.

B. *Porter's five forces*. It is an industry analysis model used to determine the intensity and attractiveness of a market and make a qualitative evaluation of a firm's position in the industry. The forces considered are the threats of substitute products, established rivals, and new entrants, and the bargaining power of customers and suppliers. This framework is particularly useful when used at the line of business industry level, that is, the market where similar products are sold, but it can also be applied to foreign country industry analysis with some appropriate modifications and caveats.

C. *Porter's six barriers to entry*. It is a widely used technique to identify and evaluate barriers to new market or business entry. The six major factors that are sources of barriers to entry are as follows (Porter 1980):

1. Capital requirements for new entrants for production facilities, start-up costs, and inventories
2. Economies of scale of existing competitors having achieved unit cost efficiencies not possible for new entrants
3. Higher costs of doing business related to access to channels of distribution, price discounts, and promotions necessary because of existing competitor long-term relationships and contracts controlling the best channels of distribution
4. Product differentiation through exceptional product design, features, quality, and brand identification and customer loyalty
5. Switching costs incurred by customers of existing rivals if they were to buy from new market entrants
6. Government policies that limit or prevent new competitor entry through license requirements, access to materials, and permits

D. *Porter's national diamond model*. It is a structured approach to examine the sources of competitive advantage from a country perspective and evaluate a company's ability to operate efficiently in a foreign market. The four components evaluated in this model are as follows (Porter 1990):

1. Demand conditions, referring to the key determinants of buyer behavior, pricing, advertising, and specific product attributes
2. Factor conditions that are both home-grown resources and highly specialized local resources
3. Related and supporting industries, that is, clusters of industries critical to the operation and growth of a given market
4. Strategy, structure, and rivalry that are determinants of national industry performance, which drive competition, innovation, and competitive advantage

E. *Porter's four corners analysis*. It is a useful technique to evaluate competitors and generate insights concerning likely competitor strategy changes and responses of different competitors to moves other competitors make, and determine competitor reaction to environmental changes and industry shifts. The four corners of this type of competitive analysis are the areas of future goals, management assumptions, current strategy, and capabilities and competencies (Porter 1980).

F. *The delta model.* It is a model used to assess a company's ability to outperform an industry over the long run through one of three distinct positions in it and develop competitively based corporate strategies. The strategy options are (a) best product, which means low cost or product differentiation; (b) total system solution, which reduces customer costs or increases their profits; or (c) system lock-in, which means competing through competitor lock-in, competitor lock-out, or proprietary standards (Hax and Wilde 2003).

G. *Ansoff growth matrix.* The Ansoff growth matrix helps a company decide its product and market growth strategy. The rows of the Ansoff matrix are existing and new markets and its columns are existing and new products. The elements of the matrix suggest that a business growth depends on whether and how it markets new or existing products in new or existing markets. That is, it helps to set the direction for business strategy by evaluating the four options: market penetration, market development, product development, and diversification (Ansoff 1957).

Advice on the use of techniques and tools: *First, go to the original sources of the tools used in SDF and really understand what they can do, how to apply them, and their limitations. Then, be eclectic in the selection of techniques and tools to use in a project. It does not mean that because all these techniques and tools are discussed in Chapters 6 through 9 and in this chapter that strategic forecasters must use all of them in each and every project. Experience and cost/benefit considerations dictate the selection of techniques and tools to use in grounding the analyses and evaluations preceding the development of scenarios and forecasts in different types of projects.*

10.5 SOUND PRACTICES AND SDF TEAM CONTRIBUTIONS

Successful new market or business entry requires that the analyses, evaluations, and execution are done appropriately within a complete, sound, and closely aligned set of processes.

A. *The SDF viewpoint.* From an SDF team perspective, the main success factors for creating reliable long-term forecasts for this type of project decisions are as follows:

1. Selection of the right new market or business entry strategy and appropriate entry mode
2. Objective assessment of megatrends and subtrends, the operating environment, the company's SWOT, and competitive activities
3. Thorough forecasting due diligence and verification and validation effort
4. Reasonable and well-grounded assumptions on thorough due diligence, screening, and testing
5. Feasible scenario development and analysis, which is the only long-term evaluation tool; all other tools being present or short-term oriented
6. Ability to align processes to the entry strategy, goals, and objectives
7. Implementing and managing effectively the marketing and sales functions

B. *The industry expert perspective.* Elements important to the success of new market or business entry projects which industry experts shared in our discussions include the following practices and factors:

1. Experience in international business transactions, marketing, advertising, and promotions as well as in foreign accounting and reporting practices
2. Identifying the right market, understanding the industry's life cycle, and understanding the foreign country and customer's buying habits and behavior

3. Conducting sound core market analysis and distinctive value proposition and possessing the capabilities and competencies required to implement successfully
4. The degree of technical innovation introduced in the foreign market through the project, anticipating new players in the industry, and having a plan to counter the impact of their entry
5. Assumption set, scenario, and forecast critical sanity checks and external advisor feedback, input, and guidance to increase confidence in the forecast
6. Reasonable and realistic estimates of market size, growth, and share subjected to sanity checks and accurate estimates of entry implementation costs and human resources
7. Selecting investing options to give appropriate control to the new entity whether it is outside the domestic market or in a new business area
8. Good partnerships, representation, agents, and distributor selection in foreign markets and screening and selecting best material and product sourcing options in the local market
9. Large scale of entry and economies of scale to obtain cost efficiencies and effective processes in planning, sourcing, making, delivering, and servicing products
10. Closeness to the company's own product portfolio experience and presence of complementary assets and capabilities
11. Early order of entry over lagging competitors and effective competitive response plans to offset the impact of new market entrants
12. Defining the most impactful performance measures, closely monitoring the management dashboard, and building an EWS and a balanced scorecard

C. *SDF team contributions.* The most valuable SDF team contributions to successful new market or business entry acknowledged in discussions with clients, senior management, and key project stakeholders include the following:

1. Bringing long-term orientation, criteria, and focus to the project and lending valuable experience in the entry mode assessment and entry strategy selection
2. Moving away from short-term forecast extrapolations to creating long-term forecast perspective to develop well-considered forecasts used to make decisions
3. Introducing discipline through complete process along with forecasting function ownership and management, and forecast realization planning capabilities and competencies

Other, often-stated SDF team members' value-adding support activities provided to new market or business entry project teams include the following:

1. Comprehensive assessments of project uncertainty and risks, which reduce time to decision substantially and result in better decisions
2. Validating data, information, assumptions, models, and results; verifying availability of company capabilities to execute this type of projects; and assessing forecast implications and risks
3. Modeling the entire project structure and its implementation and enabling project stakeholders to see the effects of changes in controllable project drivers and random events
4. Help in project managing a successful transition of assumptions and forecasts between project phases and supporting the development of the feasibility study, the business case, and the project implementation plans
5. Providing independent, critical, and objective evaluations of the proposed new market or business entry business plans and performance measures and targets
6. Planning capabilities enabling the company to respond to competitive actions and unfavorable market conditions with effective FSR plans

11.1 THE CURRENT STATE

The two ways M&A and JV projects usually get started are by an internal idea originating with senior management, Business Unit heads, and Business Development organizations or by proposals submitted to senior management by external parties through various channels. Projects coming out of the first set of channels are motivated by factors such as implementing strategic goals and objectives, enhancing the company's competitive position, and improving the financial picture of the company. Opportunities presented through the second set of channels are mostly motivated by financial engineering or profit to the idea originator. A simplified version of the current-state process to which M&A and JV projects are subjected is straightforward and applicable to many cases: A quick strategic fit assessment of the target or the potential JV partner company is usually done by the Business Development organization and then a project rationale is advanced. The Finance group is brought in to do a cursory financial analysis for the standalone and the merged company or the ongoing business and the JV operation's financials. Sometimes, alternatives to the proposed project are briefly considered and a feasibility study is put together to decide whether to move forward to the business case stage.

Organizational and cultural compatibility are not evaluated carefully but determined by senior management through their interactions with the target or the partner company's management team. The forecasting group is normally brought in late in the process or not at all and demand forecasts are created by the project sponsor, the Finance organization, or external advisors relying on historical financial data and dubious assumptions and adjustments. Ad hoc and abbreviated revenue, cost, pricing, and competitive analyses are performed using charts to prove a positive value creation project. In the case of M&A projects, the due diligence is performed by the Business Development or the Finance group in conjunction with a major accounting firm and outside advisors. The main focus of this due diligence is to verify and validate the existing business value by the information presented to the acquiring company. Company ownership, revenue composition and revenue growth, and legal obligations are the elements always investigated. In JV projects, there is no due diligence process per se, but the Finance group performs some rudimentary tests of the sources of revenue and the growth assumptions and determines the value of partner contributions.

In the business case phase, the financial forecasts are sometimes updated using demand forecasts created by Tactical Forecasting groups. Tactical Forecasting groups, however, are not experienced in long-term forecasting used to make strategic decisions that require a broad skill set in problem solving and scenario development and planning. Inflated estimates of synergies get generated by the organization spearheading the project and a business case is developed and a go/no-go decision is made based on project's net present value (NPV) estimates over a 10- to 15-year period plus a remainder value. Business plans that include operational, marketing, human resource, and financial plans are created and an implementation team is formed to execute those plans, despite foundations built on shaky grounds. The merger with or the acquisition of the target company takes place or the JV company is formed and then the painstaking integration process begins. Notice, however, that the only involvement of forecasting groups in M&A and JV projects is, if at all, limited to generating demand forecasts and rarely to helping develop estimates of future synergies.

In summary form, the main features that characterize the current state of M&A and JV projects include the following systemic shortcomings:

1. Lack of experienced project teams in parts of the M&A and JV formation processes that require external advisor intervention that is driven by motives different than project objectives
2. Limited availability and use of experienced SDF expertise and reliance on expedient rules and trend line growth extrapolations

3. Bypassing of thorough internal and competitor evaluations; megatrend, subtrend, industry, and political, economic, social, technological, legal, educational, and demographic (PESTLED) analyses; and project risk assessment due to severe time constraints
4. Negotiations driven by events of the moment and not based on actual valuations of positions and assessment of trade-offs and offers and counteroffers
5. Project implementations proceeding with limited organizational assessments and sanity checks on the project's ability to generate or extract synergies
6. Decisions made on single-point forecasts and no good project dashboard and early warning systems (EWSs) to monitor performance and with trivial competitive response and value realization plans to put into effect once risks appear on the horizon

11.2 M&A AND JV CHALLENGES AND ISSUES

The main source of M&A and JV project issues and challenges is the multitude of decision problems to be addressed, starting with determining the extent of the strategic fit of the project, namely, determining how the M&A or JV project supports the current or a new company strategy and what it brings to the firm's set of competencies and capabilities. The second difficult question to address is how to determine the best possible alternative at that point in time and verify that the proposed M&A or the new JV project is the best among different options to achieve long-term growth objectives. A third-order issue that is commonly left unchecked is to determine the fit between the company's product portfolio with that of the target or the partner's product line and how it compares with that of competitors. Here issues such as product attributes, maturity, upgrades and extensions, new product introductions, and cannibalization are not fully addressed. Also, the evaluation of the revenue stream is not evaluated critically and objectively in terms of examining the sources of revenue, its composition, stability, or the impact of external factors, pricing, and advertising on revenue growth.

Determining how much value can be created by the project is a major challenge for a number of reasons such as the following:

1. Strategic, cultural, and operational fit may not be as strong as expected to make the project implementation successful.
2. Major initiatives commonly attract competitor reactions, which are, often, far more potent than the impact of the M&A or JV project initiated by the company.
3. Getting agreement on assumptions and scenarios to develop to get reasonable forecasts is difficult when the interests of the parties diverge from the start.
4. Willingness and ability to frame and manage target and JV partner company expectations that are overshadowed by a desire to be accepted and close the deal quickly
5. Application of standard tests and criteria to subject forecasts, which is difficult due to lack of expertise to make them fit the context and the specifics of the project
6. The multitude of uncertainties and risks involved in these projects and the absence of canned prescriptions on how to identify and manage them effectively
7. Questions such as what are the risks of not doing the project, what could go wrong when doing it, and what can derail the project go unanswered
8. The extent of senior management involvement, dedicated resources, and other key stakeholder project support, which is often far less than required
9. Project sponsors who are unable to assess objectively human and other resource requirements to make the project successful and if they are available internally
10. Issues arising from incompatibility of organizational structures, corporate cultures, management styles, corporate policies, and systems

Project confidentiality, protection of sensitive information, and need-to-know constraints imposed on the project make it difficult to administer and produce resistance from individuals who are asked to contribute but cannot be told project specifics. Also, when the decision to proceed with a project is made with senior management predispositions, and under brutal time constraints, the project team is unable to determine which factors are controllable contributors to project success. Why? Because these constraints suppress certain evaluations and consideration of scenarios impacted by different factors. In this situation, it is hard to tell how the project will affect postimplementation operations in a standalone or merged scenario. Furthermore, in the presence of internal diverging interests, goals, and objectives, coordinating activities with a target company or potential partner prove difficult to accomplish.

In the absence of a well-advised strategy, M&A projects are for the most part justified by financial engineering schemes, and as a result, analyses and evaluations are performed by Finance organizations not qualified to do strategic decision forecasting. In many instances, JV projects are considered as pooling of resources schemes to get into new markets or new business. That is, they are profit motivated and therefore the purview of Finance groups. Because of such prevailing views, few companies have experienced SDF teams and complete, sound, and tested processes to put projects through end to end. Also, there are no clearly articulated goals and objectives at the start of projects, but they get generated as the process gets under way and results of ad hoc analyses come in. On several occasions, a number of M&A or JV proposals may be on the table and the first key question is which project is the best to pursue, but it is not clear what criteria one should use to decide among several projects in the pipeline. Is it the project with the highest NPV, the one with best strategic fit, the one with close cultural and management style fit, the one easiest to implement, or the one that positions the company to take advantage of new opportunities sometime in the future? In the absence of clear strategic goals and objectives and screening criteria, this question cannot be answered.

To structure and implement successfully M&A and JV projects is a formidable task that requires a confluence of unique skills and expertise to be brought in. In practice, inadequate and less than well-qualified human resources are brought into the project teams. Noticeable is the absence of end-to-end project process development and management from strategy review and strategic fit to objective opportunity assessment, to strategic decision forecasting, to critical financial evaluation, negotiation, and project implementation. In most cases, forecasting groups are brought in late in the process, if at all, with their role reduced to generating forecasts based on other groups' interests and assumptions, and updating, tweaking, and massaging forecasts until they conform to the client's objectives and wishes. There are no in-depth evaluations of company and competitor strengths, weaknesses, opportunities, and threats (SWOT), megatrends and subtrends impacting the industry, and PESTLED analyses. There is no forecasting due diligence per se other than verifying that the financial historical sales and cost data are true. And, in most cases, revenue forecasts are simple extrapolations of past growth rates normally 10–15 years out.

In every M&A project, there is undue influence by investment bankers and external advisors and experts and, for no good project reasons, the focus is on a speedy execution. In JV projects, on the other hand, a great deal of influence and pressure comes from senior management, wishing that a partnership project be executed because informal agreements were reached at high levels. At the same time, project risks are not really evaluated and ways to handle them are not developed. Project uncertainty and risk analyses are trivialized, risk management or contingency plans are rarely in place, and competitive response plans are missing entirely. Instead, they are left up to implementation teams to extract or create synergies to make up for gaps due to faulty assumptions and risks materializing. The lack of internal forecasting function management talent to manage the many activities of internal and external collaboration, coordination, cooperation, and communication (4Cs) is evident in all but a few projects. Also, project negotiations are determined by the negotiating power of the parties involved, an exercise in show of strength, and only on a few occasions, win–win arrangements are pursued. Furthermore, internal politics seem to dominate decisions and activities, even war room decisions, and exiting M&A and JV projects halfway through the process is viewed wasteful and inappropriate, but going full speed to execution is considered right.

Need for customization—Examples of differences in focus and criteria: *There are several types of M&A and JV projects, and their nature dictates the variety of approaches, analyses, and screening evaluations to be undertaken by project teams as well as the screening criteria to be used.*

- *In vertical acquisitions, the primary motive is control of certain production inputs and the main criteria are ease of implementation, relationships with other buyers, and exclusivity of factors that create barriers to entry through the acquisition. Here, control, cost reduction, and synergy evaluations are very important.*
- *In horizontal acquisitions, the driving reason is to increase market presence or to achieve economies of scale in production, distribution, and marketing. In such projects, strategic fit and cultural, managerial, and technology platform compatibility, and ability to integrate the target company are essential. Increased market share, new market entry, and consolidation assessments are vital.*
- *Concentric acquisitions, in which the acquiring and the target company are related by basic technologies, production processes, or markets, occur for the acquiring firm to move to a closely related business activity. The main criteria here are expected synergies from sharing resources, ability to enter a market with higher returns, and diversification through employment of combined resources.*

The definition of JV varies according to the laws of different countries and it carries different forms of collaboration arrangements with it from a loose association to a consortium, to a single business activity, to a permanent partnership. The motivations of domestic and international JVs are often different and some of the reasons for a foreign JV are to reduce the risks of entering a different market, to gain knowledge of the local legal and business environment, to bypass restrictions related to operating in strategic sectors, or to gain access to natural resources and cheaper labor. Diversification, speed to market, proactive or reactive competitive response, economies of scale, and directing the evolution of the industry are also common motives. Therefore, because of differences in the motives and nature of M&A and JV projects, customization of approach used by strategic forecasters in evaluating these opportunities is required.

Another major issue associated with JV projects is that, sometimes, partner contributions are made in-kind and the valuation of in-kind contributions involves an additional layer of complexity and negotiations with unsteady reference points. At other times, partner intellectual property protection considerations dominate over the need for information disclosure and, as a result, decisions are made with crucial information missing. Another unique issue encountered in certain JV arrangements is the motive of one of the partners using the JV as the means to assess the other partner company's capabilities and competencies and treating it as a potential acquisition target. Lastly, the difficult issue of getting a consensus forecast and negotiating an agreement is a major challenge when the number of JV partners increases to more than two. The complexity increases exponentially with the increase in the number of partners.

11.3 SDF APPROACH AND PROCESSES

SDF teams play a major role in M&A and JV projects in practically every aspect when they are brought in and join the project team at the start of projects. The reasons for the SDF team members' ability to add value in these projects are traced back to the following factors:

1. Thorough understanding of the company's strategy and goals and objectives, modeling the company operations, and maintaining the link of value creation to project execution and to strategy
2. Factual appreciation of the competitive and external environments and how new opportunities fit into that context
3. Organizational positioning as a business partner that provides internal expertise and support to project stakeholders through their links, relationships, and partnerships developed over time
4. Keen awareness of the analyses and evaluations that need to be performed and the information, skills, and qualifications required to support them
5. Assessing project uncertainty and corporate risk tolerance, articulating the rationale for and structuring M&A and JV projects, developing strategic posture, and helping develop a project value realization plan
6. Conducting independent, critical, and objective project assessments, validations, and verifications, and providing sanity checks

A model approach used by experienced SDF teams to make valued contributions in M&A and JV projects is shown in Figure 11.1. This systematic approach starts with verifying customer needs and validating product and technology assessments, the project rationale, and the project definition, and ends with the project postmortem analysis and report. This model creates the basis of the process SDF team members follow to perform the necessary analyses and evaluations. Notice that activities 1 and 7 are iterative and affect each other.

The SDF approach shown above only outlines the steps and direction of the process. The outputs from each step serve as inputs to the next process step and to other project evaluations, as insights for follow-up evaluations, and as indications to pursue certain lines of investigation. However, Figure 11.1

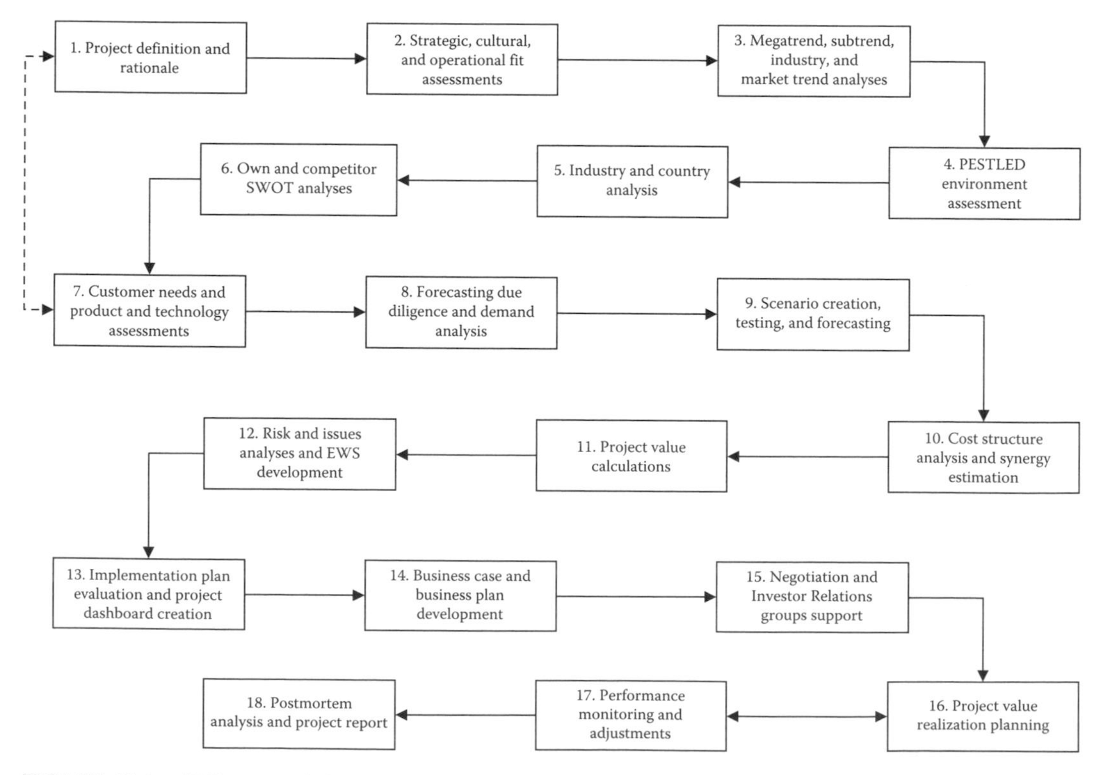

FIGURE 11.1 SDF approach for M&A and JV project support.

does not show the supporting and sustaining elements of the process, which include factors such as the following:

1. Understanding the scope of the project, determining informational requirements, and establishing what is known and unknown, knowable and unknowable
2. Determining levers and factors within the control of the company and those not controlled by it but by external factors
3. Assessing the degree of corporate risk tolerance versus the uncertainty and risk that come with the proposed project and strategic posture
4. Translating strategic goals and objectives into a sustainable project rationale and SDF goals and objectives and developing relationships with key stakeholders
5. Modeling the standalone, the resulting entities of M&A or JV projects, and how they impact company operations in the postimplementation period

The steps outlined above and which are involved in executing the SDF processes in this type of projects are the following:

Step 1. Prior to engaging in the project definition and articulation of project rationale, the SDF team establishes strong functional links and close personal relationships with the project team, all planning groups involved in the project, and key stakeholders. This step is probably the most important in determining the contributions SDF team members are able to make to the project. What flows out of this step is team consensus on clarity of project scope and level of involvement, preliminary screening criteria, determining resources needed to execute the project, identifying qualified internal and external experts, and creating the SDF team's goals and objectives in the project.

Step 2. What comes out of the fit assessment phase first is a unified project team view concerning consistency of the own company and target or partner company's strategic goals and objectives. Then an assessment is made of the degree to which the strategies and the cultures of the parties involved in the project are compatible and if a productive relationship is possible with minimum levels of friction. This involves an examination of the management structures and philosophies, policies and practices, guiding values and principles, and operational and technology platforms, all of which are crucial in determining if the project is likely to succeed.

Step 3. Before market trend analyses take place, a review of current megatrends and subtrends and their impact on the industry and the company is undertaken to create a reference point and context. Upon completion of this element of the SDF approach, the project team has obtained the industry perspective and has a good sense of how megatrend-related influences are likely to impact the project in the future. After the various trend analyses, uncontrollable external influences are identified and determinations are made of their impact on the company and the project and how to address them.

Step 4. The industry and country analysis step has different degrees of significance in the opportunity evaluation according to the type of M&A and JV projects. It is essential, however, in the case of foreign acquisitions and international alliances. Two things are determined after this step:

1. The industry structure and the legal and business operating environment, government policies and tax implications, strategic and structural barriers to entry, distribution channel availability, and likely marketing and sales costs after project implementation
2. The company's ability to manage risks due to political stability, social trends, business practices, tax policies, access to resource inputs, lack of qualified labor availability, and power of labor unions

Step 5. A company and key competitor SWOT analysis is done, and once this activity is finished, SDF team members step back and perform an evaluation of the findings up to this point. When a truly independent, critical, and objective evaluation is performed, the project team has identified the major gaps and issues to be addressed and has determined the project success requirements vis-à-vis the existing company's capabilities and competencies. Subsequent to this step, SDF team members are in a good position to proceed to the more granular assessments of customer needs and product attributes and technology features, applications, and uses. The result of these assessments and the findings of market research are used as validations of the product and technology positioning and market potential. But, more important is the development of a comprehensive set of assumptions vetted by the project team and which is subjected to further scrutiny in the process phases that follow.

Step 6. The forecasting due diligence part of the SDF approach is different from the corporate due diligence effort that seeks to uncover any hidden information concerning ownership, financial reporting, monetary or contractual, obligations, and legal liabilities. Instead, the goal of the forecasting due diligence is to ensure that all possible intelligence and insights are validated and brought into the development of forecasts in order to assess project uncertainty and risks accurately. The insights created in this stage serve as crucial input to the next step and the deliverables of this activity are

1. Sanity checks performed on assumptions made to this point and appropriate revisions.
2. Making explicit and clarifying many of the implicit assumptions made by key stakeholders.
3. Identification of risk factors related to expectations, wishes and desires, and implicit and explicit assumptions.

Step 7. The development of distinct and plausible scenarios enables SDF team members to model and describe the transition from the existing to the future, post M&A or JV project implementation state, and generate a set of forecasts. Key outputs from this step include the following:

1. Hypothesis testing and measurement of the strength of drivers, investigation of interrelationships, and assessment of feedback effects
2. Identification of the weak links in the scenarios and possible controls, project support, and intervention needed to ensure progress along the projected path
3. Unit and price forecasts and marketing and advertising expenditures needed to support the revenue forecasts, which serve as inputs to financial models
4. Assessment of the implications of the different forecast scenarios for company human and financial resources vis-à-vis availability of such resources
5. Identification of future-state risks and their sources, evaluation of the magnitude and likelihood of their occurrence, and planning ways to manage them

Step 8. The cost structure analysis and synergy estimation step is a joint effort whereby SDF and other team members provide their evaluations in their respective areas of expertise. At the end of this process phase, reliable revenue–cost and cost–price relationships are developed, possible sources of synergy are identified, and synergies are quantified. This information represents a synthesis of revenue and cost projections, which is the main input to the calculations of project valuation that is captured in the project's NPV calculations. Notice, however, that other inputs that include factors such as the valuation of intangibles, comparable transactions, deal premiums, terminal value, competitive bids, and negotiating position influence the final project valuation.

Step 9. The other factors mentioned last in Step 8 are not determined until the completion of the negotiations, signing of the agreements, and the project implementation is under way. Some initial estimates, however, enter the risk and issues analysis that is a continuation of the uncertainty and risk assessment performed in the scenario development and forecasting step following project valuation.

One objective of the risk and issues analysis step is to, once again, identify and understand questionable assumptions, capability and resource gaps, lack of implementation management talent, and integration issues. The second objective is to use that understanding to create an EWS to communicate the threat of risks as they appear in the horizon. Having come this far, the project team is then beginning to develop an implementation plan, which is then subjected to an independent, critical, and objective assessment with the SDF team participating. Here, the role of SDF team members is to validate the consistency of information and assumptions used in the implementation plan with current market realities. The recommendations of the SDF team are incorporated in the updated implementation plan, which is one of the inputs used in the development of the project business case.

Step 10. The business case is a living document, which captures all the data and information used, the analyses and evaluations performed, intelligence and insights created, and financial evaluations. It is all brought together in one place and the results are summarized in the business case recommendation to either proceed with the project or not. The draft implementation plan and the findings of the business case are translated into a new set of business plans for the standalone and the merged or JV company. By now, informal understandings, exchanges, and negotiations become formalized in the negotiation stage, and the role of the SDF team is to provide support to negotiators related to assumptions, scenarios, and demand forecasts; market research and competitive analysis; project support requirements and risks; and identification and valuation of trade-offs.

Step 11. Upon completion of the negotiations phase, legal agreements are finalized and so are the implementation and business plans. As the process unfolds, implementation is under way and the project value realization plan is finalized. This plan consists of the forecast realization part, the competitive response plan, and the project risk management plan. However, the involvement of the SDF team does not always end here. In many cases, it performs, along with Finance, the monitoring of project performance pertaining to revenue forecasts, competitive reactions, and validity of assumptions. Performance monitoring involves variance analysis that requires a good understanding of the assumptions and scenario that generated the forecast used in valuing the project as well as familiarity with real-time, actual market developments. Since the SDF team is in the best position to do this, it investigates deviations from expectations and proposes support adjustments, value realization plan initiatives, and forecast fine-tuning.

Step 12. Before its project involvement ends, the SDF team participates in the project postmortem analysis and produces a report of how well the forecasting function and the project were executed and lessons learned so that they can be used in future projects.

11.4 SDF TOOLS AND TECHNIQUES

Critical thinking and problem-solving techniques and many of the tools and techniques used by SDF team members in M&A and JV projects, such as system dynamics modeling, are discussed in Chapters 6 and 7, but some specific techniques that deserve another mention include the following:

A. *Measuring uncertainty and strategic posture.* Adapted to M&A and JV project needs, this is a useful technique used to assess the level of uncertainty in the industry at a given point in time ranging from a clear future to few scenarios to a range of scenarios, to ambiguous future. Once the level of uncertainty is determined, the company's strategic posture can be defined. That is, a decision is made whether the company will play a leading role in shaping the industry's future or follow the leader and adapt to the future or reserve the right to play, which means not investing early on but just enough to stay in the game. After defining the firm's strategic posture, different strategic moves are examined, which means that strategic portfolio actions are evaluated that are appropriate for the current situation. Possible actions include the following:

1. No regret moves that generate returns regardless of industry outcome
2. Options used when the right to play posturing is chosen
3. Big bet moves, which under some outcomes may be very profitable but lead to large losses under other outcomes

The importance of this analysis is that it can guide SDF team members in the selection of the forecast approach, models, and scenarios used. In the case of clear future level of uncertainty, a single revenue forecast for an M&A or JV project may be adequate but in the alternative futures uncertainty level, discrete scenario forecasting is required. When a range of future industry outcomes is identified, SDF team members look for a few key factors that can define the range, but there are no discrete scenarios and the outcome can be anywhere in this range. Here, the approach is to focus on reasonable forecast scenarios and understanding trigger events that can point to which scenario the market is headed to. Forecasting in the occasion of an ambiguous level of uncertainty is impractical because it is impossible to identify a range of outcomes or the scenarios that could lead to the outcome being in some range. In the case of an ambiguous level of uncertainty, experienced SDF team members help the project team reach a decision by investigating ways of scaling down and measuring uncertainty and guiding the analysis in terms of risk that is a measurable uncertainty. Why? Because risk is a conscious choice of actions and not determined by fate; risk is a measurable uncertainty. Thus, instead of dealing with unknown probabilities of events and outcomes, that is, uncertainty, the SDF team's focus shifts to approximate probabilities, key events that are likely, and determining outcomes.

B. *Identification of black swans.* The term black swans refers to events of extreme impacts that after the fact are predictable, but before their occurrence nothing from past experience would make them seem possible. It is also a way of SDF team member's thinking and challenging conventional wisdom that allows for rare and seemingly impossible things to happen in the future. It helps focus the strategic forecasters and the project team on clarifying what they know and do not know so that they do not underestimate the uncertainty originating in things that are unknown or unknowable. The application of this tool is useful in all M&A and JV forecasting projects, but it is especially valuable when faced with ambiguous levels of uncertainty (Taleb 2007b).

C. *Evaluation of alternatives matrix.* This tool consists of a matrix that in the first column lists all the elements to consider in the evaluation, and the proposed projects and other alternatives currently open to the company across the first row. The idea here is to get consistent total scores for the alternatives so that they can be ranked accordingly and in a more objective manner. Common elements evaluated across opportunities include strategic, cultural, and technology fit; project NPV; risk impacts; implementation costs; and synergies. To make the evaluation of alternatives more meaningful, in addition to the assessed value ascribed to each element, a weight is assigned that signifies the importance of that element to the company to come up with a score for each alternative.

D. *Decision tree analysis.* Decision trees are a method to identify alternatives, lay out options, and investigate possible outcomes associated with each option selected. Hence, the analysis of decision trees helps to choose between the options available and to create a balanced perspective of project costs and benefits and risks and rewards. Drawing decision trees for M&A and JV projects involves the following steps:

1. Start a decision tree with what needs to be decided and draw a small box representing this on the left of the paper. Boxes indicate decisions and circles uncertain outcomes.
2. Draw out lines to the right for each alternative solution, writing the solution along the line.

3. At the end of each line, consider the results. If the result of making a decision is uncertain, draw a circle. If the result is another decision that needs to be made, draw another box. Write the decision above the box or the circle.
4. Starting from the new decision squares on the diagram, draw out lines for options that could be selected; from the circles, draw lines representing possible outcomes and make brief notes on the line explaining its meaning.

Repeating this procedure until all the possible outcomes and decisions originating with the first decision are drawn is a well-structured approach to spell out the sequence of events that need to take place to arrive at an outcome analysis. This facilitates analysis of the timing and the likelihood of occurrence of certain events and decisions. Therefore, decision tree analysis is an indispensable aid in creating plausible, distinct scenarios of future states in M&A and JV projects.

E. *Measuring the moat.* The term, "economic moat," was coined by Warren Buffett, and for our purposes, it refers to a competitive advantage that is difficult to copy or emulate and that creates barriers to competition from others in the industry. Measuring the incremental moat created by an M&A or JV project entails evaluations of management skills, capabilities, and competencies; patents, brand identity, technology, and buying power; and other factors that impact operational efficiency. Measuring the incremental moat is most valuable in the negotiations of the residual value assigned to the project valuation, but also useful in the development and selection of scenarios.

F. *Strategic gap analysis matrix.* This tool is primarily used to identify the gaps that an M&A or JV project is expected to fill, but it is also used to determine strategic fit and areas of overlap in strategic areas. The rows of this matrix are strategic elements such as organizational capabilities, technical competencies, cost structure, and product differentiation. The columns show the current state, the desired future state, the current strategic gaps, and the benefits and value of closing these gaps for each of the strategic areas examined. A complete strategic gap analysis is useful in determining the relative negotiation positions of the parties involved in a transaction, in supplementing the economic moat analysis, and in estimating residual value growth. SDF team members also use it to evaluate the incremental moat created by a project, to develop future-state scenarios, and to better assess project risks.

G. *SDF strategic management model.* The model shown in Figure 11.2 was developed specifically to identify and model the links between operational performance, forecasting, strategy formulation, and strategy implementation. In this representation, the evaluation of operational results drives the external environment analyses, and the evaluation of market reaction signals, which affect the corporate financial plan, business and operational plans, and serves as a reference point for the new strategy valuation and implications assessment. The usefulness of this model is that it enables SDF team members to trace impacts from operational results to strategy formulation and implementation and back to operational results. That is, it is a tool to determine and measure different interventions to the current state of company operations in the event they are warranted.

Operational results and megatrend, industry, market trend, and SWOT analyses, as well as feedback from the options analysis and strategy formulation, influence the SDF team's future-state forecasts and options that feed the corporate financial plan and the business development plan. Note that the business development plan has feedback effects on the SDF future-state forecasts and drives the business and operational plans. Business development plans, including M&A and JV projects, drive the assessment of uncertainty and the adaptive strategy realization plans, but are subject to feedback effects from them as well. There are direct causal and indirect and feedback effects from factors in this model that are captured in the market signals and market reaction evaluation. The linkages captured by this model are among the following factors: options analysis and strategy formulation, uncertainty assessment and strategy realization plans, strategy valuation and assessment of its

communicating the appearance of risks in the horizon and helping in the project value realization and the risk management efforts.

11.5 SDF ANALYSES AND EVALUATIONS

Around the world, the business operating environment is becoming progressively more difficult and strategic decision forecasting even more challenging because as Yogi Berra put it, "the future ain't what it used to be." Increasingly, a company's ability to compete is based on capabilities, competencies, and knowledge, and this is what the SDF team brings to M&A and JV projects. It does that by knowing the sources and understanding the types of information to search for the various analyses and evaluations it performs in the different segments of different strategic projects. As soon as the SDF team is brought into the project, it reviews the corporate strategy, does a quick situational analysis, identifies the informational requirements for the project, and determines what is known and knowable and unknown and unknowable. It then assesses corporate risk appetite and forms a fist impression of the level of project uncertainty and potential strategic posturing. With those activities completed, the SDF team is ready to undertake the step of the process validating project rationale.

A. *Project rationale and structure.* To assess the M&A or JV project rationale, SDF team members examine the background, reasons, and project sponsor's expectations that led to the proposed project. Then, the scope of the project and its goals and objectives are clarified, which lead to the proposed project being created and its structure being defined. For M&A projects, understanding the advantages of different types of acquisition (vertical, horizontal, or concentric) is essential to define and structure the project, while for JV projects, clarification of the type of partnership arrangement sought is necessary.

The consistency between the motives for undertaking the project and the objectives to be met is evaluated in light of what is taking place in the marketplace. Opportunity costs of alternatives are briefly considered in conjunction with the project team and the question of what happens if the project is not undertaken is addressed. Feedback of project sponsor and senior management is obtained, and at the end of this step, the project team presents a clear and unified project structure and rationale across the organization.

B. *Fit assessments.* The first part of project fit assessment is in the strategic area, namely, consistency of proposed project vis-à-vis corporate goals and objectives. Strategic fit assessment also involves validating the compatibility of the own, target, or potential partner company's risk tolerance levels and strategic posture as well as verification of the own, target, or potential partner company's strategic goals and objectives. These assessments entail comparisons not only between stated visions, positions, goals, and objectives but also with expectations that are not always explicit, clear, or stated upfront.

Cultural fit assessment is the second part, and it requires an evaluation of the consistency or harmony between the own and the target or potential partner company's philosophy, values, policies, and practices. This is the most important element of fit assessment because it determines the extent to which organizations can function together with a minimum amount of friction. It also determines the changes needed in policies and practices and their feasibility and cost. In most cases, organizational fit is part of the cultural fit assessment where the compatibility of the own versus target or potential partner company's management structures takes place.

The third part of the fit assessment is that of operational fit where the own versus the target or potential partner company's technologies and systems compatibility evaluation occur. Technology fit includes platforms and licenses, versions, and releases, and systems fit extends across major corporate human resource, financial reporting, and other management systems. A complete operational fit assessment also involves an appraisal of the own versus the target or potential partner company's key processes and procedures. After this fit assessment is finished, the project team has a good idea of the likelihood and the effort that will be needed to make the project a success.

C. *Trend analyses.* The trend analyses performed in M&A and JV projects are of megatrends, subtrends, and market trends. We mentioned earlier that megatrends are defined as major forces in human development that cause shifts in thinking that have broad and defining changes on nations, industries, and companies. As such, they are universal and determine the current state and the development of the future of government, business, and consumer states globally. A subtrend, on the other hand, is a lower level, more specific influencing force or tendency whose impacts are more localized.

The evaluation of megatrends begins at the economy- or country-wide level, proceeds to the industry level, and ends at the product, service, and technology levels of the own company and the target or potential partner company. The evaluation of the interactions between and the feedback effects among megatrends is as important as the effect of individual megatrends and subtrends. Market trend analysis is the evaluation of the direction and the causes of consumer or user tendencies at the total market and market segment levels that determine demand. The result of the trend analyses performed is the creation of intelligence and insights to reset the company vision driving the M&A and JV projects and implement plans correctly, thereby setting the stage and direction to transform the way business is done to take advantage of megatrends, subtrends, and market trends.

D. *Environmental assessment.* The environmental assessment performed by or for the SDF team in M&A and JV projects varies according to the specifics of the opportunity, but it evaluates key PESTLED elements. It is a continuation of the trend analyses performed, which seeks to identify environmental factors impacting the company's operations and determine to what extent they can be taken advantage of and what changes are required to do that. PESTLED analysis is the next level analysis beyond the impact of megatrends and subtrends on the company and includes an investigation of their impact and operational issues.

Other external environment issues such as licensing, resource procurement, and supply chain constraints are also investigated in this step vis-à-vis external environment developments. The environmental evaluation is traditionally the first step of a strategic analysis, an ongoing effort to identify and manage the probable effects of external forces on the company's survival and growth. Upon completion of the evaluations in this step, SDF team members are in a position to determine how external environment factors are likely to affect the industry as a whole, the company, and specific product lines in future years.

E. *Industry analysis.* It is an analysis of conditions in the industry at a particular time and a market assessment tool to appreciate its complexity using Porter's five forces model. Industry analysis attempts to answer the question of what are the key factors for competitive success and identify the important structural features of your industry via the five forces of the industry analysis model. Again, the focus of SDF team members in this analysis is to

1. Better understand the forces at work and the own company's and the target or the JV partner company's industry position relative to other alternative entities.
2. Make informed judgments about future developments.
3. Create reasonable assumptions related to future industry developments.

Industry analysis begins with a project team's review of products and markets, skills, and competitors contained within the industry; it is followed by the industry structural analysis per se and concludes with identification of key success factors. Sometimes, industry analysis requires an evaluation of country dynamics and benchmarking of best-in-class companies and practices. Done properly, it is very useful because it determines the competitive rules and strategies to use to position the company in the industry via M&A and JV projects so that it can serve a niche market or offer products that give it an advantage over competitors.

F. *Own, target, and partner company SWOT analyses.* These analyses are among the first steps in planning the SDF team's activities in M&A and JV projects, which, in turn, help the project team focus on key project issues. In this step, the SDF team sifts through huge amounts of information and summarizes it into strengths and weaknesses that are internal issues, and opportunities and threats that are external factors. The analysis begins by clarifying and assessing stated project goals and objectives and identifying internal and external factors that help or could detract from achieving those objectives.

SWOT analysis is a project team's effort that helps to determine if the project objectives are really achievable, given the findings of the analysis, and if they are not achievable, the project needs to be restructured and objectives reset. While conducting the SWOT analysis, alternatives to the proposed project may be uncovered which the company is capable of exploiting, and, by understanding its weaknesses, the project team can better manage project threats. The impact of this analysis is far more potent if it includes an evaluation of the target or the partner company as well as the SWOT of the new entity created by the project.

The findings of the SWOT analysis are evaluated for their effect on the fit assessments performed earlier and the determination of capabilities and competencies help to determine the existence of or the possibility of creating synergies. Another result that flows out of a complete SWOT analysis is reference points for sanity checks of some explicit and implicit project assumptions and the identification of risks. This intelligence is then used to create and validate distinct plausible scenarios and develop future-state projections. Finally, insights obtained through SWOT analysis are used in the creation of the project value realization plan.

G. *Customer, product, and technology assessments.* These assessments are derived from the market intelligence gathering and evaluation effort. They, in turn, form the foundation of and drive the forecasting due diligence and guide the demand analysis and forecasting process. Notice, however, that market intelligence gathering and evaluation is an ongoing decision-driven, cross-functional activity that considers customers and products and reflects on the opinions of industry experts and decision influencers.

First, it considers customer needs, customer profiles, and market segment characteristics and then product, service, or technology line characteristics and attributes. It also draws on the findings of the megatrend, PESTLED, SWOT, and industry analyses and focuses on the voice of the customer. Then, the effects of new products, extensions, and enhancements are evaluated as well as those of geographic and channel expansions. Lastly, product and technology lifecycle position and management, sales and marketing channels, pricing policies, promotions, advertising, and customer retention programs are assessed.

Effective company customers, products, and technology assessments require reviews and evaluations of industry studies, market research, customer satisfaction surveys, and benchmarking studies. They also require evaluations of earlier competitor, target and partner company experiences in similar projects and operations, marketing and sales, new product and technology development, customer satisfaction, and financial support provided to projects. Here, SDF team members seek different sources and types of data, ideas, and intelligence at different stages of product or technology life cycles considered: In the early stages, market intelligence is crucial; in the growth phase, customer intelligence is most important; and in the maturity phase, competitive intelligence is most valuable.

H. *Demand analysis and forecasting due diligence.* In this phase, SDF team members revisit informational requirements to support the M&A or JV project, identify sources of intelligence, and determine what can be known and what cannot. Then, the data and information gathering, verification, and validation get under way, and the forecasting due diligence begins in earnest to answer the why, when, where, and how questions related to demand. Demand analysis and forecasting due diligence build on earlier evaluations and, depending on the nature and the particulars of the project, they include validations and confirmations, such as

1. Validation of company and target historical, cross-sectional, and market segment unit and revenue data and analysis of revenue composition and growth by product and market strata of own, target, or potential partner company.
2. Review of advertising, promotions, pricing, and discount policies and practices and validation of estimated historical price responses, price sustainability evaluation, and projections.
3. Detection of outliers versus external factors that impacted sales over the past five years and identification of events that caused the outliers.
4. Determination of events in the next 5–10 years, which are outside the company's control, their likelihood and timing of occurrence, and potential effects.
5. Confirmation of the environmental factors that the industry and the company are likely to be influenced by in the next 5–10 years.
6. Review of the technology evolution analysis, likely future developments, and impacts on other technologies and future business.
7. Turning expectations and implicit assumptions into explicit assumptions, validation of earlier used assumptions and results, and sanity checks of analyses so far.
8. Preliminary modeling of the company's current operations and future states at a level sufficiently detailed to capture key linkages and model the new entity operations and measurement and testing of relationships and feedback effects.
9. Independent validation of first-order total market size estimates and uptake projections used in the initial stab at revenue projections.
10. Reevaluation of risks identified in the SWOT analysis and the uncertainty evaluation and agreement on what constitutes acceptable risks.
11. Reassessment of the forecast implications for productive capacity and financial and human resources.
12. Evaluation of different project support levels, hypothesis testing, and "what-if" analyses to produce insights for the next phase in the SDF process.

The results of the forecasting due diligence and demand analysis are used as reference points and benchmarks to compare with expected outcomes, to validate new assumptions, to determine the extent to which project goals and objectives are achievable, and to restructure the project or reset objectives and strategic posture. The recommendations of the SDF team on consistency of project objective vis-à-vis the analyses and evaluations in this step are communicated to senior management and in some instances form the basis of a decision on whether to proceed or not.

I. *Scenario development, testing, and forecasting.* In M&A and JV projects, scenarios are narratives about how the company's future will unfold, but not actual predictions. They are plausible descriptions of how pertinent external factors could develop, interact, and create new opportunities and threats. They are exercises that enable the project team to envision the future and create the basis for system dynamics models to be created to generate forecasts. Scenario development and planning is of particular importance primarily because project success depends not only on company performance, but also on the performance of the acquired company or the JV entity created by the project. Their purpose is to challenge current thinking and get insights into future possibilities by recognizing and charting uncertainty in order to investigate factors that may determine the company's future. Scenario development and planning enhances project team communication, collaboration, and learning and helps to obtain agreements and arrive at decisions more efficiently.

There are three levels of driving factor changes addressed in SDF-created scenarios, which are integrated into one coherent description of future states, namely, the following:

1. The macro arena that describes external environmental changes, that is, megatrends, PESTLED developments, and industry and market evolution
2. The middle level that shows exchanges in the transactional environment, that is, the effects of interactions of customers, suppliers, and competitors with the company
3. The operational level that describes the links between corporate strategy, company operations, competencies and capabilities, and project goals and objectives intended to create a new future state.

Models or scenarios first? *Model development comes first when there is some certainty of future outcomes and data are available and because scenario development is an extension of models to determine the effects outside the model. However, when there is a lot of project uncertainty and causal models are not feasible, scenario development is the first and only option. Notice that testing of scenarios takes place after computational models of storylines are created, which show how different levels of driving forces result in alternative future states.*

J. *Cost structure analysis and synergy estimation.* After providing the demand and revenue inputs to financial models, the main role of the SDF team is to identify, project, and validate synergy estimates and provide competitive intelligence and cost benchmarks to the Finance organization responsible for cost analysis. Additionally, it conducts analyses and reality checks of unit cost estimates for the combined entity, in the case of an M&A project, or for the JV company. It also validates the components, estimates, and reasonableness of upfront investments and ongoing capital requirements.

The Competitive Analysis subteam helps SDF team members to assess historical costs and cost efficiency management and monitor the analysis of the own company cost structure as well as those of the target, the combined, and/or the JV company. The acquisition and target integration or JV implementation costs are estimated to arrive at combined entity or JV company cost projections. The main focus of the SDF team here is to determine or validate the likelihood of current or future cost advantage, its sources, and sustainability. Another SDF team support's function is to help in the financial due diligence area and articulate the conditions, requirements, and scenarios for cost leadership.

Synergies take different forms, but the most common and easier to estimate are revenue enhancements, cost efficiencies or reductions, and decreased future capital requirements. There are, however, incremental costs to obtain synergies and their sources and impact magnitudes on customers need to be considered as well. The value added by SDF team members here is to help the project team in creating synergy harnessing scenarios and modeling synergy creation. But this is not the end of its involvement because the identified and estimated potential or expected synergies are subject to negotiations, and here the SDF team provides competitive and industry averages and norms, and performs specialized analyses and evaluations.

K. *Project value calculations.* In most cases, the value created by an M&A or a JV project is the sum of the NPV created over a 5- or 10-year period, the remainder value in perpetuity, various acquisition premiums, expected synergies, and tax impacts. However, other elements enter the picture and include the negotiating power of the parties involved, competitive bidding fears, price sweeteners, and other special considerations. And, sometimes, SDF team members conduct or participate in analyses, sanity checks, and evaluations of those elements. They also play a big role in the synthesis of revenue and cost analyses, modeling, scenario development and forecasting, and resulting value projections.

SDF team members revalidate NPV calculations derived from distinct scenarios and help in the development of consensus on the most likely view, perform sanity checks on the synergy estimates, and provide intelligence on industry practices, averages, and norms. However, their most important analyses in this step are as follows:

1. Evaluation of the project residual value, which requires identifying future events, factors, and scenarios that result in continuous future revenue growth, stagnation, or decline for the target or the JV company
2. Identification of risks in achieving project goals and objectives and the impact associated with each of the other valuation components
3. Creation and valuation of trade-offs and assessment of terms and conditions, which ultimately determine the real value created by the project

L. *Risk and issues analyses.* The uncertainty and risk analysis of feasibility studies is to a large degree related to project goals and objectives and the deal structure. In this step of the SDF process, the risk analysis done is related to factors around project valuation and the company's ability to implement the project successfully, although there is some overlap and reevaluation of the first category of risks. This reexamination engages the entire project team and involves a closer assessment of strategic objectives, cultural and management compatibility, and congruence of project goals and objectives vis-à-vis competency and capability issues. It also involves reexamining the sources and nature of business, financial, and operational risks; the technology platform and human resources and financial reporting systems; and second-level sanity checks of assumptions, models, forecasts, and their implications.

The risk analysis and determination of uncontrollable event and black swan occurrences includes the effects of negotiated terms and conditions of the deal and the various components of project valuation. And, it revalidates assumptions concerning the availability of qualified managers to implement and manage the integration of the target and the running of the JV company. This type of risk analysis also entails evaluations of the consistency of present and future objectives and expectations of the parties involved and the company's ability to create and execute a business plan that extracts or creates new synergies to levels expected in the negotiated deal. Furthermore, it anticipates the first-, second-, and third-order industry and competitor reaction to the project.

Management and resolution of risks and issues is the other part of this phase, which requires the attention of the SDF and the project teams to the following items:

1. Definition of measurements and the creation of KPIs to determine the extent to which project goals and objectives are met
2. Creation of a management dashboard to display the selected indicators of underlying realization of project value
3. Development of appropriate risk sharing, mitigation, shifting, and management options and executable risk management plans

M. *Implementation plan evaluation.* The steps of the SDF team support for M&A and JV projects shown in Figure 11.1 appear to be sequential, and to some degree, they are, but at times, some steps are performed in tandem or in the reverse order. Usually, the implementation plan evaluation step takes place earlier in the process, but in many instances, it is delayed until after valuations take place and proceeding to project implementation appears very likely. In any case, the important element of this evaluation is an independent, critical, and objective evaluation of the key implementation plan components, which are as follows:

1. The implementation process, agreed to timelines, assigned resources, milestones and handoffs and specified deliverables
2. Senior management involvement and support, the adequacy of qualified personnel, and budgetary support for project execution
3. Retention agreements of key target personnel and secondments to the integrated or the JV company and their sustainability
4. The tracking of progress in different implementation phases and the estimated total cost for each phase

The project implementation evaluation is where the last set of checking, screening, confirming, verifying, and validating all the factors that impact the project value created, and the likelihood of achieving the project objectives takes place. It is an evaluation of the project management capabilities of the implementation team, but more importantly of the degree to which 4Cs can be effected in the own company, the standalone or integrated target company, and/or the new JV entity. And, it is the last and crucial piece of the evaluation that makes all the difference in the world.

N. *Business case development.* A compact definition of business case is attributed to the *Financial Times*, which describes it as "an explanation of how a new project, product, or other opportunity is going to succeed and why people should invest money in it." It is the place where the SDF team's analyses and evaluations come together and overall project team's consensus is reached. The business case builds on the project feasibility study, all the analysis and evaluations of the SDF team, and the sanity checks performed throughout the duration of the project. The business case is also a living document where key information, assumptions, models, and results and their implications are captured and it is owned by the project sponsor. The functions of the business case are to be the following:

1. The repository of the documentation of all analyses and assessments that determine the project value beginning with the results of a strategic gap analysis and the articulated project rationale
2. The source of guidance and a decision-making tool by discussing the costs and benefits of doing the project as well as not doing it
3. The basis of senior management approval to authorize capital expenditures and implement the project

The analyses performed and the support SDF teams provide in business cases are discussed in more detail in Chapter 19.

O. *Negotiations support.* In SDF, negotiations support is a key support function step of the process and is usually provided indirectly through the project team leader, but often in direct discussions with company project negotiators. The support starts with SDF team members evaluating the needs and pressures of project participants to conclude the transaction in order to determine their negotiation power. SDF team members are the source of invaluable environmental, industry, and competitive intelligence and insights. The findings of the analyses and evaluations up to this point are used in creating the company's opening positions, fallback positions, and walkaway positions as well as determining those of the other side.

Real-time support for the negotiations team and senior management from SDF team members is delivered through making available actionable competitive intelligence and insights and providing industry practices, norms, and averages and applications. It is also useful in the identification and quantification of trade-offs valued by the other side, that is, finding things to be given up in exchange for things the company really values. The SDF team's support in negotiations is the subject of discussion in Chapter 17.

P. *Project value realization plan.* A project realization plan is a map of actions required to ensure that project goals and objectives and expected values are achieved. It consists of three basic elements such as

1. The competitive response plan, which, after recognizing the competitive and industry reaction to the project, prescribes a set of actions to offset or counter that reaction.
2. The risk management plan, which is a set of preestablished rules and actions to follow and which is activated when the EWS communicates the need for intervention.
3. The forecast realization plan, which guides company interventions through controllable factors and adjusts the forecast drivers and support levels to required levels in order to achieve the expected performance.

The development of the project realization plan starts with determining in the preimplementation period the company controllable factors as well as the events and factors that external entities might be able to influence, such as government organizations, industry associations, and industry experts and opinion influencers. Then, future uncontrollable events are identified and their likelihood of occurrence estimated. This enables the creation of a risk management plan with options open to the company, including insuring against certain risks, risk elimination, mitigation, and sharing. For this plan to be effective, specific personnel is assigned specific functions at predetermined points in time. A more detailed discussion of future-state realization (FSR) planning can be found in Chapter 18.

Q. *Performance monitoring and variance analysis.* Ordinarily, the involvement of the SDF team in M&A and JV projects ends with the creation of the project value realization plan. But, often, its involvement continues until the success of the project is judged to be well under way to being achieved. The key drivers of the company business have been identified in the demand analysis stage and they are useful in creating project performance measures that are influenced by those drivers. Thus, the charge of the SDF team in this step is to identify which variables to monitor and define corresponding measures to be compared with actual data that feed the management dashboard and the EWS.

At times, involvement of SDF team members in variance analysis is necessary to help the project sponsor to

1. Determine what negotiated terms and conditions are not met and understand the nature of the problems and issues causing deviations from expected performance.
2. Recognize what uncontrollable factors and events have occurred and how they impacted project performance.
3. Identify and describe new scenarios of external factor influences.
4. Help the new entity management team deal with the new situation and communicate progress and developments to the wider organization.

Having done all these, the SDF team is now ready to undertake an end-to-end analysis and evaluation of how well the project was done.

R. *Postmortem analysis and project report.* Postmortem analysis, also known as post project review or project audit, is the process of looking back and performing a critical assessment of how the project was handled after it is completed. The purpose of SDF team's participation in the postmortem analysis is to do the assessment and document in a project report how well the key elements of its own processes were done, and to share the findings of the analysis throughout the company. This is done by uploading the report to the company's Intranet database and creating guidelines and procedures to be put into effect in future projects.

The postmortem or project review process: *The SDF postmortem analysis is performed by the manager of the SDF team, and in order to judge how well functions were performed and issues handled, a number of components are evaluated for each step of the process and include the following:*

- *Communication, cooperation, coordination, and collaboration (4Cs) within and between the SDF team, the project team, and the wider organization*
- *The level of senior management and project sponsor involvement and support*
- *Team member's integrity, flexibility, and professional conduct in all internal and external transactions*
- *The level of expertise in their areas and technical competence of team members and understanding of the company business and its strategic goals and objectives*
- *Prior project team member's experience relevant in analyses and evaluation for M&A and JV projects*
- *Whether the right informational requirements were identified and sound models, techniques, scenarios, and tools were used and how due diligence findings was turned into insights*
- *The suitability and completeness of processes and procedures, whether adjustments were required to address project specifics, and whether cost–benefit considerations drove decisions*
- *Extent to which policies, processes, and procedures were followed and the manner in which they were executed*
- *Examining how things got done with an independent, critical, and objective eye, and determining what was done right and what was done incorrectly*
- *What issues, problems, and conflicts arose within the own company and with the target or partner companies and how they were resolved*
- *How timely, relevant, and constructive was stakeholder's input, guidance, and feedback to information shared and information requested*
- *What lessons were learned from the team's experience and target or the partner company's experience during the project that could be applied to future projects*
- *What parts of the SDF processes, procedures, and practices must be changed to enhance process efficiency and increase support levels*
- *What significant mistakes were made or issues handled incorrectly that are not to be repeated in future projects*
- *How the SDF team's execution of their broad area of responsibility in the project can be improved and specific tasks made more efficient*

11.6 SOUND PRACTICES AND SDF TEAM CONTRIBUTIONS

M&A and JV project success requires a confluence of the right objectives, sound processes, unique experiences, and appropriate analyses in all functional areas involved. Naturally, the significance of each varies according to the particulars of the project, but following is a set of sound SDF team practices and success factors that are repeatedly highlighted in postmortem analyses and in discussions with project sponsors or team leaders.

11.6.1 Sound Practices and Success Factors

Factors that external experts, sponsors, and project teams consider important in the predeal consummation phase cover every project phase and reflect particular needs. But the differences among those factors suggest that each project should determine critical success factors appropriate for it.

Case 1. A set of critical success factors from a business development group perspective includes the following factors:

1. Well-defined and articulated corporate growth strategy and congruence with project goals and objectives
2. Complete strategic, market, and operational gap analysis and clearly stated rationale, goals, objectives, and reasonable project scope
3. Appropriate assessment of strategic, cultural, organizational, technology, and operational fits
4. In-depth historical and market segment demand analysis and forecasting due diligence
5. End-to-end forecasting process management, sharing of findings of evaluations, all-around communication, and inclusion of client input and project team feedback
6. Well-balanced analyses and evaluations, and a tightly knit and defensible business case to arrive at consensus decisions efficiently

Case 2. A different set of factors considered most important comes from the perspective of the formation and management of a JV company:

1. Sufficient senior management involvement and support and dedicated qualified managers to implement the project
2. Detailed modeling of the before and after JV company formation operations
3. Critical evaluation of alternatives to the proposed JV project and selection of the best option
4. Megatrend analysis, evaluation of the external environment, and SWOT analysis to determine consistency of project objective with the reality of the situation
5. Management of internal and partner future performance expectations, ongoing partner contributions, and support to the JV entity
6. Effective project structure; sharing of costs, benefits, and risks; and the negotiation of a win–win and sustainable arrangement
7. Obtaining efficient financing of the JV company based on the merits of project economics demonstrated by the forecasting due diligence
8. Effective, step-by-step coaching of senior management and the Investor Relations group by the head of the forecasting group

Case 3. The source of the preimplementation success factors that follow is an industry expert advising senior managers

1. Focus on customer needs and congruence of meeting those needs with the proposed project goals and objectives
2. Creation of well-reasoned and balanced assumptions, extensive testing of their implications, and updating as new evidence comes in
3. Timely evaluations and incorporation of feedback from project stakeholders and internal and external experts
4. Independent, critical, and objective assessment of project rationale, deal structure, and implementation team capabilities

5. Reasonable future position scenarios, sound revenue and project value estimates, identification of major project risks and issues, and assessment of their implications
6. Obtaining a good part of the initial negotiation positions and reasonably good terms and conditions

Case 4. The factors considered essential by finance experts in coming up with sound project valuations include the following:

1. Experienced project team members and qualified forecasting experts dedicated at the start and being active participants throughout the project
2. Validation of proposed project rationale and goals and objectives and assessment of several alternatives and options to close strategic gaps identified
3. First-rate situational analysis and conceptualization of the problem to be solved, complete and correct information and data, critical assessments, and detailed models of pre- and postimplementation operations
4. Validation and verification of unit sales and pricing data, in-depth demand analysis and due diligence, and evaluation of project objectives feasibility vis-à-vis due diligence findings
5. Creation and validation of balanced assumptions, multiple levels of sanity checks, and identification of risks and potential impacts
6. Development of well-thought-out and plausible scenarios, testing of ideas and hypothesis, and simulation of "what-if" scenarios
7. Modeling of operations with sufficient level of detail to feed financial models, highlighting key value drivers, and ranking their impact
8. Accurate revenue and cost projections and reasonable estimates of costs and benefits of potential synergies
9. Upfront agreement on performance measures, management of dashboard indicators, intervention signals, and timing of actions to offset threats of risks
10. Practical negotiation plans, creation and valuation of desired trade-offs, evaluation of terms and conditions, and negotiations support

Case 5. Sound practices from experts in project implementation are captured in the following set of success factors:

1. Qualified human resources dedicated to the new entity assigned before implementation begins to shape the business plan
2. Clear project goals and objectives to drive the creation and execution of the implementation plan and guidance on how to use forecast solutions correctly
3. Complete and reasonable project assessments and implementation processes and procedures motivated by cost–benefit considerations
4. Sufficient number of skilled managers to implement the project effectively, deal with transition issues, and execute the new company's business plan
5. Sound project implementation plan with clear delineation of responsibilities, deliverables, and handoffs, and new entity business plans
6. Reasonable assumptions and achievable forecasts of project value created and management of synergy expectations to realistic levels
7. Senior management involvement, support, and commitment in all project execution phases and insistence on project management competence
8. Creation of win–win arrangements and competent, critical, and objective identification and resolution of implementation issues
9. 4Cs and information sharing internally, with the target or partner company, and with external advisors

Case 6. The set of sound project implementation practices articulated by M&A and JV project advisors includes the following factors:

1. Development of trade-offs valued to the other side to create win–win and agreed to forecast-based, and not target-based, expectations
2. Sound integration planning and execution linked to sound forecasts, driven by reasonable scenarios and assumptions
3. Clear definition of the responsibilities, timelines, deliverables, and format of handoffs in execution of the implementation plan
4. Strong senior management leadership, commitment, involvement, and support and issue escalation processes and intervention in cases of unresolved conflict at operational levels
5. Well-drafted, achievable business plans for the new entity that address and guide operations, marketing and sales, human resources, and finance functions
6. Definition of good performance measures, close monitoring of operational results, and understanding of the forces behind observed forecast deviations
7. Creation of a project value realization plan, which includes extraction and creation of synergies, adjustment of controlled forecast drivers, a competitive response plan, and a risk management plan
8. Complete postmortem analysis, sharing of learning throughout the company, and recommendation of changes to implement in future projects

11.6.2 SDF Team Contributions

The contributions of SDF teams that participated early on in M&A and JV projects overlap with the sound practices and success factors mentioned above. What explains this is the sharp focus of SDF team members on maintaining support excellence in every step of the SDF process and their close relationships and communications with project team members and other project stakeholders. The sources of SDF contributions are postmortem review sessions and include actual deliverables or help provided to project teams:

1. Clarification of project rationale, goals, objectives, and expectations and validation of inputs
2. Descriptions of how megatrends, industry, and market trends and PESTLED developments are likely to impact the company's future
3. Identification of black swan scenarios and help in preparing contingency plans
4. Distillation of insights from competitor, industry, and SWOT analyses to determine achievability of project objectives
5. Explanation of the modeling concepts of the own and new company operations, identification of key demand drivers under the company's control, and creation of demand, price, and other revenue-impacting assumptions
6. Demand analysis, forecasting due diligence, and identified and quantified forecast risks that formed the basis for scenario-based revenue and cost projections
7. Introduction of well-structured processes, discipline and rigor in the analyses and evaluations, and sharing of tools and techniques that helped project team members do their jobs better
8. Validation of the reasonableness of target or potential JV partner company's projections, sanity checks of revenue and cost forecasts, and assessment of their implications
9. Description of reasonable and plausible scenarios, identification of inputs and support required to generate accurate forecasts and corresponding valuations, and performing revenue forecast sensitivity analysis and "what-if" simulations of distinct scenarios
10. Forecasting function project management excellence and guidance provided to project stakeholders on the application of forecasts

11. Identification and measurement of the value of trade-offs and what the company may get in return, development of negotiation positions, and evaluation of terms and conditions
12. Creation, testing, communication, and updating of a common project-wide assumption set, and identification of project risks and actions to manage those risks
13. Development of performance measures, a management dashboard, and an EWS of M&A and JV project forecast performance
14. Coaching of senior management, investor relations, and external affairs on key elements of the forecasts, major assumptions used, project risks, and risk management issues
15. Helping to condition and manage target and partner company expectations and determine customer and competitor reaction
16. Validation of the impact of the project on company competencies and capabilities and prediction of the project impact on the R&D, NPD, and Portfolio Management activities beyond financials
17. Development of an actionable response plan to counter risk occurrence due to competitor or industry reaction, incidence of uncontrollable events, or false scenario assumptions
18. Sharing of lessons learned in the project postmortem analysis sessions and right and wrong handling of major issues that can serve as guiding principles in future projects

12 Product and Technology Licensing

Licensing of products, services, and technologies is the assignment by the licensor to the licensee the rights to produce and sell goods and services, apply a brand name or trademark, or use the licensor's patented technology. Licensing is governed by agreements that are legal contracts between the licensor and the licensee that cover a wide range of business arrangements and define the rights and obligations of the parties. Licensing agreements are important strategic decisions, and in today's highly competitive environment, they are a part of the growth strategy of many large companies. For example, in the US pharmaceutical industry, licensing is necessary to ensure revenue and profit growth, and today almost half of large pharmaceutical company revenue is derived from licensed drugs. In the technology sector, licensing is a business strategy that generates substantial revenues and, to enhance profitability, expertise has been developed to manage it from the idea to qualification to incubation to commercialization down to licensing.

The majority of licensing agreements have standardized common elements that include the following:

1. The scope of the agreement, which deals with exclusivity, transfer of rights, territorial restrictions, and distribution rights
2. Time schedules and milestones specifying the length of the contract, product to market dates, and renewal options and conditions
3. Financial terms and conditions that include advance payments, revenue sharing or royalties, minimum payments, and how payments are calculated
4. The licensor's ability to monitor and control the licensed article's quality and market performance
5. The licensee's obligations to maintain quality standards and commercial performance, protect intellectual property, and control patents and trademarks, advertising, and inventory requirements
6. Exit terms and conditions for either party to terminate the licensing agreement and the penalties that apply

One element of licensing agreements that receives a great deal of attention from both sides is the financial arrangement and negotiations around this aspect of licensing projects. Hence, a lot of emphasis is placed on the development of the data, assumptions, evaluations, and long-term forecasts that underlie the revenue and cost projections and drive the value creation and the negotiations around it. The scope of the discussion in this chapter is limited to the commercial aspects of licensing agreements that involve transfers of rights for goods, services, and technologies. The different variants of project ownership transfer agreements, such as build–transfer–operate, build–operate–transfer, and other variants are a distinctly different topic, which belongs to the subject of project finance discussed in Chapter 15. Licensing projects are a clear case where the strategic decision forecasting (SDF) team is best qualified to provide the support needed to structure value creation and win–win projects.

12.1 THE CURRENT STATE AND CAUSES OF FAILURES

Licensing is a tool to achieve strategic growth objectives, and as a result, its role has expanded significantly and the number of licensing deals has been increasing over the past ten years, while focus shifted from intellectual property deals to value-creating projects. Increased product and technology licensing activity has produced increased competition, which now requires more complex deal structures, processes of evaluation and financing, shifts into new research efforts, and moves into new markets.

In some industries, such as pharmaceuticals and information technology, substantial progress has been made in the overall handling of licensing projects. However, in other industries, the current state of licensing has not reached high proficiency levels and one of the first observations is that while there is an increased need of licensing to achieve strategic growth objectives, the objectives of licensing projects are limited to the immediate financial benefit type. Some observers attribute this to the fact that the need for higher level of deal sophistication is not met by early participation of competent strategic forecasters and a corresponding increase in project team skills, training, and experiences.

Creating effective licensing partnerships is a challenge because of the increased competition for good partners and while more product introductions are expected to produce higher earned licensing revenues, in some cases they do not meet that expectation. In most cases, there is no clear understanding of how licensing affects the company's future and the key decisions that need to be made to create successful projects. The perspectives of licensor and licensee are viewed as negotiations space boundaries because the transfer of rights issues and strategic implications for both parties are not given sufficient attention. Licensing project processes tend to be incomplete and inadequate, and the roles and responsibilities of project participants are ill defined, overlapping, and often confusing. And, even though more products and technologies are licensed, the linkage and coordination with portfolio management activities remain weak and ineffective. The causes of failures are many, and in this section, we present the findings in licensing projects of different companies and highlight their causes.

Case 1: Absence of internal capabilities. There is a lack of strong internal capabilities in structuring and evaluating licensing projects, which necessitates the use of external advisors to provide deal evaluations and implementation guidance. Also, licensing projects suffer from lack of personal attention of senior managers, insufficient cooperation and clear communication, and poor monitoring of progress on project objectives and the health of partner relationships. The widespread shortage of talent and experience in deal making, project implementation, and alliance management areas extends to managers with the right mix of strategic forecasting, problem solving, and project management skills, which are so important in licensing deals. In most projects, the forecasting group is given a very limited role to provide data and forecasts and revise them as needed to get the transaction done. As a result, licensing projects are usually not viewed with open and innovative minds and partnerships fail to meet the objectives of both parties.

Case 2: Unharmonized priorities. Project evaluations are done under stress conditions by team members lacking required skills and objective perspective and with limited input from other functional area experts. This problem is compounded by the risks of new product and technology development, unclear and unharmonized goals and objectives, poor project implementation, cultural clashes, and poor partnership management. The result is that almost half of licensing alliances of this company do not meet the initial partner expectations. Licensing projects take a long time to complete because of disagreements over the value of the deal, which is traced back to differences in assumptions, views of the operating environment, and long-term forecast methods and models used. As a general rule, current-state licensing forecasts are based on static statistical models, analogs, and optimistic market research data and assumptions. Scenario development is often reduced to looking at baseline, optimistic, and pessimistic forecasts created from suspect assumptions. And to give the impression of rigor in revenue forecasting, thousands of simulations are performed to define a range of possible outcomes.

Case 3: Lack of partnership processes. A major problem in licensing projects is that long-term financial forecasts are generated by each of the parties involved without the benefit of a disciplined, well-structured, and objective process of valuing partner contributions in the context of unknown environments and management of the project. The successful partnership elements that are generally missing from handling licensing projects include the following:

1. A well-coordinated and communicated portfolio management and licensing strategy linked with strategic forecasting processes
2. Appropriate selection and screening of partners, expectations, existing relationships, and project management skills
3. Strong strategic forecasting group internal and external functional alliances, personal relationships, and high levels of cooperation and collaboration
4. Complete licensing processes and procedures, which define the responsibilities of team members and the role of the forecasting team
5. Having a market research strategy, developing good product profiles at the outset of the process, and producing supporting evidence

Case 4: Strategy and coordination limitations. The problems in this case have to do with the core project team's poor understanding of the corporate and project strategy and goals and objectives. Their inability to define and communicate project processes and success factors, and their weak skills in the forecast project management part are also responsible. Other observed limitations include the following:

1. Conducting the screening and creation of alternatives required to develop the feasibility study recommendations on an ad hoc basis
2. Not managing the due diligence effort so as to uncover gaps, problems, and risks
3. Not ensuring the availability of qualified resources, experts, and external advisors to provide analyses, trade-off valuations, and support to negotiations

Further limitations are noted in informing, guiding, or updating the project sponsor and senior management and failing to oversee the supporting teams' evaluations and subject them to sanity checks. Managing the development of inputs for the financial analysis and project valuation is often assigned to the Marketing and Finance organizations, which control the data and information flows internally, with the partner company, and with external advisors. There is little coaching of project participants and licensing subject matter experts on the intricacies of strategic forecasting for these projects or ensuring a smooth transition of responsibilities between different phases of the project.

Case 5: Project team limitations. Problems and issues related to project team member limitations identified in this case include the following: Project team members focusing exclusively on the specific tasks assigned, not seeing other possibilities, and failing to collaborate with other subteams and ensure quality and timeliness of handoffs. They do not raise issues, concerns, problems, risks, and gaps or report concisely and communicate findings of analyses, results, and their implications effectively or provide actionable recommendations. Forecasters and other internal experts do not fully address the specific issues and areas of concern to project clients, and their proposed solutions are weak as are the cost–benefit considerations and the defense of their positions. More significant gaps, however, are the project team's excluding suitable experts from the team, limiting the participation of external advisors and not managing their conduct as members of the project team. Project team inability to define scope of advisor engagement and their responsibilities, include them in subteams as needed, and learn from their experiences is very common as is not protecting sensitive information or using advisors wisely knowing that they are very expensive.

Case 6: Wrong perspective and skills. Technology licensing is a category of projects that in addition to strategic forecasting and financial evaluation skills, decisions require expertise in the application of tools to assess not only internal and external environments, but also technology evolution, strategic partner actions and reactions, and the impact of terminating agreements. That expertise is missing. Licensing projects are broad scoped, but needs and issues vary by the type of project, the specifics of the situation, and the skills of participants. In this case, instead of being viewed as strategic projects, they are viewed as financial engineering undertakings. As a result, accepted strategic forecasting applications have not been developed or methods and processes created that help to assess projects, reduce the time to decision, and manage future uncertainty and risks. Thus, licensing projects are a product of strategic initiatives that start with good intentions, but are then fraught with conflicts of interest as the terms of the deal take center stage.

Case 7: Participant positions not understood. Win–win licensing deals require that the different positions of the licensor, the licensee, the producer, the user, and the customer perspectives, which vary across types of projects, be well understood and taken into account. In this instance, however, the position and interest of the dominant project participant is allowed to determine the focus and attention in the project. Partner expectations are not subjected to sanity checks, and this leads to poorly defined projects, unreasonable valuations, drawn out negotiations, and costly, short-lived projects. Additionally, in some projects, clarity of purpose and strategic intent, which are the starting point of successful projects, are missing. Also, a silo organizational structure and parochial interests do not allow for a unified ambition and capabilities to be developed. Hence, a common long-term objective does not materialize and resources are not leveraged to reach the strategic objectives.

Case 8: Limited qualifications. Due to limited availability of internal competencies and capabilities, ascertaining that partners possess the right qualifications is done on an ad hoc basis and partner selection decisions are based on personal relationships of senior managers. As a result, important project success factors go unchecked, such as the future partner having a good alliance record and reputation, similar project vision and sense of urgency, and the ability to pay for the licensed product. This problem is complicated by inability to establish a potential partner's efficiency of decision-making processes, the wherewithal to provide adequate support for implementing the deal and market and sell a product aggressively, and the capacity to manage the entire life cycle of the alliance so as to maximize deal value. Wrong partner attributes explain why licensing projects of this case have a history of poorly handled deals and project teams do not learn from experiences with former partners who failed to provide strong support needed to implement the deal. Also, because many stakeholders favor internal projects at the expense of the partner's product and because they have convoluted processes, they are unable to see the strategic value of licensing alliances.

Case 9: Organizational challenges. Organizational structure failures and missing licensing expertise result in product management groups leading licensing projects, and forecasting expertise is brought in the process late and with limited scope. Hence, functions such as detecting inconsistencies in strategy, establishing project team qualifications and preparation, conducting a thorough forecasting due diligence, and managing the alliance formation are not given proper attention. Negotiations are sometimes conducted without reliable data and forecasting team support, and little attention is given to uncertainty and risk assessment and management. In addition, liaison support is absent, while monitoring project performance is limited to tracking and reporting actuals. Because evaluations for most licensing projects are financial engineering driven and negotiated by strength of position and not by good forecasts, external environmental factors often do not enter project valuations. Further, since licensing agreements do not assign proper responsibility in managing forecast realization and competitive response planning, it is difficult to ensure that project objectives are realized. Consequently, long-term forecasts used for project valuations are subject to negotiations and not rigorous analysis, sound assumptions, and scenario modeling and forecasting. As a result, forecasts are based on simple and static models, heuristic methods, and optimistic market research results.

Case 10: Project exit issues. The product and technology licensing process is made worse in this case by sole focus on intellectual property protection and definition of rights, clauses, timelines, fees, royalties, and not on the viability of the alliance. Also, the heavy influence of senior managers is a difficult hurdle to overcome in structuring effective exit clauses and termination of licensing agreements. Industry norms and analogs are used extensively in forecasts, but they are often misleading because they come from different project team experiences, product uniqueness, and operating environments. Another problem is that since negotiated forecasts rely on market research and customer satisfaction surveys, they do not address effectively the differences in the own versus competitor product attributes, while the product user positive feedback is inflated. These issues are magnified by divergence of licensor and licensee interests, which develop over time, the absence of reliable performance criteria, and early warning systems (EWSs) whose functions are vague or poorly defined.

Case 11: Wrong project approach. Due to the licensing project rationale being motivated by financial results and occasionally by enhancing the company's capabilities and competitive position, alternatives to licensing are not examined. Why? Because they may need longer payback periods or could involve higher levels of uncertainty relative to the company's risk appetite. Also, the costs of acquiring product rights are underestimated because licensor and licensee forecasts do not take into account the product upgrades and improvements necessary over product cycles to manage forecast performance and the product lifecycle management costs.

The timing and the effects of licensing a product on customer experience, especially when licensing to a joint venture (JV) company and related risks and issues, are not fully understood and reflected in valuations. Licensor critical questions such as how the product fits the partner's portfolio, how interested the partner is to develop and can exploit the product to its full potential, and how the different parameters affect the value of the product are given a light treatment. Licensee issues given equally light treatment are questions such as what is the potential value of a product to the company, do potential product uses fit the firm's R&D portfolio, what are the risks and costs of developing the product, are there synergies to be extracted, given that they could be impacted by unforeseen events and uncontrollable factors?

Case 12: Deal complexity. Deals are more complex today than five years ago, especially the multiparty alliances that are used to speed up innovation. However, there has not been steady development of skills in strategic forecasting, alliance formation, and licensing implementation to ensure corporate strategy, alliance, and portfolio management integration. A major need that goes unmet is to continuously assess the value of new products and technologies that require balancing marketing with value propositions; this is a difficult balance to achieve. And since licensing is a key component of corporate strategy, investors look at licensing deal structures closely because they are aware of the difficulties in managing the complexity, uncertainty, and urgency of product and technology partnering and acquisition strategies. For these reasons, the Business Development group emphasizes the need for ongoing licensing opportunity identification and evaluation, and alliance project management skills. However, the reality is that these qualifications are missing in this project team that, eventually, learns the skills needed on the job. Additional complications arise when there are multiple partners involved in the R&D, manufacturing, and distribution and marketing areas. That is, the involvement of several participants increases project complexity exponentially.

Case 13: Lead and communication problems. Product and technology development consortia are becoming accepted, but the commonly take silo approach to licensing alliances does not work well. Why? Because in most projects, the Business Development and R&D groups do not drive alliance strategy and because Marketing and industry advisors create partnership opportunities. Licensing deals do not always consider investor reaction to the alliance created by the project, and when insufficient attention is paid, these projects become liabilities. This happens because licensing targets are given the same due diligence treatment as acquisition targets, namely, watching for window dressing. Another important issue is that the project lead organization assumes overall project responsibility but fails to facilitate communications with senior management and other stakeholders. Also, since there is no real forecasting process management, the lead organization oversees the project team's progress and assumes the forecasting and project negotiator roles. As a result, they are unable to negotiate quickly win–win arrangements and report convincingly to the outside world the true value of a deal.

Case 14: External effects on forecasting. Licensing has resulted in increasing importance of core technologies, but it has also increased the interdependence for joint supply of product components and shorter product life cycles. The major issue here is that licensing core competencies have not kept up and the effects of technological change on the licensing parties versus competitors and the industry as a whole are not well understood. Also, identifying and verifying the technology user needs and the effects of new technology on user performance are difficult to assess in the midst of rapid technology changes. And, assessing causality and the interaction of technology trends with megatrends, subtrends, and political, economic, social, technological, legal, educational, and demographic (PESTLED) changes is harder to incorporate in the technological risk assessment and revenue and cost forecasting under any approach. Furthermore, it is difficult to understand and assess the effects of key technology licensing parameters and their implications for the value of licensing deals. This challenge is further accentuated in the absence of inadequate project support and competitive response and forecast realization planning.

12.2 SDF APPROACH, PROCESSES, AND EVALUATIONS

12.2.1 SDF Approach and Processes

The SDF team's approach and roles it assumes are structured to provide maximum support to the five phases of the licensing process it impacts, which is shown in Figure 12.1. Note that the roles of the SDF team depend on which side of the project it lends support to, but its contribution is maximized when the analyses, forecasts, and evaluations it performs take into account the perspective of the other side. For that to materialize, SDF team members need to have access to and work closely with counterparts in the potential partner company from the start of the project.

The five project phases that require SDF team support define its scope of involvement, the approach, and many of its activities, but the bulk of the SDF team's involvement is in the commercial aspect of licensing. However, since the commercial aspects impact the licensing implementation and require monitoring of project performance, SDF team members play important support roles in those areas as well. The commercial aspects of licensing projects of SDF team involvement are shown in Figure 12.2.

After strategy and project strategy review, the next SDF team activity in a typical company's licensing process, which is shown in Figure 12.3, is an environmental scan followed by an

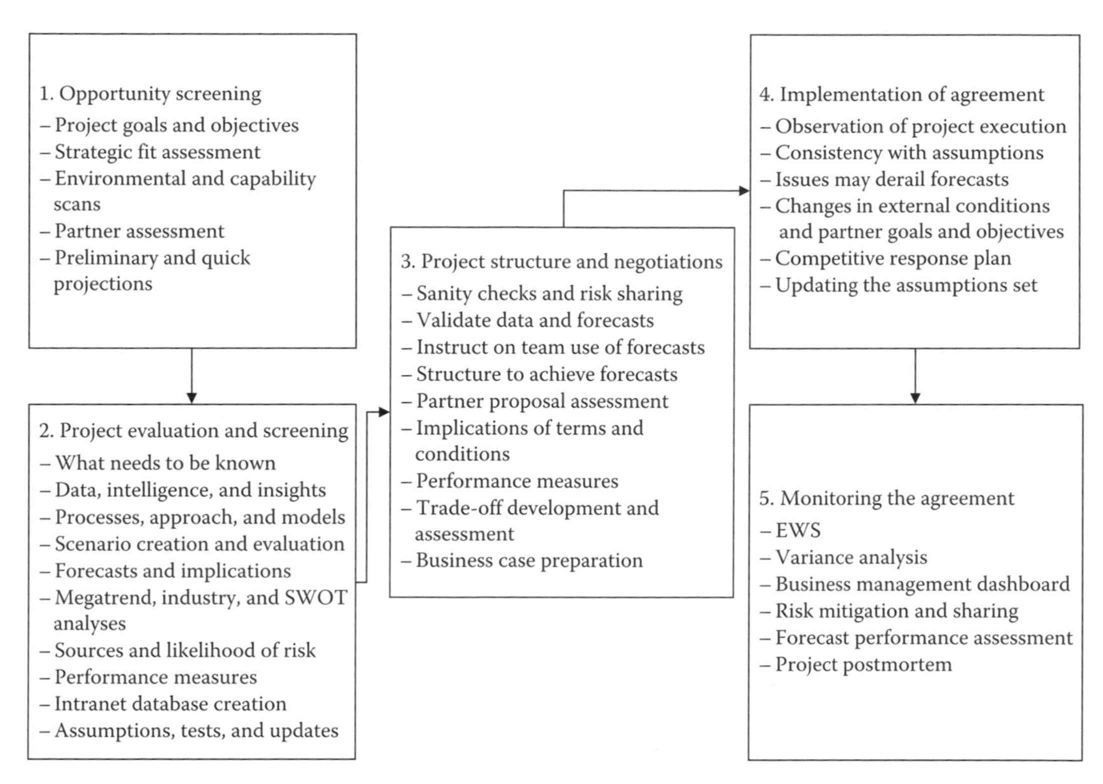

FIGURE 12.1 SDF team approach and roles in licensing projects.

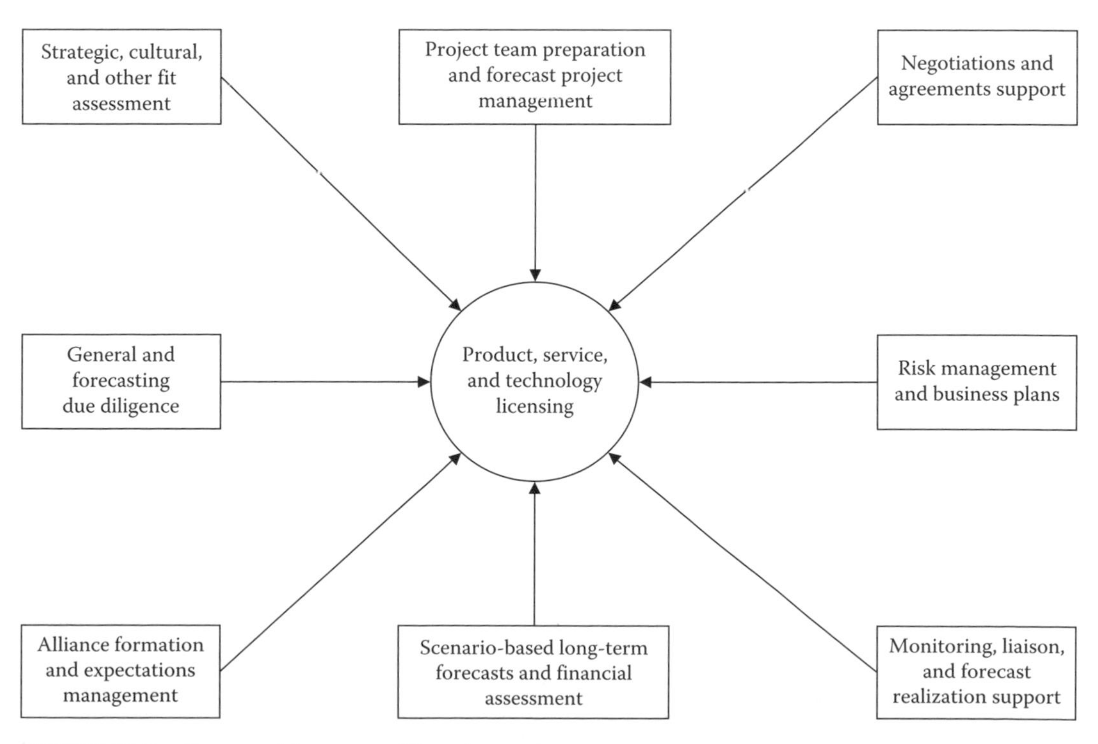

FIGURE 12.2 Commercial aspects of licensing projects.

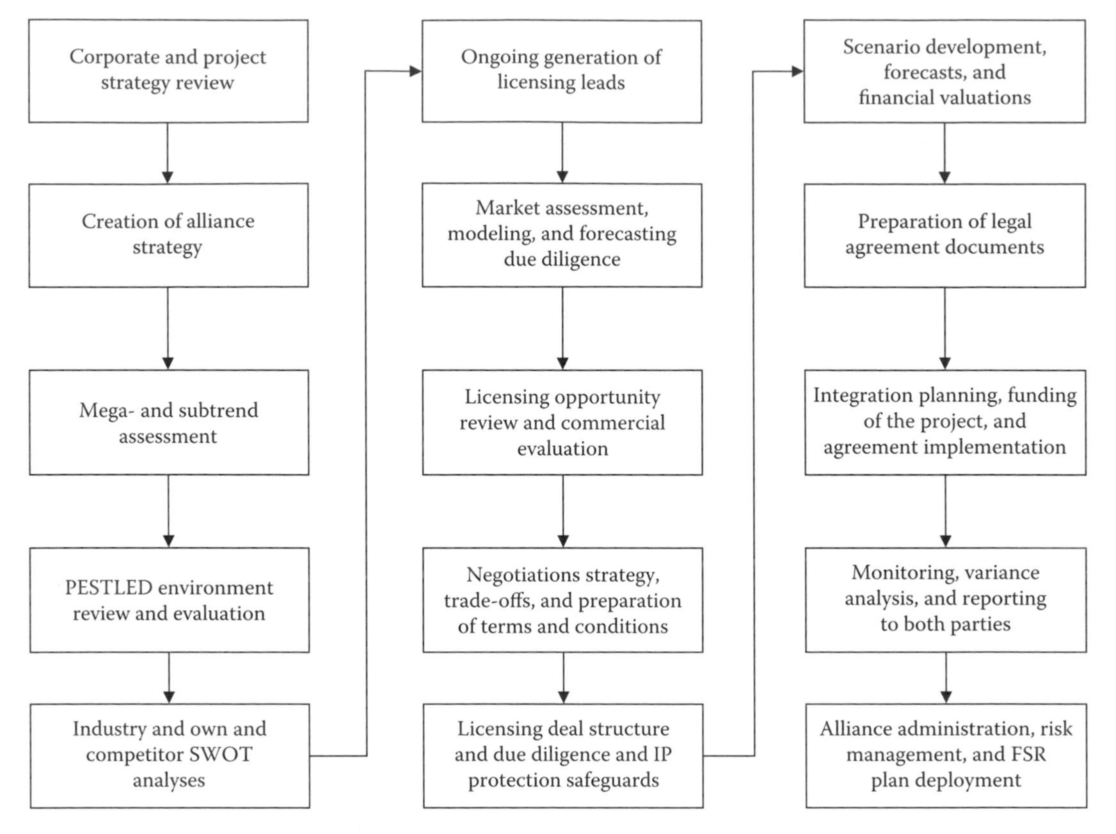

FIGURE 12.3 Typical licensing project process.

initial market assessment and forecasting, and the development of project rationale and project objectives. The screening of potential partners comes next and the preparation of project materials takes place, such as a feasibility study and desirable deal structures and terms of the deal. Alternatives to the proposed project are considered and ranked according to company selection criteria.

Once a deal structure is contemplated, negotiations begin in earnest, while the due diligence effort is under way and sanitized long-term forecasts are developed to support financial valuations. At this point, funding, intellectual property protection, and implementation issues are settled and legal agreements are created. Monitoring of deal performance and management of the alliance begin after deal closing and implementation. Notice that since the SDF process and the generic licensing project process overlap, the support provided by SDF team members covers the majority of the project team's needs.

The SDF team is best qualified to generate sound project forecasts due to its central organizational positioning, strategic decision support expertise, and its internal and external links and relationships. The support elements in the commercial aspects of licensing projects include the following:

1. Clarifying or defining the rationale for licensing vis-à-vis other growth options and the rules and criteria for partner selection based on competitive analysis, evaluation of past performance, and industry experiences
2. Designing a sound project structure by understanding how licensor and licensee interests could diverge over time, harmonizing those interests, and helping to manage alliance issues to achieve realization of the desired future state

3. Explaining the appropriate role of senior management, the project team, and key stakeholders in the forecasting process to reduce negative influences and to enhance high levels of communication, coordination, cooperation, and collaboration (4Cs)
4. Helping to assess the impacts of rights, clauses, timelines, fees, royalties, and intellectual property protection; the data, information, and the common assumption set driving the project; and exit clauses and termination of agreements; and evaluating their implications
5. Making sense of and bringing transparency to the licensing forecasting process by introducing a well-structured analytical approach, eliminating unsuitable industry norms and analogs, and testing the reliability of market research results
6. Creating meaningful performance criteria and reliable management dashboards and EWSs to judge deal performance and manage competitor reaction to the project and the risks that could materialize in the forecast horizon

After the rationale for a licensing project and strategic fit are validated, the SDF process in the evaluation stage requires participation in generating background information, leads, and deal sourcing. Helping in internal preparations means involvement in licensing reviews and creation of agreements that cover data, models, forecasts, and evaluations used in the project. Also, a mini-due diligence is conducted at the feasibility study phase with SDF team support to quantify expected synergies, unintended negative effects among business units, and resources needed to support the project team. In the mini-due diligence stage, the SDF team identifies and clarifies expectations and growth goals and objectives and validates basic assumptions.

Once the initial evaluation and the feasibility study are done, information exchange with the potential partner takes place. Then, the due diligence to validate data, assumptions, and forecasts leads to the licensing project's evaluation from a technical and commercial perspective. Deal structures are advanced and negotiated and legal agreements are drafted, and once the financial assessment is completed and sanity checks are performed, a business case is prepared with a recommendation to proceed or not and corporate approvals are obtained.

Often, potential licensing partners are projecting themselves as preferred partner, but experienced licensing teams create R&D and Business Development outposts to get access to better alliance opportunities. Such teams are supported by SDF team members and focus on the business aspects of the deal structure. Part of the process here involves addressing issues such as exclusivity levels, trademarks, geographic area or market segment, types of income, and payment type. The SDF team support activities are shown under the six major processes:

A. *Preevaluation phase.* In this phase, SDF team activities encompass assessing project consistency with the corporate, R&D, business development, and product portfolio management (PPM) strategies, the fit of the proposed licensing project, and its consistency with market realities. Once the positions and interests of the various participants are understood, project and SDF goals and objectives are developed and screening and selection criteria are advanced. SDF team members brief the project team on forecasting issues and requirements and help to identify alternative opportunities, reassess strategic fit, screen proposals, and develop a feasibility study. The SDF team is the steward of the forecasting process, helps in the project due diligence function, and ensures all-around communication of its findings. It reviews earlier legal agreements, performs sanity checks, and helps manage the roles and evaluation activities of consultants, investment bankers, and external advisors. Additionally, it consults project stakeholders on forecasting issues, provides inputs and recommendations for corporate approvals, and ensures proper transfer of forecast responsibilities across project phases.

Why would SDF teams be allowed to engage in activities normally performed by other organizations?

1. *They are the central keepers of competitive intelligence, data, and earlier evaluations.*
2. *They can verify that assumptions are consistent with the terms and conditions in licensing agreements.*
3. *They are able to obtain maximum benefit from consultants and external advisor experiences.*
4. *They are in a good position to identify events and assess the implications of important factors for project success.*
5. *They have the ability to screen and validate inputs to decisions impacting the project's commercial aspects.*
6. *They can gauge the support levels needed to execute the project successfully.*

B. *Licensing evaluation.* In this phase, the primary focus of the SDF team is on the situational analysis, grounding of the evaluations, and performing sanity checks. The commercial assessment elements include potential sales, expected development costs, lifecycle considerations, level of exclusivity, territorial licensee reach, and deal structure. Here, the SDF team helps to obtain and maintain sponsor and team support, to conduct a thorough opportunity assessment, and to develop sound forecasts and future-state scenarios. Also, it identifies project risks and major issues, assesses forecast implications, and provides inputs and insights to the financial analysis and project valuation. Further, it facilitates the creation and proper evaluation of alternatives to the proposed project, manages the identification and valuation of synergies, and assists in the development of the project business case. The methods and techniques used are similar to those of new product development (NPD) discussed in Chapter 9.

C. *Prenegotiations.* Prior to negotiations and development of the business case, SDF team members help the project team perform a preliminary screening of the proposed deal and conduct the forecasting due diligence. They help to quantify expected business unit synergies, perform sanity checks, identify resources needed to execute the project successfully, and develop the SDF team's support plan for negotiations. The progression of SDF team activities involves a second review of the own and target company sales trends, pricing, and advertising. It also involves evaluation of competitors and external environment, checking existing or developing new assumptions, and updating initial sales forecasts. An evaluation of recent deal information takes place to identify biases and address cross-cultural issues. Also, the parameters influencing the deal value are examined, and the value of the licensing transaction is estimated through financial models. Then, the partner forecasts and supporting rationale, assumptions, and documentation are obtained and evaluated.

D. *Negotiations phase.* In this phase, SDF team members assist in managing the creation of a consensus out of forecasts based on different partner assumptions. They also help in creating and valuing initial, fallback, and walkaway negotiation positions; perform analyses and provide inputs to the negotiation team; and help create a good negotiation plan. They create negotiation scenarios to test partner claims and positions and develop and value trade-offs to exchange for concessions. Also, during negotiations, they provide support and insights into structuring the deal, create trade-offs identify financing requirements based on licensing project's economics, and assess the implications for negotiated terms and conditions.

E. *Project execution.* In the execution phase of the licensing process, the SDF team helps the implementation team understand the forecast scenarios, their implications and risks and advises on the proper use of forecasts, and the conditions for creating an efficient implementation plan.

It assists the project team to ensure that the right governance structure is put in place to manage the licensing venture and helps the implementation team to execute the project as envisioned in the forecast scenarios used in the decision. It also helps to manage the implementation, integration, and consolidation of operations and plays a major role in the development of the project business plan. Further, because of being the central owner of data, models, and assumptions, it assists in synergy identification, creation, and extraction planning. The SDF team also helps with conflict resolution and with 4Cs and continued senior management support.

F. *Postimplementation.* During this period, the SDF team lends its expertise to the liaison subteam to ensure effectiveness of monitoring and managing project performance. In some projects, the SDF team monitors new entity performance, interfaces with the steering committee and counterparts in the partner entity to assess or adjust performance targets and see the competitive response and forecast realization plan put into effect when conditions warrant. Because of their understanding of the customer's and key competitors' capabilities and competencies, senior SDF team members serve at times as steering committee members in JV companies created to bring licensing products or technologies to market. They also play a key role in the project postmortem analysis, disseminate the commercial area findings, upload lessons learned in the Intranet database, and provide recommendations for future project and SDF process improvements.

When the licensor is the technology-producing agent and the licensee is the technology-using entity, SDF team members work with counterparts in the partner company to create a common, acceptable technology forecasting process to help the project team and their key elements involve the following:

1. Identifying and defining market needs for the new technology, applications, user benefits, and possible new applications
2. Identifying and evaluating new technology-based products and services beyond the technology part of the project
3. Helping to define the desired licensing project type, evaluate the nature of technology advances needed to meet user expectations, and assess project uncertainty and risks
4. Creating plausible, distinct scenarios of future states and development stages of technology and estimating the useful economic life for that investment
5. Determining how technological innovation will change the competitive environment of the industry and developing plans to counter competitor responses
6. Evaluating the own company and competitor performance, benchmarking best-in-class project support, and forecasting market adoption rates for the new technology
7. Evaluating the speed of new technology innovations and technology adoption differences by geography, user preferences, and new versus current technology costs
8. Describing scenarios of how technology will invent the future beyond the project implementation and understanding how disruptive technologies will affect the business, the industry, and the users
9. Determining the length of technological leadership sustainability through external environment, financial, and other resource costs and benefits evaluation
10. Helping the R&D organization determine what particular technology to focus on developing, which is a problem similar to the one encountered in the NPD area

12.2.2 SDF Analyses and Evaluations

The SDF team analyses and evaluations performed in licensing projects are in direct support of the key project plans, namely, the strategic plan, the due diligence plan, the negotiations plan, the business plan, the financing plan, the implementation plan, the competitive response and forecast realization plan, and the risk management plan.

A. *Preparation activities.* In the preparation stage, SDF team members determine what information needs to be known and what is known to develop long-term forecasts, what factors are under the control of the company and the partner, and what factors are not. They also define the criteria to use in screening products to be licensed, strategies, and potential partners. SDF team members also help screen licensing partners and build internal consensus. The screening performed enables creating opportunity profiles through their analysis, initial assessment, and the feasibility study.

B. *Project rationale clarification.* In the project definition phase, the SDF team clarifies the purpose and rationale of licensing through an evaluation of the costs and benefits of the product being licensed and checks the consistency of the project rationale with corporate strategy and objectives. The internal consistency of project goals and objectives is verified and what the company and the potential partner want to get out of a specific licensing project is confirmed. Then, the consistency of project objectives with historical data, benchmarks, and competitive intelligence is verified and a rough screening of alternatives being considered takes place.

C. *Strategic objective clarification.* SDF team participation in clarifying strategic plan objectives is required because of the analysis and the sanity checks it performs using market, company historical data, and competitive intelligence. SDF team members evaluate the major external environment developments and issues and help the project team explore different opportunities to licensing. They create functional, professional, and collaborative forecasting links internally and externally and help set achievable project objectives based on market realities, corporate risk tolerance, and strategic posture. Also, they monitor the implementation of strategic plans to ensure realization of objectives because they impact forecasts and provide recommendations to keep project plans on track.

D. *Product portfolio evaluation.* After the corporate strategy review, the SDF team conducts a current product portfolio evaluation in conjunction with the PPM group. It reviews the state of company's R&D situation, investigates product development pipeline issues, and assesses the company's strengths and weaknesses in the various product commercialization areas. Then, it is in a position to assist the project team in making a sound licensing strategy, creating an alliance formation process that includes screening potential partners, and understanding the implications of partner requirements. Thus, a good portfolio review helps to manage partner expectations and create the environment to maintain a positive chemistry, which is so important in alliance formation and management.

E. *Defining the scope of due diligence.* Significant effort is expanded by SDF team members in the licensing due diligence to define the purpose, process, and types of due diligence needed for the commercial aspects of the project. The due diligence includes the following:

1. The own company forecast due diligence
2. The partner's general due diligence
3. The partner's forecast due diligence

Energy is also devoted to selecting due diligence participants, developing recommendations, and producing the due diligence report. SDF team members create a project-specific licensing due diligence process and sometimes use consultants to establish initial partner contact. A key element of licensing success is data, data format, and quality, and the SDF team ensures that the reports provide a detailed outline, cover confidential and nonconfidential data, and address all information requested. In the company visits that are planned and corporate presentation material that is exchanged, the SDF team evaluates the data and information received and identifies, creates, and tests assumptions about the potential size of the product's market.

F. *Deal due diligence.* The SDF team's due diligence focuses on understanding the value of the deal from the perspectives of both parties and evaluates the deal potential as if the product is part of licensee's product portfolio. To that extent, in addition to assessing how large the market is, how much market share the licensed product can obtain, and development and marketing costs, the SDF team helps evaluate the potential value of the licensed product under different scenarios. It validates processes, assumptions, data, models, and scenarios used. It performs sanity checks and assesses the forecast implications for resources needed to execute the project well. It also assesses the impact of delay in product development and identifies the risks to each party. That is, it helps the project team determine a range of expected deal values for the licensor and the licensee in preparation for support the SDF team provides in negotiations.

G. *Revenue analysis.* In the product revenue due diligence phase, an external and internal environment analysis is performed. It includes an evaluation of megatrends and subtrends, PESTLED and industry factors, and internal and external strengths, weaknesses, opportunities, and threats (SWOT) analyses. The product attributes are evaluated vis-à-vis competing products, the key market drivers are identified, and their interrelationships are measured. Then, an evaluation of total market and market share of the licensed product takes place based on evidence of the sources of market share. Unit pricing assumptions and considerations are tested and advertising and promotional requirement estimates are developed. Ability to execute on the various aspects of bringing the product to market as planned and gaps are identified, and an SDF solution and recommendations are developed. The underlying assumptions of unit and sales forecasts are individually and collectively evaluated, and the implications for those forecasts are assessed.

H. *Analytical modeling.* The work of SDF team members in the analytical modeling area begins with benchmarking around licensing best practices from industry and competitor experiences. Then, the effort moves to intelligence creation and developing the key common assumptions set that includes creating, testing, and updating total market size; product uptake; and market share assumptions. The key variables impacting the value of the licensing opportunity are identified, and the impact of different parameters and model inputs are evaluated with special attention paid to interaction and feedback effects. Also, the sales force, advertising, and marketing expense requirements are identified in order to determine product acceptance and uptake.

I. *Cost and pricing analysis.* The product pricing due diligence focuses not only on the product or brand, place, and promotion elements, but also on product effectiveness, enhancements, and value created for the user. Market share estimates are reassessed before profit maximization exercises begin to make sure that stable, competitive prices are put into effect and return-on-investment (ROI) targets are established. The viability of skimming, prestige pricing, penetration pricing, extinction pricing, and price differentials and discounts are also investigated. The cost due diligence is usually a Finance organization function and deals with developing cost of goods sold (COGS) and selling, general, and administrative expenses (SG&A). However, the analysis of SDF team members in this area involves ensuring that all cost components are included, reasonable, and consistent with those of competitor products. Product development costs, sales and marketing, upfront investment requirements, milestone payments, royalties, and transfer payments are also validated and verified. Then, product liability insurance and financing costs questions are answered and future capital contributions, unexpected outlays, and transfer pricing issues are investigated. Lastly, issues and potential tax implications are assessed by the project team.

J. *Scenario development.* To identify situations in which the licensing opportunity is attractive, the SDF team engages in scenario development and planning with models that capture deal revenues and costs. Extensive hypothesis testing and "what-if" and sensitivity analyses take place, and model simulations are performed to reduce decision uncertainty and arrive at a most likely forecast view quickly and efficiently. The critical negotiation elements are identified and fleshed out in order to value potential trade-offs. Throughout the analytical modeling effort, the focus is on ensuring data quality and reliability and scenario model adequacy to provide reliable inputs to Finance that builds the project financial model. The net present value (NPV) analysis and risk-adjusted NPV estimates, the internal rate of return (IRR) comparison with alternative projects, and the Monte Carlo simulations are performed to identify the greatest impact variables on deal value. They also investigate alternative commercial terms that may make the deal more attractive and look at alternative deal structures from the partner's financial viewpoint.

K. *Financial model inputs.* The SDF team's inputs to financial models are derived from a number of evaluations, which, often, begin with a critical assessment of existing or partner forecasts and underlying assumptions. Licensing revenue projections are scrutinized from the company, the partner, or the alliance perspective, and the rationale of investment and expense requirements is checked against prior experiences and industry benchmarks. Also, partner contributions are subjected to analysis and valuation in order to arrive at reliable value measures for input to financial models and the development of a sound valuation to support the business case.

L. *Negotiations support.* SDF team analyses and evaluations for negotiations and legal agreements begin with the evaluation of the competitive position of the licensing project parties and identifying the crucial elements in the negotiations process. The development of negotiation positions is based on the creation and valuation of trade-offs based on alternative scenario forecasts. To help the project team create a more effective negotiations environment and a win–win arrangement, the SDF team quantifies prenegotiation project risks, evaluates the negotiated terms and conditions, and assesses postnegotiation risk sharing. SDF inputs to the legal team are highlighted areas of uncertainty, forecast risks, and environmental factors, which impact the value of the deal and may need to be addressed through terms and conditions in legal documents.

M. *Implementation support.* An experienced SDF team extends its involvement in licensing projects to the deal implementation phase and helps develop common goals and objectives and identify resources to manage the licensing partnership. Having a good understanding of the negotiated terms, the team is in a position to help Product Management monitor effectively the performance of the launched product and help in the creation of alliance business plans and targets. Also, by managing the assumption creation, data flows, and information exchange, SDF team members prevent misunderstandings, help resolve conflicts, and help to get internal team commitment in the implementation phase. Experienced in management dashboard and EWS creation and performance measures, they help the project team to develop these systems and provide the links for data inputs to populate them.

N. *Risk management support.* SDF team support begins with assessment of project uncertainty and risks and includes the own and partner company SWOT analysis and investigation of industry trends. This support and in-depth competitive analysis enable SDF team members to identify factors and events outside the control of the partners in a licensing deal, which leads to the identification of forecast risks and possible timing of their occurrence. Quantification of forecast risks and impacts on project value are other functions performed by the SDF team along with helping the project and implementation teams to develop and deploy forecast realization and competitive response plans. Here, the SDF team plays a major role in creating options and moves to manage risks and in developing contingency plans.

12.3 SDF TECHNIQUES AND TOOLS

The tools and techniques used by SDF team members in forecasting for licensing products and services are similar to those used in NPD. However, technology forecasting tools are more specialized and require a broader spectrum of skills and expertise. Technology is the application of science, engineering, and art to meet human needs, and technology forecasting includes the method and processes used to predict the future state of technology, its uses, and user adoption. Technological forecasting is a subset of futures research and refers to the process of predicting the characteristics and timing of technology advances. It is an integrative modeling approach of using data and information to predict technological advances.

Technology and technological forecasting require a good understanding of the interaction and feedback effects of customer needs, technical systems, market dynamics, and megatrends and subtrends. As such, it requires inputs from diverse sources and methods to create and validate future-state technology forecasts. There are three sorts of technological forecasting tools and techniques used by SDF team members to predict technology innovations and future states: quantitative, judgmental, and integrative general tools and techniques.

12.3.1 Quantitative Technological Forecasting Techniques

These are statistical models that attempt to deal with complex interdependencies and forecast technical capabilities not devices. They include the following widely used models:

A. *Trend extrapolation models.* Assuming that the steady stream of technological innovation will continue into the future, these models make future inferences based on past experiences. They include (a) statistical curve fitting of a technical parameter over historical data and projections, (b) limit analysis whose idea is that technological improvements eventually reach some absolute limit or constraint, and (c) simple and multivariate correlation used when a technology is a precursor to another or when innovations in one can be adopted by the follower technologies.

B. *Growth or S-shape curves.* These are models where technologies follow the path along the invention, the introduction of technology and innovation, the diffusion and growth, and the eventual maturity phases.

C. *Envelope curves.* They are growth curves with stringent constraints that can be overcome by new technological capability applications.

D. *Substitution models.* These models imply that the progression of events leading to technological change is inevitable. These models are based on three premises:

1. New technologies are alternative and competing ways of satisfying needs.
2. Once substitution begins, it will continue to completion (cannibalization effect).
3. Substitution of new for old is proportional to the amount of the old adjusted for natural market growth and additional applications.

Note: *The quantitative technological forecasting techniques are techniques also used in tactical forecasting; that is, they are short-term methods. To be used in long-term technological forecasting, they must be supplemented by or integrated with strategic forecasting methods and techniques discussed in Chapters 6 and 7 and the qualitative methods discussed here.*

12.3.2 Judgmental Technological Forecasting Techniques

These are qualitative methods that benefit a great deal from the input and judgment of competitive analyses and SDF team evaluations. They include the following techniques:

A. *Monitoring or innovation tracking.* This is a purely competitive intelligence gathering activity, which requires SDF team members to stay informed, anticipate technology changes, and scrutinize the different phases of discoveries and innovations before they become widely known. The areas of technology monitoring and tracking include the following:

1. Concepts, ideas, suggestions, and plans as well as research proposals and grants
2. Scientific findings and laboratory reports
3. Announced field trial results and commercial introductions

B. *Network analysis.* It is an extension of innovation monitoring used in exploratory forecasting, which focuses on identifying opportunities that may result from applications of current R&D efforts. It is also used in descriptive forecasting whose essence is to determine what research results are necessary to realize a certain capability or system.

C. *Scenario forecasting.* This type of technological forecasting attempts to describe the future of technology improvements and innovations based on system dynamics modeling. It determines those drivers that are subject to a company's control, some key assumptions, underlying trends, and environmental conditions. These scenario models are then used as flight simulators to determine what happens as inputs to them change or what the inputs should be to achieve certain outcomes.

D. *Morphological analysis.* This is a prescriptive forecasting technique used extensively in the automotive industry that uses a process that makes assumptions regarding future customer needs or wants and then evaluates possible systems, technologies, and innovations to satisfy them. To do that, it uses a matrix, which is called the morphological box, where technology parameters are listed on the first column and the alternative ways of achieving each of those parameters are defined (Zwicky 1969).

E. *Relevance trees.* They are variants of network analysis and are used to study technological goals and objectives and choose appropriate and specific R&D projects. This method requires the SDF team to work with R&D and engineering groups to determine the most appropriate path of the relevance tree by placing in a logical and hierarchical order the objectives, broad methods, processes and procedures, performance measures, sales and cost estimates, and describing the resulting research projects (The Futures Group 2004).

F. *Cross-impact analysis.* This is an extension of the Delphi method, which uses a matrix to assign systematically confidence to forecast events and technology driving factors by including dates of occurrence and probability of each event and factor occurring. Like the Delphi method, this technique uses a panel of technologists whose opinions are summarized and fed back to the group, which then generates a new set of judgments. The key elements of this technological forecasting technique are as follows: There is balance in the composition and viewpoints of the panel experts, anonymity is maintained in opinion gathering and distribution, there is iterative balloting to moderate outlier positions, and reasons are given about specific judgments to reach consensus (Lipinski and Tydeman 1979).

12.3.3 Integrative Technological Forecasting Techniques

The four most widely used integrative technological forecasting techniques and tools are the interrogation model, strategic market research, bibliometrics, and theory of inventive problem-solving (TRIZ) forecasting.

A. *The interrogation model.* This technique involves preparing questions and interpreting the answers in four steps: Interrogation for the need of technology, examination of the underlying causes, interrogation for the relevance of technological innovation, and interrogation for the reliability of the new technology or system (Martino 1983).

B. *Strategic market research.* This technique uses analytical tools, expert opinion, and user interviews to visualize the opportunities created by changes in company technology, markets, and customer needs. The elements of this process are intended to

1. Ensure consistency of technological change and marketing and sales strategy.
2. Identify, validate, and evaluate emerging market needs and user preferences.
3. Help assess the total market value of new technology and the impact on related technologies.
4. Evaluate the requirements and risks associated with entry into new markets through a new technology.
5. Create a basis of knowledge to forecast adoption rates of the new technology.

C. *Bibliometrics.* This technique is also known as research profiling and is an adjunct tool to technological forecasting that scans large databases and captures information indigenous to patterning and content of relevant literature. This analysis uses counts of publications, citations, and patterns to measure and interpret technical innovations and their effects. The basic premise of bibliometrics is that the data elements it captures are useful indicators of R&D activity, innovation, and states of future technology. The value of bibliometrics lies with its ability to show the progression of knowledge and its application in a specific area of technology and how it can be used to determine the shape and the location of S-curves (Narin 1976).

D. *TRIZ technological forecasting.* TRIZ is the Russian acronym for theory of inventive problem-solving invented by Genrich Altshuller and it is a great tool used in making strategic forecasting decisions that drive R&D and new technology and product development projects. While some technological forecasting tools deal with technology parameters, TRIZ forecasting deals with the structures and conditions that make the realization of the parameters possible. As such, it is a practical technique to predict innovation and a technology forecasting tool based on a systematic, logical analysis of patterns called the Laws of Evolution.

Notice that the essence of TRIZ forecasting is to identify changes needed to move a technology, system, process, or product to the next position on a logically predetermined path of evolution. To accomplish that, the 40 TRIZ inventive principles and the 8 patterns of how systems develop over time are exploited one by one to establish where a certain technology is coming from and create ideas and scenarios about where it will evolve in the future. For more on the TRIZ technological forecasting technique, the Law of Evolution, and applications, see Eversheim (2009).

12.4 SUCCESS FACTORS AND BENEFICIAL PRACTICES

12.4.1 Project Success Factors

There are a number of common beneficial practices and success factors in licensing projects, but the product particulars and the deal structure really define what contributes most to the success of a licensing project. Factors considered important to deal success are the definition of the product being licensed, creative product development, use of reliable data and insights, and people with expertise that are essential. Then comes the win–win deal structure, which includes appropriate product pricing, well-negotiated points of agreements, good financial terms, buy-back rights and deal exit terms, and intellectual property control. The success factors are shown by the following functional area cases:

A. *Due diligence.* The success factors in the due diligence area are establishing appropriate initial contact with the potential partner and screening, using experienced SDF team members or external consultants in evaluating forecasts, and evaluating the licensor report. To do that, the licensor report should provide detailed outline of reference material and address all data and information requested. Another key element of the licensing due diligence is data format and quality and appropriately planned company visits to assess corporate presentation materials. Also, testing the assumptions about the size of relevant markets and the partner's ability to deliver the product as specified in the agreement is crucial to understand the particulars of deal structure and the project valuation. Furthermore, verifying the intent of licensing agreements, enforcement of contractual compliance, and how contract and cultural conflicts will be resolved is needed to assess the forecast risks and the likelihood of realization of project financials. Last, and most important, is the validation of the extent of management support and internal coordination and cooperation of stakeholder groups.

B. *Competitive advantage.* The factors critical to the success of a licensing project from a competitive advantage perspective are the following practices:

1. Understanding the strategy of competitors and beating them without engaging in licensing partner competition
2. Minimizing the number of project stakeholders and ensuring coordination, collaboration, and communication in the commercial area
3. Creating strong partnerships and preparing for long gestation licensing projects
4. Building and maintaining alignment of strategic forecasting with other functions
5. Focusing on licensing implementation to ensure deals succeed after deal closing
6. Determining the degree of cultural fit, managing relationships, and adhering to the licensing alliance process
7. Maintaining the partnership's health by focusing on three areas: the broad project goals, the tools and structures to develop a profitable product, and the means to measure progress
8. Creating models, metrics, and techniques for successful partnerships and turning them into a corporate competency for growth

C. *Understanding the deal.* From the perspective of understanding the deal, the following practices are considered success factors in licensing projects:

1. Focusing on understanding the value of deal from both parties' perspectives to see if it is a good deal and a win–win arrangement
2. Evaluating the deal potential as a good addition to the licensee's product portfolio
3. Determining how large the market is and how much market share can the product obtain in the next 10–15 years
4. Identifying all development, testing, and marketing and selling costs, and determining the potential value of the product under different scenarios
5. Evaluating the impact of delaying product development and the risks to each party
6. Ascertaining balance between the value to the licensor and the value to the licensee to ensure to project viability

D. *Senior management view.* From a deal-making standpoint, the practices and factors leading to success are as follows:

1. Articulating a clear and compelling rationale for a project, senior management team experience in licensing projects, and having done comparable deals
2. Employing appropriate industry norms, standards, and benchmarks, and looking at analog product performance
3. Using sound competitive analysis, forecasting methods, external advisor assessments, and independent opinions

4. Seeking partners whose culture, needs, and strategy fit closely to those of the company and good relationship management in all activities
5. Delineating the roles and responsibilities of the licensing partners and spelling out the alliance governance details in contracts
6. Managing partner expectations, having conflict resolution mechanisms in place, creating competitor response options and moves, and helping deploy risk management plans effectively
7. Effective marketing and sales management in the postlaunch period, monitoring performance, and close partner cooperation

E. *Process perspective.* The process quality factors that are the main success factors in licensing projects include the following:

1. Using the expertise of SDF teams to create integrated processes with product management teams in order to create sound forecasts
2. Limiting the number of project team participants to one person from each stakeholder partner organization
3. Having a clear understanding of partner's roles and responsibilities, and a complete and effective process, and developing and testing of assumptions jointly with the partner
4. Creating a dedicated implementation team to produce joint marketing and promotional programs with the partner
5. Communicating the value of the alliance internally, to the partner, and to customers
6. Coordinating with the partner project activities, schedules, and plans; creating performance indicators to monitor progress; and having a common risk management plan in place

F. *Forecast perspective.* The forecast-related factors considered essential for project success are the following:

1. Forecast team leadership in the commercial area, consulting clients and senior management, and presenting effectively forecasts and recommendations
2. Bringing discipline, rigor, processes, and transparency to the modeling of the commercial aspect of the licensing project
3. Helping clarify the company in- and out-licensing strategy and develop realistic goals and objectives, defining the licensing decision problem, determining informational requirements, and creating the knowable information set
4. Researching comparable deals, industry norms, standards, and benchmarks, and evaluating several analog product performances to determine the most applicable
5. Modeling the licensing process and determining key relationships and feedback effects and defining plausible future-state scenarios
6. Performing the forecasting due diligence to validate processes, assumptions, data, models, and scenarios and ensure adequacy of resource support for the project
7. Accurate assessments of megatrends and subtrends impacting the industry and a thorough environmental (PESTLED) evaluation
8. Distilling the SWOT analysis to confirm the availability of critical capabilities and competencies to execute the project successfully
9. Clarifying and validating the vision, purpose, and long-term direction of the firm vis-à-vis megatrends and the competitive environment; creating and validating the assumptions; and updating and maintaining a record of changes and their rationale
10. Identifying the forecast risks and the critical issues in the deal and assisting in the creation of a risk management plan

12.4.2 Beneficial Practices

The effects of beneficial licensing project team practices contribute to project success equally as the success factors identified earlier. These beneficial practices include the following:

A. *Competency viewpoint.* Good licensing projects consider the following practices as highly contributing to the success of the deal: Accurate situational and grounding analyses are crucial as is sound project valuation to identify, screen, choose, and negotiate deals appropriately. But they should be tailored to the specific project needs. Managing the scope, expectations, and the terms and conditions of the deal to maximize value and developing and valuing contingent strategies are required sound practices. However, developing core competencies to get a preferred partner status and using predeal valuation terms and conditions to drive postdeal success are central to long-term success. Also, establishing realistic and achievable commercial goals with a partner and monitoring closely project performance are vital along with leveraging controllable factors to achieve expected results.

B. *Technology licensing.* In technology licensing projects, several elements are crucial to success, such as understanding current and future user needs, developing products that meet them, and creating value and integrating technology subject matter expertise with SDF capabilities. They are a must to ensure forecast relevance, validity, and acceptance. Technology transfers must be implemented holistically and with high 4C levels because they involve hardware, software, people, processes, information, organizational structures, and synchronization of plans, functions, and behavior. The ability to find the right balance of interests and costs and benefits to participants is important, but development, validation, and updating of insightful and reasonable assumptions and realistic forecast scenarios are more important. Also, correct evaluation of the impact of licensing and technology transfers on portfolio management activities, company resources, different business units, and potential synergies is imperative.

Licensing experts also view the following technology licensing aspects as making the difference: formulating a partnering strategy and executing flawlessly, quickly, and efficiently and understanding the effects of megatrends, industry trends, and the mechanics of technology discovery. Also, sound scenario forecasts that help determine the range of potential value creation for each technology are crucial. Just as significant are the partner's intent to want and the ability to achieve superiority in the licensed technology, knowing and screening partners, determining controllable factors and events to reduce uncertainty, and having a process to monitor the agreement.

C. *Licensor versus licensee perspective.* Product licensing experts make a distinction of success elements for the licensee and the licensor parties. The licensee success elements are determined by the potential value of the product to the company over the license period, the extent to which the new product fits the company's R&D strategy and its product portfolio, and the level of synergies and the ease to be extracted. From the licensor's perspective, the elements that determine success are as follows: the timing of the licensing, that is, should the product be licensed now or later; the partner's ability to exploit the product to its full potential; the degree to which the product fits the partner's R&D and product portfolio; and the partner's interest to develop the product further. Also, understanding how different assumptions affect the forecasted value of the product, the risks and costs of developing the product, and the upfront payment, milestone fees, royalty rates, and equity requirements is crucial to project success.

D. *Relationship management.* The beneficial practices in relationship management aspect of licensing deals are often overlooked and have undesirable consequences. Internal high 4C-level practices are a necessary core competency because licensing alliances are an important part of company growth strategy. Projects and processes must also have effective partner communication channels to build strong partnerships, manage partner needs and expectations, and communicate the importance of licensing projects. Identifying and emphasizing the key drivers of team building and frequent dialog to ensure effective project implementation are important. However, the competitive advantages in licensing are the organizational and resource management capabilities and actionable competitive intelligence. Also, partner competencies in identifying needs and developing, marketing, and distributing products, and use of their capabilities to manage outsourced market research, validation of forecasts, and sales and marketing activities are important project success elements.

E. *General beneficial practices.* When licensing is viewed as a strategic tool for growth, the choice of alliance is guided by matching strategic objectives, and deals are made to close strategic and product portfolio gaps. Licensing projects should build on current strengths, enhance strategic objectives, and generate new opportunities, but a comprehensive project, structure, and overall cost of the deal due diligence is crucial. Also, of vital importance are creating effective processes and structuring responsibilities, risks, and rewards, objective evaluation of assumptions, well-considered forecast scenarios, measurement of project risks, and development of effective contingencies and response plans to counter threats to the alliance's business plan.

F. *SDF team contributions.* The majority of the SDF team's unique contributions in licensing projects is in the commercial area and varies according to project needs and the individual SDF team members' experience. However, other areas are also beneficiaries of SDF team best practices, which include the following:

1. Clarifying project objectives and participant's responsibilities and creating the organizational links needed for effective project evaluation and implementation
2. Providing rigorous sanity checks and scrutiny of strategic objectives, links to financial models, and preparation of negotiation positions and trade-offs
3. Describing clearly plausible deal scenarios that form the basis of forecasts and future-state realization (FSR) plans
4. Articulating the pros and cons and the impacts of doing and not doing the deal on the company and the partner
5. Convincing senior managers of the forecast reliability to reach consensus effectively and reduce time to decision through effective presentations that are backed by evidence
6. Creating a management dashboard and an EWS to monitor and report project performance based on comprehensive measures
7. Providing a well-supported recommendation to proceed or not along with documentation of the rationale and forecast use caveats
8. Developing a joint competitive response and forecast realization plan with partners to ensure that the expected success indeed materializes
9. Assessing the strength of the company's existing position and common ground with the potential partner and scrutinizing the expected value of the deal
10. Implementing moves and options to manage the critical issues with the partner as the project is implemented and risks appear on the horizon

Word to the wise: *Major benefits accrue to forecasters who study the causes of licensing project failures and the success factors for the following reasons:*

- *Knowing the particular needs, issues, and challenges of this type of projects*
- *Finding ways to fill gaps and limitation in their processes and methods*
- *Learning from the failures and the successes of others*
- *Getting involved early on and knowing enough about controllable factors to resolve issues under the company's power*
- *Distilling lessons to apply to particular problem areas and prepare properly for future projects*

13 Corporate Reorganizations and Turnarounds

In everyday parlance, corporate restructures, reorganizations, and turnarounds (CRRTs) are terms used interchangeably and refer to events and processes by which a company is transformed to address the financial difficulties it is experiencing under the current structure, operations, and management team. CRRTs involve changing its ownership, operational, or legal structures to better organize operations, improve product quality and customer service, and enhance their financial performance. There are internal and external CRRTs, those focused on fixing short-term problems versus those addressing problems that require long-term focus. They can also be strategic, operational, financial, and environmental change, or shareholder value driven.

CRRTs can also be proactive, reactive, or involuntary and involve changes in the existing organizational structures and management teams, in the firm's business definition and corporate strategy, and in consolidations and reductions in size and scope of operations. In other instances, they entail changes in the way transactions among existing business units are conducted, in operating agreements with customers and suppliers, or in recapitalization of the corporation and rearrangement of its assets and liabilities. However, the majority of restructures are reactive and the main reasons are debt and financing reengineering to deal with specific crises, which, in turn, require changes in other areas of the company's operations.

The focus of this chapter is on strategy-driven CRRT projects intended to alter organizations, functions, products, and financials to adapt to external environment changes and to reaffirm the company's reason for being in business. Hence, our discussion does not address projects dealing with short-term operational and financial issues. The strategic focus on CRRT projects is a fertile ground for applications of strategic decision forecasting (SDF) solutions, methods, techniques, and tools. It is because they require a thorough understanding of causes that brought about the need for restructuring, modeling, and assessment of the impacts of new strategic initiatives on corporate resources. SDF teams are best qualified to help CRRT project teams better understand controllable and uncontrollable factors, to create objective turnaround scenarios, and to identify levers that help the most in the realization of the project goals and objectives and the corporate long-range plans and, indeed, the company's survival.

13.1 CURRENT STATE AND CAUSES OF FAILURES

CRRTs have been increasing over the past quarter century, and with capital markets becoming globally integrated, they have become a common tool to manage financial performance. The prevailing current approach is short-term focused and based on overall simplistic and naïve assumptions. It consists of evaluating immediate financial problems, developing a plan to resolve them quickly, reshuffling the organization, restructuring debt, or tweaking some other financial lever. However, because the complexity of reorganizations is underestimated and the shortage of qualified project leaders is so severe, their track record has been less than satisfactory and, in most cases, downright failure.

CRRTs are more often than not based on opinions, expectations, wishful thinking, untested ideas, and simplistic modeling of operations to get new cash flow forecasts due to cost reductions. In the prevailing state, SDF teams are ordinarily excluded from CRRT project teams and evaluations of company changes on customers are not given high priority. Most of them are not proactive projects but rather immediate and short-lived reactions to operational underperformance and financial distress. The prevalent mindset in the current CRRTs is to address long-neglected strategic, competency, and operational performance gaps and issues with short-term fixes alone. That is why they are more painful, take longer to execute, and have lower success rates, unlike the preemptive CRRTs, which are long-term focused and a far less common occurrence. Various causes of CRRT failures are shown under the categories below.

Disconnected plans. There is a weak link between company strategies, business plans, and performance in companies forced to reorganize and this affects unfavorably all three when the main focus is on changing the terms on company assets and liabilities due to current financial difficulties. CRRTs are driven by a need to make financial obligations equal the cash flow stream, but in most instances fail to meet stated objectives beyond the current and next quarter financial results. Why? Because of the following reasons:

1. All focus is on the action side and not on an action–reaction–action planning framework needed to address long-term problems and solutions.
2. Long-term forecasts are not created using distinct and realistic scenarios of company operations models.
3. Success realization plans to ensure that restructuring has met its objectives consist of doing nothing more than what was done over and over previously.

Wrong attitude. As a general rule, CRRTs are viewed as corporate financial engineering solutions and the preferred solutions in these projects are cost decreases through headcount reductions and balance sheet reengineering. Thus, the slash and burn attitude is widespread in financially troubled companies where priority one is to do a liquidity analysis and follow up with debt consolidation and restructuring. In such cases, companies reorganize using an abbreviated process that includes identifying the financial problems, making the decision at senior management levels to restructure, creating a reorganization plan, and implementing it in 30–90 days. In those cases, the uncertainty involved in the discounted cash flow calculations is ignored or dealt with very superficially.

Expedience reigns. In the current state of CRRT implementations, there is heavy concentration on short-term, expedient solutions sold to creditors and investors as low-hanging fruit opportunities that cannot be missed. Many companies in need of reorganization undertake them with partial appreciation of their complexities and the myriad of issues to be addressed prior to plan execution. Some CRRTs involve drastic changes in contractual relationships and implied trust-based dealings with suppliers, creditors, and employees, and they result in unexpected negative results because the use of short-term measures to fix long-term problems results in the inability to contain the effects of negative reactions. And, because focus is directed to the short-term issues, there is only financial modeling of the conceived CRRT project in order to measure financial impacts, while other effects are ignored. Demand analysis, forecasting due diligence, external environment assessments, and analysis of feedback effects are overlooked. A common forecasting approach followed by most CRRT project teams is to start with short-term financial forecasts and extrapolate to mid-term and then to long-term futures. These are deterministic forecasts created not on some tested scenarios, but on optimistic assumptions characterized as unrealistic after the fact.

Short-circuit processes. In these CRRT cases, the commonly used truncated process is to start with evaluating the nature and sources of current problems and come up with options to resolve them. The next step is to create a reorganization plan that is communicated to the employees, creditors, suppliers, and other external stakeholders such as Wall Street analysts or regulatory agencies. Subsequently, the plan is put into effect and all employees are evaluated, duplicate functions are eliminated, and those underperforming are terminated. This process leads to a top-down reshuffling of the organization starting with senior managers all the way to hourly labor employees. In the last step, the restructuring and consolidation of debt and recapitalization take place.

Process oversimplification. Here, the problem with current CRRT processes exemplifies how naïve expectations can be when CRRTs are dealing with financial instead of systemic issues responsible for poor performance. The complaint from companies that tried this approach and failed is that the condensed restructuring process driven by financial distress focused only on a few elements:

1. Containing cost increases due to high marketing and advertising costs, labor inefficiencies, and out-of-control operating expenses
2. Determining the major financial problems and concentrating the sales and marketing efforts on market share gains
3. Negotiating the financial details of the CRRT plan with creditors and suppliers and obtaining their agreement
4. Putting into action tactics and moves to eliminate losses and sharpen focus on profitable areas

Incorrect choice of instruments. Short-term instruments commonly used are viewed as choice weapons, but are mostly ineffective because they lack the benefit long-term view and forecasts based on proper internal and external environment evaluations. These instruments consist of the following:

1. Streamlining internal organizations, reshuffling responsibilities, and modifying creditor and supplier relationships and the terms of trade
2. Making small investments to update production, sales and marketing, and distribution facilities
3. Undertaking activity-based costing to reduce costs in underperforming areas
4. Rationalizing the company's product line and market presence and optimizing its product and asset portfolios in accordance with new market realities
5. Restructuring its debt, which means reducing, renegotiating, and changing the terms of loan agreements to improve liquidity, and recapitalizing, which involves changing the debt–equity mix by trading bonds or equity with investors

Restructures are common. Restructuring is now a routine management responsibility not only in the business sector, but also in educational institutions and government agencies with similarly meager success rates. The instruments used to affect CRRTs vary by the nature of the business problem and some of them induce drastic changes, while others require less severe adjustments. The more difficult but effective strategic and long-term approaches are not used often primarily because of the urgency to meet financial obligations and lack of experienced turnaround managers. They include projects of a different nature, such as the following:

1. Changing the corporate culture, which is a difficult long-term undertaking and requires strong senior management leadership, commitment, involvement, and unwavering support
2. Refocusing the company by changing its strategy, goals, and objectives and modifying its internal and external policies
3. Changing the company business definition, which means entry into new markets and businesses or exit from existing geographies and operations
4. Restructuring of company assets, which entails sales or purchase of valuable assets, such as spin-offs and mergers, in order to enhance their performance and increase their value
5. Modifying the legal ownership structure of company holdings, which are accomplished through M&A activities in order to maximize shareholder value

TABLE 13.1
Reasons for CRRT and SDF Team Support Provided

Motive for CRRT	SDF Team Support When Warranted
Address financial difficulties	Investigate demand issues, assess revenue implications, identify risks, address PPM adjustment issues
Fix short-term problems	Evaluate quality, pricing, and customer support impacts; create forecasts; determine competitor reactions
Fix long-term problems	External environment assessment and support strategy to address chronic and systemic issues, create scenarios and forecasts, and identify competitor reaction risks and plans to deal with them
Debt and financing reengineering	Assess impacts on quality, pricing, advertising, customer support, product portfolio rebalancing, and lifecycle management support
Human capital optimization	Determine impacts of megatrends and external factors provide competitive intelligence, insights
Organizational realignment, streamline R&D, and NPD	Perform megatrend, industry, and SWOT analyses; assess effects on NPD, quality, launch support
Process improvement and total quality management	Assess external factors, competitive intelligence, insights, benchmarks; modeling support
Take advantage of new opportunities	External factor analysis, SWOT insights, sanity checks, scenario planning, forecast implications and risks
Optimize PPM	External factor and market analyses, SWOT insights, rebalancing scenario forecasts, risks, evaluation of implications
Get more control over the business	External factor analysis, strategy evaluation, modeling, sanity checks, scenario planning, forecasts, implications and risks
Develop capabilities and competencies	External factor, industry, market, and SWOT analyses, competitor insights, benchmarks, sanity checks
Implement strategic and tactical benchmarks	External factor, industry, market, and SWOT insights, benchmarks, modeling of operations, scenario planning, risk assessment
Obtain economies of scale	External factor, industry, and SWOT analyses; impact on pricing, modeling, sanity checks, and competitor reactions
Adopt new technology platforms	External factor, industry, and market analyses, technology forecast evolution, scenario planning, risk assessment, strategy evaluation
Frugal or clever (Jugaad) innovation	External factor, industry, and market insights, market research and validation, scenario planning

Note: CRRT, corporate restructures, reorganizations, and turnaround; NPD, new product development; PPM, product portfolio management; SWOT, strengths, weaknesses, opportunities, and threats.

SDF due diligence related to customer demand, pricing, and competitor issues is performed with the purpose of identifying the root causes of problems before generating final forecasts based on well-thought-out and balanced scenarios. Validating cash flow forecasts and assessing the level of uncertainty and risks associated with these forecasts is next in the sequence of the SDF process. In proactive CRRT projects, SDF team member's upfront involvement and participation are needed to conduct evaluations of proposed alternatives and new opportunities and to help keep the projects on track. When turnaround projects entail portfolio rebalancing, the SDF team's focus is to assess the impacts of rebalancing activities and provide insights into competitive and market reaction. The ultimate SDF due diligence objective is to support continuous process improvement in the analytic and evaluation support areas in all turnaround types of projects.

The SDF approach and levels of support provided depend on the reorganization mode senior management wishes to pursue in a given project. The various restructuring and turnaround modes

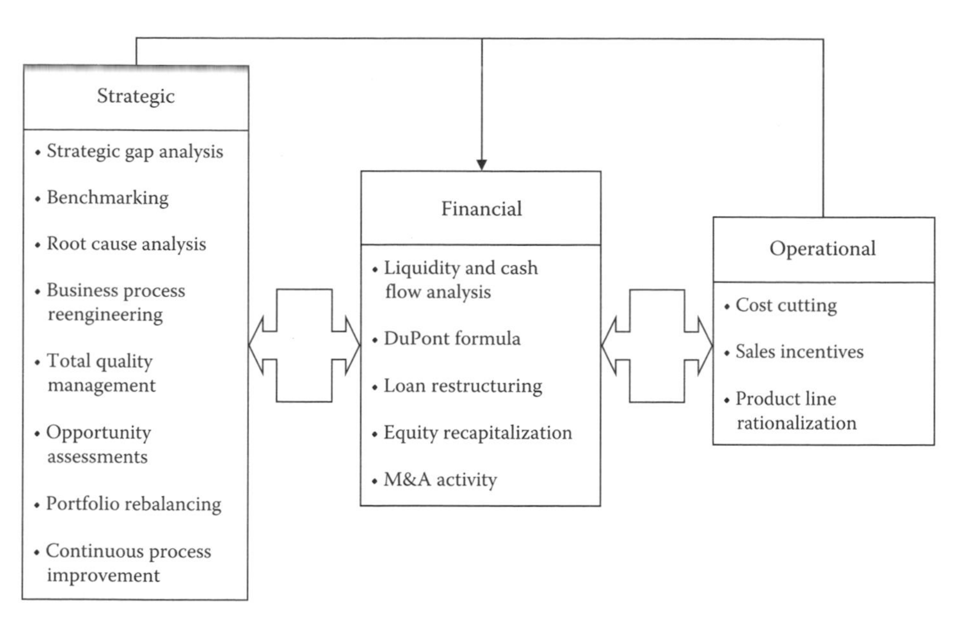

FIGURE 13.1 Restructuring and turnaround modes and instruments.

and common instruments used are shown in Figure 13.1. Again, strategic CRRT mode projects are strategy driven, long-term focused, and more difficult to execute. But they are more successful because they help address root causes of problems, explore new opportunities, and make changes that have long-lasting effects. Financial mode CRRT projects are ordinarily focused on fixing critical problems immediately in order to avoid disastrous consequences, although financially driven M&A projects could take longer to implement than other instruments in this mode.

Operational mode CRRTs take more time to implement than most of the financial type except when the sales incentive and advertising expenditure change instruments are used. Notice, however, that the CRRT and SDF processes are highly intertwined and impacting each other.

Strategic restructuring of organizations and processes occurs after an in-depth review of corporate strategy, goals and objectives, and policies, and identification of the root causes of problems. It is affected through the instruments shown in Figure 13.1 and requires professional management of corporate revitalization processes, rigorous and structured analyses, and objective evaluations to achieve the reorganization goals and objectives. As is the case with other strategic decisions, turnaround plans can proceed after a complete review of strategic goals and objectives and are not based on an operational results review alone. Determining why a company or an organization is not meeting its goals and objectives involves using a basic template that includes the following steps:

1. Management review of the current situation and problem identification and evaluation across all functional areas
2. In-depth demand analysis to identify sources of problems related to product quality, attributes, pricing, and customer satisfaction
3. Evaluation of megatrends, subtrends, PESTLED (political, economical, social, technological, legal, educational, and demographic) developments, industry trends, and market structure changes
4. Analysis of root failure causes through a series of investigations to determine why there is a problem in order to address the underlying causes
5. Development of long-term strategic and restructuring plans resulting from the turnaround management analyses and evaluations

6. SWOT analyses to determine congruence of reorganization goals and objectives with the company's ability to execute plans and meet those objectives
7. A realistic implementation plan, strongly supported by senior management and executed by experienced turnaround managers, and effective management of expectations

In all successful turnaround projects, business process reengineering (BPR) drives many effective changes through modifications in core processes, which result in changes in organization structure and personnel changes, technological platforms used, corporate policies and procedures, and decision support structures. And, by process reengineering, we do not mean the downsizing usually associated with it, although reengineering may or may not result in that. Strategic CRRT processes viewed as consistently producing positive results that affect SDF processes, analyses, and evaluations begin with preparing for reorganization by building a competent team, creating customer support objectives, and developing a strategic purpose.

The next activities in the strategic CRRT project processes are mapping the current processes, creating process chart illustrations and activity-based models, followed by modeling of operations and simulating activity-based costs. The next team activity is the evaluation of current processes, which is done by analyzing the model results and identifying disconnects and value-adding activities. Using competitive intelligence and industry knowledge to benchmark key internal processes and practices with appropriate standards is necessary in designing, creating, and validating the new processes intended to drive turnaround changes effectively. Having validated the new processes, the creation of organizational and functional prototypes takes place, which enables simulation of transition plans and development of a reasonable implementation plan. However, to roll out the plan with reengineered processes requires initiating training programs, and sustainability of turnaround performance requires an ongoing improvement effort, which, also, involves continuous progress monitoring and measuring and periodic performance reviews against plan targets.

The analyses and evaluations performed by SDF team members participating in turnaround projects are tailored to address the sources of problems faced by the company and answer questions by two complementary approaches: employing a variant of the strategic management model, discussed in Chapter 11 and illustrated in Figure 11.2, and identifying and focusing on the six key problem areas shown below.

A. *Strategic management model.* In a summary form, the strategic management model approach begins with a team evaluation of operational results and benchmarking activity in key areas. A comprehensive evaluation of the external environment and a strategic gap analysis follow in order to identify and define problems to address through a CRRT project. The desired future state of company performance is articulated in different alternatives, transition paths, and decision option plans. The impact of proposed plans on other initiatives and on the corporate business and financial plans is evaluated with the uncertainty and risks surrounding cash flows taken into account. An analysis of incoming and outgoing market signals and evaluation of investor reaction takes place and a CRRT strategy and implementation plan are formulated.

B. *Focusing on key problem areas.* This approach involves SDF team participation and focusing on the following six areas and going through the process of identifying and addressing root causes of problems:

1. *Strategy-related gaps and limitations.* The problems commonly encountered in this area include wrong strategy, goals and objectives, poorly communicated strategy and lacking focus in execution, erroneous corporate policies or in conflict with strategic goals and objectives, and half-baked reorganization plans with unclear or conflicting goals and objectives and not strategy driven. Lack of continuous innovation, productivity gains, and product differentiation strategy are major problems to overcome. More damaging, however, is not reviewing major past strategic decisions to

appreciate the consistency and reasonableness of objectives and performance target setting, project management requirements, and lessons learned about strategic decision implementations. In this area, the main role of the SDF team is to ensure consistency of project objectives with market realities, perform sanity checks, and provide competitive intelligence and insights to the project team.

2. *Organizational issues.* Many times, problems that surface in different areas have their roots in organizational issues such as the following:

a. Wrong organizational structures and unaligned or missing functions not supporting achievement of high performance
b. Wrong corporate culture, not conducive to making changes until things get out of hand
c. Lack of quantity, quality, motivation, and incentives of key staff and employees

Other thorny issues are the duplication and overlap of business functions that lead to noncooperation, confusion, and conflict. Also, the mismatches of CRRT tasks and competencies are a big concern rarely addressed through internal talent and not until a reorganization is about to get started. However, the issues SDF teams are primarily concerned within this area are its organizational position and governance structure, the functional links with other planning organizations, and high communication, coordination, cooperation, and collaboration (4C) levels need to execute a CRRT project successfully.

3. *Management hurdles.* These are major limitations and inefficiencies, which are overlooked or neglected and manifest themselves in other problems. They include the following:

a. Management ineptness, lack of vision, short-term results mindset, and internal politics and conflicts
b. Inefficient monitoring of employee's performance, low labor productivity, and superficial support for continuous improvement programs
c. Senior management not understanding the role forecasting plays and the skills required to identify and manage uncertainty and risks
d. Internal and external communication ineffectiveness, unclear messages, and misinterpretation of reorganization plan objectives and execution details
e. Lack of proactive reorganization planning, risk aversion, and slow corporate demise due to fear of drastic change consequences
f. Shortage of experienced turnaround project managers along with lack of competent, large project implementation management talent

The areas in which the SDF team helps to address management hurdles include the following activities:

a. Assisting in proactive planning to enhance the chances of project success through modeling and scenario development and planning
b. Creating forecasting and project evaluation processes needed and defining appropriate performance measures
c. Identifying CRRT project uncertainty and risks and determining consistency of senior management risk tolerance with risk levels involved in the proposed project
d. Testing the validity of assumptions used against competitive data, industry standards, and benchmarks
e. Performing analyses and evaluations to determine the feasibility of undertaking a CRRT project successfully

4. *External predicaments.* These refer to areas that are more often than not given superficial attention, but further complicate difficult situations and the management of problems.

This is so because they are viewed as outside the company's control or may be treated as givens and include the following:

a. The impacts of factors such as megatrends, subtrends, PESTLED developments, and regulations on company operations and performance
b. Unexpected, unrecognized, or misdiagnosed changing consumer tastes and preferences, and new market conditions, discontinuities, and shifts

Other complications arise from avoidance of dealing with external factors because they are considered uncontrollable or due to technology advances and platform changes that catch companies unprepared to adjust to them and take advantage of potential applications. And since no black swans are identified because this attempt is considered wasteful and the probability of such events is so small, they are not given serious attention.

In this area, SDF team support includes the majority of analyses and evaluations common to strategic projects, namely, the following:

a. Performing megatrend, subtrend, industry, PESTLED environment, and SWOT analyses
b. Determining controllable events and extent of control, uncontrollable events and likelihood of occurrence and impacts, and identifying black swans
c. Assessing risks in CRRT initiatives and helping the project team in contingency planning
d. Modeling CRRT problems, developing scenarios, and creating forecast solutions, including long-term technological forecasts
e. Determining the support and resources needed and helping create a CRRT expectation realization plan

5. *Knowledge gaps and process troubles.* This is a pervasive problem area, especially in companies that do not support TQM and continuous improvement programs. The gaps and limitations include missing, incomplete, wrongly communicated, broken down, or difficult to implement processes, and practices. Beyond broken internal processes, there is little learning from competitors and absence of serious benchmarking efforts. There is, also, internal and external coordination ineffectiveness, which results in inappropriate management dashboards being created and wrong performance measures and indicators being monitored. These adverse consequences are made worse by ineffective migration from the short-term to long-term forecast horizons, lack of focus, and poor implementation plans. Furthermore, setting unrealistic initial performance targets in one area without equivalent targets in complementary areas, such as high-revenue forecasts and low-cost projections, is common. Also, having no uncertainty and risk assessment of financial projections performed and having no risk management plans to execute in the event cash flow projections are irresponsible practices.

The support of the SDF team in the process and knowledge areas comes primarily in the form of the following activities:

a. Creating sound processes and introducing best practices, benchmarks, and appropriate industry practices
b. Reducing CRRT project uncertainty to its residual component and performing evaluations on risk management plans, moves, and actions
c. Taking ownership of the forecasting function management across project phases and management of the assumption set driving CRRT project plans
d. Helping the project team develop realistic project targets without negative effects on customer demand using competitive intelligence and insights
e. Creating a company Intranet database to house data, assumptions, plans, and useful information and institutionalizing CRRT project learning

6. *Operational problems*. These are common occurrence events that come up because of lack of attention in monitoring performance and addressing problems as they come up. They include the likes of the following:

a. Product attributes that do not meet customer needs, inferior product quality, and inefficient pricing vis-à-vis competitor offerings
b. Poor customer service record over long periods and decline of customer satisfaction leading to customer defection
c. Failures of PPM, that is, inadequate NPDs, ineffective sales, marketing, and advertising plans, and unsuccessful product upgrades and extensions
d. Out-of-line operating costs due to production process inefficiencies, unmanaged input costs, poorly trained labor, and lack of management experience
e. Spending versus cash flow imbalances due to unrealistic assumptions, expectations, plans, and equal attention to the revenue and expense streams
f. Supply chain issues ranging from order entry, materials management, production interruptions, capacity limitations, and distribution channel problems
g. Absence or failure of existing accounting control systems to detect problems as they arise and the means to course correct
h. Partial understanding of operations and ineffective systems modeling to identify drivers and feedback effects and drive financial projections

SDF team support in dealing with operational issues, indicative of systemic problems, begins with a current customer needs assessment, product evaluation, review of market research and determining how customers would be impacted by the CRRT project. Other elements of SDF team involvement include the following:

a. Having a major role in the development of reasonable assumptions set to feed financial planning models
b. Performing in-depth industry and SWOT analyses to assess the company's ability to operate profitably under the status quo and the desired restructured scenarios
c. Modeling company operations, identifying controllable key drivers, interactions and feedback effects, and how they impact the bottom line
d. Performing sanity checks on expectations, implicit and explicit assumptions, models, and results of evaluations
e. Monitoring progress through the project dashboard and communicating risks via the EWS

In addition to the strategic management model, SDF team members use many of the problem-solving techniques and tools discussed in Chapters 6 and 7. The following tools are widely applied in CRRT project support:

A. *Modeling of current and future operations*. The SDF team approach to replicate the functioning of a company is heavily influenced by systems thinking in creating conceptual and detailed company operational models. Systems thinking is a structured approach to problem solving and a way to view company operations from a broad perspective. This perspective includes assessing organizational structures, performance patterns, and feedback cycles in its operations, rather than focusing only on the specific components of company operations. Also, SDF team members attempt to explain what is happening between the components over time that keeps the system going. Going from conceptual to computational models involves use of flowcharting to connect all the pieces of the system that make up the company operations, the direction of causality, and the feedback effects. This approach allows the SDF team to create dynamic models that fit different CRRT modes, produce stock, and flow estimates, and they allow for hypothesis testing and "what-if" simulations.

In the preparation stage for a CRRT project, three SDF activities take place:

1. Defining SDF project goals and objectives that are consistent with those in the current strategic plan
2. Identifying what is needed for SDF to achieve its objectives in support of the turnaround effort
3. Developing a plan to maximize the SDF team's contribution to the success of the turnaround project

Note that the SDF analyses and evaluations are in support of and center on the CRRT project process, are adjusted to meet the project team's needs, and are always focused on keeping the project on track. The commonly performed SDF activities, analyses, and evaluations that evolve around the general process outline of the strategic management framework and include the following sequence of activities in the CRRT process stages.

1. *Preparation stage*. In the preparation stage, SDF team members help the project team review past turnaround efforts in the company, research industry and competitor turnaround initiatives and experiences, and provide insights and learning from earlier projects. The goal is that by the end of this step, three things are accomplished by the project team:

a. Articulating a first and broad restructuring vision, obtaining stakeholder inputs, designing a preliminary turnaround process, and creating a draft reorganization plan
b. Coaching and preparing stakeholders for reorganization problem solving, which is expected to result in turnaround and improved performance
c. Presenting an implementation blueprint prototype with embedded continuous improvement for senior management review and approval

2. *Goals and objectives definition*. In this step, the SDF project goals and objectives are defined and shared with the project team. Armed with the results of the preparation stage assessments, the feasibility of the turnaround project objectives is given its first test and the three process elements to be done are as follows:

a. An in-depth review of corporate and organizational strategies, goals, and objectives to determine flaws and inadequacies and ensure their consistency with market realities and the objectives of the turnaround plan
b. A strategic gap analysis of key areas identified in need of improvement and a comparison of company versus industry and competitor performance to give an indication of the extent of changes needed
c. A conceptual model of company operations and computational business segment models developed to assess in rough terms the value of filling the strategic and operational gaps that have been identified

3. *Definition of needs*. The next order of business is for the project team as a whole to identify what is needed to achieve the turnaround objectives in terms of senior management involvement, project support in human and financial resources, communication channels, and reporting progress on team member's activities and deliverables. While these requirements are being determined, SDF team members identify what is needed in terms of process development for SDF solutions, analyses and evaluations, data and models, external expert support, and other requirements. Then, they create a plan to maximize contribution to the success of the reorganization project.

4. *Alternatives to proposed project*. In this step, the project team with support from SDF team members creates, describes, and evaluates alternatives to the proposed turnaround project and selects the one most likely to come to fruition. In parallel, it analyzes current processes, benchmarks

performance in key functional levels, and learns from the best in the business. The current process mapping activity takes place and tentative performance measures are created with input from key stakeholders, but not target levels; they are created when a better understanding of the CRRT models is obtained. During this phase, root cause analyses are performed and a better understanding of possible solutions emerges. Then, the project team identifies drivers of performance that it can control and factors that are considered as uncontrollable are noted and given special consideration in the risk management efforts that follow.

5. *Grounding of SDF evaluations.* While the process mapping activities take place, the grounding of the SDF analysis and the due diligence effort begin in earnest. First, an assessment of the impact of megatrends, subtrends, industry and market changes, and PESTLED environment developments on company performance takes place. Extensive industry and SWOT analyses are carried out to verify and validate company capabilities, competencies, and ability to differentiate its products. Also, the key assumptions of the turnaround plan are tested and validated. The objectives of assessing the feasibility of turnaround objectives in a future-state environment, identifying major threats, and determining the potential impacts of risks are the major accomplishment of the SDF due diligence, which takes place in every project.

6. *Validation of information and findings.* Once the SDF team completes the external environment assessments, the SWOT and industry analyses, and the validation of data, models, scenarios, and other important pieces of information, the project team is in a position to create and validate the final turnaround assumptions and test the project objectives one more time. In the meantime, the results of the analyses so far are used to assess potential impacts of external factors and those of the contemplated changes on customers, product demand and revenue, and future company performance.

7. *SDF team participation.* At this point in the process, the SDF team better defines its participation, recalibrates the role it will assume, communicates its future goals and objectives, and provides the rationale for its plan. It then proceeds to develop or obtain market studies, competitive intelligence, data, models, and tools to conduct the analysis and evaluations that need to be performed. Then, it establishes the factors critical to the success of its participation in the turnaround effort and revisits the measurements of performance developed earlier. Having done that, it can further tailor the analysis to the specifics of the company's current situation and strategic needs.

8. *After the assumption set.* By now, the assumption set is firmed up and shared across all stakeholders and agreement is obtained on the set of assumptions used to model the restructured operations and to capture and evaluate all the costs and benefits involved and qualitative implications for assumptions examined. Then, models of relevant parts of company operations are developed based on business process mapping and linked together to form a corporate model of operations. An appropriate management dashboard is created and the performance measures to feed the dashboard and the EWS are revisited and finalized.

9. *Tests of reasonableness.* The SDF team, with input from the project team, develops, describes, and tests the reasonableness of three distinct and plausible restructuring scenarios. It then conducts a hypothesis testing and "what-if" analysis, and a number of simulations to obtain a range of outcomes for each scenario. A set of three forecast future states of reorganization based on the selected scenarios is developed and the reasonableness of the forecasts and their implications for resources are assessed. At this junction, it is very important to reevaluate the impacts of the reorganization on product quality, customer satisfaction, and customer demand and to make appropriate adjustments based on their feedback.

10. *Internal and external risks.* The SDF team assists the project team to identify internal and external risks involved in each reorganization scenario and define early warning indicators needed to alert the organization of risks appearing in the horizon. When the most likely scenario is selected and agreed upon, the project team identifies the uncertainty and risks associated with the financial forecasts resulting from it. After that, SDF team members help in the preparation of plans for interventions to ensure that the expected CRRT project goals and objectives are realized.

11. *Refining the restructuring vision.* Having obtained a significant level of confidence from the analyses and evaluations to this point, the project team refines the earlier articulated restructuring vision and finalizes the reorganization process. The project team creates an updated reorganization plan with targets and embedded continuous improvement to obtain stakeholder and senior management concurrence and approval. The Investor Relations group is coached and advised on all aspects of the turnaround plan and the external communication of the plan and its expected results are shared with the investment community.

12. *Uncontrollable events.* The restructuring plan guides the implementation of scheduled actions and controllable events. However, provision must be made for cases where unforeseen or uncontrollable events occur, and usually, this is done through some kind of risk management plan. The SDF team plays an active role in the creation of a reorganization value realization plan to address occurrences that require interventions, to direct plan adjustments and fine-tuning of targets, and to develop actionable moves and tactics recommendations.

13. *Performance monitoring.* When the turnaround plan is rolled out and changes take place, it is not clear whether things move in the right direction unless continuous monitoring of established performance indicators and variance analysis take place. Monitoring performance of the reorganization plan in key areas and investigation of deviations from expected outcomes are crucial to determine the extent of progress made, the causes of unexpected changes, and if and when interventions stipulated in the value realization plan are needed. The participation of the SDF team in this activity is advisable and valuable to assess market reaction and stakeholder feedback and to gauge the magnitude of adjustments needed to the reorganization processes, inputs, resource support levels, and expectations.

14. *Project review and postmortems.* If the SDF team remains engaged in the reorganization effort, it helps the remaining project team to conduct periodic project reviews and provide its assessment of progress. However, in every project, the SDF team participates in the reorganization project postmortem review to highlight what was done well and assess what went wrong, where, why, and what could have been done to keep the project on track. The last deliverables of the SDF team are to

a. Upload the knowledge and information gained in the project to the Intranet database.
b. Provide its recommendations to senior management concerning changes that can be made to enhance the intelligence gathering, modeling, scenario development, forecasting, monitoring, and the execution of the value realization plan.

13.4 SOUND PRACTICES AND SDF TEAM CONTRIBUTIONS

A very practical and useful definition of a critical success factor shared in discussions with a senior manager with substantial experience in CRRT projects is that it is a tool to take quantifiable objectives and break them down to simple, actionable parts so that the project team can deal with them. This definition of success factors is used in discussions with project sponsors and obtaining opinions of project team leaders.

13.4.1 Sound Practices

There are several practices and factors considered important by the participants in our discussions, but the significance ascribed to each of them varies by the particular project needs, as shown in the sets that follow.

Set A. The practices and factors shared by a group of corporate reorganization project experts, which are viewed as crucial to successful CRRT projects, are as follows:

1. Absolutely clear, concrete, and strategy-related turnaround goals and objectives and performance measures, which are fully understood by the entire organization
2. Senior management commitment and support along with respect of their personal time required in major project decisions
3. Expectation of and preparation for disruptions and time needed to communicate the reorganization plan internally and sell it externally
4. Identification and understanding of proponents, opponents, and their motives and enlistment of the support and approval of internal and external allies
5. Willingness, ability, and readiness of project sponsors and team leaders to reconcile multifaceted unit objectives and conflicting stakeholder interests
6. Formation of a skilled reorganization team lead by a seasoned turnaround manager and extensive upfront preparation and planning
7. Reorganization must follow strategy; changing organizational charts alone does not affect cultural, strategic, and operational reforms needed
8. Dealing with the uncertainty involved in discounted cash flow analyses and assessment of potential impacts on reorganization goals and objectives

Set B. Senior management team must "walk the talk" in reorganization projects made it to the top of the list of sound practices coming from business unit reorganization experts. Other sound practices include the following:

1. Corporate and business unit strategy drive operations, growth, and the turnaround effort
2. Meeting customer needs, focusing on customer service, and clearly articulating and communicating project rationale and objectives, which are the main tasks
3. A well-defined project plan with clearly delineated roles and responsibilities
4. Including reorganization stakeholders and key employees in the decision process and addressing directly major employee issues before changes are implemented
5. Flawless communication, coordination, and cooperation within and between business units and swift execution of the reorganization plan by a team of experienced turnaround managers
6. Understanding that process and due diligence extend beyond financial considerations and include employee and customer reactions
7. Strategic partnerships and strong relationships with suppliers and creditors, which are essential to turnaround success

Set C. A third set of common success factors obtained from turnaround experts in different industries includes the following elements:

1. Strong commitment, constant involvement, and unwavering support by the senior management team
2. Alignment of turnaround plan goals and objectives with corporate strategy, company capabilities and competencies, and market realities
3. Assignment of experienced managers to the project on a full-time basis and use of proven reorganization methods, processes, and plans

4. Effective end-to-end project management with clear problem ownership and accountability and well-functioning organizational structures, plans, resources, and processes
5. Transparent models of business operations simulated under reasonable scenarios to test assumptions, hypothesized links, relationships and causality, benefits, and costs of the CRRT project
6. A convincing business case of reorganization benefits exceeding costs, which takes into account the uncertainty involved in reorganization cash flow projections
7. Documentation of the reorganization objectives and roles and responsibilities of stakeholders along with high 4C levels throughout the organization in all aspects of the project

Set D. To get a different perspective on what makes a difference and determines CRRT project success, a more comprehensive set of sound practices was obtained from external subject matter experts. This set includes the following practices:

1. Accessibility to senior managers, persistent commitment to the company's vision, exemplary conduct, and integrity in all dealings as well as dependable support
2. Maintaining a long-term orientation while dealing with short-term issues, validating the company's vision, and creating realistic and consistent objectives driven by corporate strategy
3. Determining how different elements of the reorganization impact the plan objectives, but keeping focus and attention to customer needs, product or technology quality, and customer satisfaction throughout the project
4. Independent, critical, and objective analysis of customer requirements as well as key stakeholder needs, evaluation of feedback provided, effective execution of reorganization plan components, and excellent end-to-end project management
5. Setting reasonable targets, focusing on the most impactful elements of the reorganization plan, making employee morale a priority, and dedicating resources to reach targets
6. Highly coordinated, consistent, and effective communication internally and externally and implementation by highly qualified leaders in the shortest time possible
7. Using a balanced scorecard approach to monitor closely progress in all areas of change to ensure that expected results are being realized over time
8. Selecting reorganization project team members prudently, building stakeholder confidence of forecasted benefits and costs, knowing how to address critical parts of the plan, and being able to monitor progress closely
9. Collaborative culture, transparency, openness, and action mindset; continuous learning orientation; and inviting input and feedback from the stakeholder base
10. Making continuous improvement and TQM to simplify functions and products, aligning processes and functions, and increasing productivity top business priorities

13.4.2 SDF Team Contributions

The array of SDF team contributions echo particular project needs and are captured in the accounts that follow.

Account A. The involvement of SDF teams in CRRT projects is considered a decidedly value-adding participation according to project team leaders who did a comparison of "without" and "with" SDF team participation. The contributions they rated as valuable or beneficial in our discussions were in several areas, starting with SDF team members soliciting, assessing, and incorporating the input of customers affected by the reorganization plan. Helping to create and validate the assumption set, modeling of operations, and scenario planning provided invaluable insights. Also, the results and the efficiency with which they were produced and inputs and advice were dispensed due to the SDF team's internal functional links, personal relationships in other planning organizations, and access to external experts were central in providing focus and keeping the project on track.

Account B. To better deal with the uncertainty of reorganizations, the SDF team first identified megatrends, subtrends, industry, and market trends that help assess future product demand and competitor plans. This gave more credence to the insights created by the external environment and internal capabilities and competencies analyses and the competitor assessment. It was also central in determining the limits of what the turnaround project could achieve at what costs. Those insights and their understanding of competitor's experiences in reorganizations helped the project team to create and test the assumption set used in the reorganization plan. In addition to that, their disciplined and well-organized processes and problem-solving approach provided a prototype for the rest of the team to copy.

The project team gained the confidence of senior management when SDF team members demonstrated how they start with a review and analysis of the current situation to create conceptual models, then go on to flowchart functions and operations, modeling parts of the company, and developing an integrated company operations model that can be simulated. The scenario development activity they lead was vital to visualize the possibilities and pitfalls of reorganization plans. Also, the "what-if" analyses done through system dynamics models enabled the project team to visualize how to go from the current state to the future state and back to the current-state scenario to determine sensible changes needed to achieve the expected levels of performance.

Account C. The revenue and cost due diligence performed by SDF team members was valuable because it uncovered gaps in the project team's awareness of turnaround plan impacts on product quality and customer satisfaction, as well as the degree of competitor reaction to the turnaround effort. However, their understanding of the industry benchmarks they introduced to the analysis and appreciation of cost levers were key to the effectiveness with which financial evaluations were performed.

Account D. A highly valued SDF team contribution was their leadership in creating a turnaround management dashboard, an EWS, and a progress monitoring and value realization plan. Their root cause analysis of factors related to demand and revenue-related issues created insights crucial to the creation of more effective restructuring plans. Also, the participation of SDF team members in periodic reorganization project reviews and the postmortem analysis was very helpful because they developed actionable recommendations for process improvements in the current and future turnaround projects. The SDF team's contributions were more valuable and effective in strategic turnaround and less so in financial reengineering projects.

14 Uncertainty and Risk Management

Risk is present in every strategic decision due to, among other factors, the long timelines involved in strategic projects. There are several definitions of risk and they vary by application and situation and so is the case with uncertainty. In fact, risk and uncertainty are incorrectly used interchangeably to describe lack of clarity, predictability, and potential for loss. Also, risk management and contingency planning are considered the same. The prevailing view is that risk is a problem that can be avoided or mitigated, while another view describes risk as a situation that could lead to adverse consequences if not addressed. Risk is also portrayed as the threat that some unexpected event, intervention, or issue will adversely affect a company's ability to achieve its objectives. Others characterize risk as the exposure of decisions to unintended losses in future projects or as what one pays in exchange for higher potential returns when the future state or outcome is not known for certain. In all definitions, the two components of risk are the losses that can be caused by some event and the probability of the event occurring.

To clarify the meaning of terms and their use in this chapter, we give the definitions as used by risk management teams. Uncertainty is inability to determine the correct future state or to predict the outcome of a decision, and it is characterized by inability to assign probability to the occurrence of outcomes. Uncertainty is broader than risk and may or may not involve risk. The measurement of uncertainty consists of defining possible outcomes and the prospect of each outcome occurring. However, in strategic decision forecasting (SDF), Charlie Sheen's description of uncertainty as "a sign of humility and humility is just the ability or the willingness to learn" is probably the most insightful and useful definition. In strategic projects, risk management is the process of identifying, assessing, and prioritizing of risks addressed by a systematic and disciplined method to assign ownership to a central focus point that can monitor and control the occurrence and impact of threats and help in the realization of project objectives. Alternatively stated, the essence of risk management is to ensure the consistency of project expectations and goals and objectives with corporate risk tolerance and build the ability to bear some risk while initiating actions to reduce the occurrence of unfavorable outcomes.

Contingencies are anticipated project-relevant events that the project team usually considers as having a low probability of occurrence, but when they crop up they have major impacts. Thus, for SDF team purposes, contingency planning is a systematic way to help identify what can go wrong in a given project and prepare with strategies, plans, and approaches to avoid, cope with, or exploit such situations. Risk management deals with threats before they have been realized, but contingency planning involves continually confronting the likelihood of an adverse event occurrence and preparing for it. In practical terms, contingency planning is part of the project strategy that deals with uncertainty by developing specific responses to possible future conditions before they occur.

Contingency planning is done through "what-if" scenario simulations, the Delphi method, and other approaches to help the project team think about major contingencies and possible responses. Scenario planning, thinking, and analysis are strategic planning methods used to make flexible long-term plans and decisions. Scenario planning is about creating and evaluating distinct, plausible scenarios of future states to bring forward surprises and unexpected gaps of understanding. They are used to order and condition the perceptions of senior managers and to make sound strategic decisions for all plausible future states. The selection of the best or preferred future comes after all aspects of uncertainty and risk management are addressed, and this is where the SDF team makes significant contributions by establishing itself as a central focal point in the uncertainty and risk management function.

14.1 THE CURRENT STATE

For most business people, there is little or no difference between uncertainty and risk, the two terms are used interchangeably, and risk equals financial loss. In cases where uncertainty and risk management are more broadly defined functions, the record of effectiveness has been between acceptable and good. These cases, however, are the exception and most of the time risk analysis focuses only on financial risk. The common approach is to deal with risks in the immediate horizon and overlook or discount longer term threats. Also, lack of experience in effective risk management leads to inertia and delays in dealing with risk issues until it is too late. As a result, instead of uncertainty and risk management being an ongoing activity, they are thought of as stop-gap measures. An expedient approach used to deal with uncertainty and risk is to adopt very conservative forecasts or pessimistic views in planning and bypass altogether engaging in risk management. Another major issue encountered in risk management practices is the lack of an uncertainty and risk management strategy and an integrated risk management plan implementation. This is partly due to ad hoc prioritization of threats to address gaps in current project processes.

In the minds of some senior managers, risk management equals contingency planning, which is done only in instances of low occurrence probability and high-impact threats. That may explain why risk management in the current state is not incorporated into every step of planning, and in some cases it is only an afterthought. Lack of turnaround expertise, silo mentality, internal politics and frictions, and organizational and functional misalignments are often blamed for the absence of disciplined and rigorous risk management. While these reasons are responsible for some, it is the lack of structures, methods or templates, and processes that accounts for the many failures to manage threats to strategy and project objectives.

A common and unsettling practice in uncertainty and risk management is the degree of the superficial environmental assessment performed. Megatrend and subtrend analysis consists of listing some general or industry-wide trends with no evaluation of how they can impact future states and specific company operations. In most cases, industry analysis is something done by the Strategic Planning group once in a while and not by project teams unless the project involves a new business, a new geography, or a new market entry. Also, shallow and biased evaluations of political, economical, social, technological, legal, educational, and demographic (PESTLED) conditions are common, which result in creating questionable assumptions that lead to erroneous conclusions. More appreciably harmful, however, is the lack of independence, criticality, and objectivity in strengths, weaknesses, opportunities, and threats (SWOT) analyses, especially in sizing the capabilities and competencies needed to deal with risks associated with strategic projects. This leads to forming incorrect conclusions and ultimately to wrong decisions and project failure.

Inconsistency of corporate strategy with project objectives is a source of risks that is not given sufficient attention as is the difference in corporate risk tolerance versus risk taking needed to achieve expected results. The presence of a silo mentality prevents the effective communication of threats and is usually the culprit of failures, but we find that other factors contribute equally to this problem, namely, poor coordination of internal and external initiatives and actions and inadequate monitoring of performance that lead to ineffective risk management. Also, monitoring problems compounded by poorly defined indicators are used in reporting performance, which masks the impact of realized risks.

In the absence of strategic decision forecasting, long-term forecasts are simple extrapolations of short-term forecasts driven by recent experiences of underlying factors. In such projects, forecast users use single-number forecasts and have incomplete information about the likelihood of risk factors associated with the forecasts. Superficial demand analysis, absence of root cause analysis, and poor due diligence are common occurrences and are responsible for the inability to uncover and articulate immediate and long-term project risks. Additionally, strategy execution and project implementation suffer from lack of process reliability analysis and expert project management, and risks go uncovered and unnoticed from project phase to project phase. Not only that, but there is no serious scenario development, "what-if" analyses, and scenario planning. More importantly, there is a lack of forecast realization plans, which should be an integral part of risk management plans.

Traditionally, the focus on financial measures in risk management entails treating risk management through Monte Carlo simulations of project strategy on net present value (NPV). That is, the ranges of NPV measure and prioritize risk and the result is that uncertainty management is reduced to picking a conservative point on those ranges. Absence of discipline in scenario planning leads to key risk factors and events not being revealed; and while plausible future-state scenarios are simulated, the end result is that the teams responsible for executing strategy or a project are not prepared to deal with threats when they arise.

14.2 UNIQUE ISSUES AND CHALLENGES

Strategic projects have varying degrees of uncertainty and risk associated with them, whether it is an action, no action, or reaction decision. Managing them is the most important responsibility because they determine the future of the company and are the most difficult to carry out because of the immense consequences associated with wrong strategic decisions. Nevertheless, making decisions under uncertainty is the most important and most difficult responsibility of senior management. Decision making, however, is complicated by the multiplicity of sources and causes of risks in the midst of continuous change in operating environments and business drivers in extended forecast periods. To begin with, due to globalization of business, continuous technology changes, changing consumer tastes and user preferences, and higher levels of competitive activity, there an increased need for sound uncertainty risk management. And to do it successfully, it requires skills and experiences in different functional areas across key stakeholder organizations that are not present. As a result, there is reliance on external experts to moderate and guide uncertainty and risk management activities, which is not always the best solution because they do not have identical interests or the same level of understanding company operations. Hence, the crucial need for competent SDF and uncertainty and risk management to make the right decisions and obtain expected results.

Each strategic decision requires its own uncertainty and risk management solution because of its own peculiar needs and circumstances. For example, new business or market entry requires a different approach and process to deal with their uncertainty and risks than do merger and acquisition (M&A) and strategic alliance projects, licensing ventures, and turnaround projects. Each of these types of projects involves different approaches, processes, plans, activities, timing of decisions, and actions. But, when risk management is considered an ongoing and costly activity that consumes a large amount of senior management time, it is often concluded that it is more practical to postpone, ignore, or simply accept risk. And, the widespread view that risk is the price you must pay in exchange for getting higher returns so just accept it is difficult to change. This explains the passive stance of decision makers when it comes to analyses, evaluations, and planning to handle threats and risks to project plans.

Risk management and, to a lesser extent, uncertainty management necessitate not only recognizing and developing strategies and instruments to counter them, but continuous adjustment to changes in order for projects to achieve desired objectives. It would be fairly straightforward to follow predetermined processes that, with appropriate adjustments, lead to certain results. But separating market signals from contradictory actual data as is reported is a big challenge, especially when there are no effective early warning systems (EWSs) in place to detect threats and provide advance notice and time to evaluate and understand operational results. Late recognition of risk is followed by a flurry of activity thought to be appropriate but, instead, causes poorly designed and executed responses. Also, for a number of reasons, the organizational inertia that maintains the status quo is difficult to overcome and course correct as conditions warrant it.

Risk management in the financial industry is embraced and managed on an ongoing basis by experienced stakeholders and line managers. This, however, is not true outside the financial industry where risk is often viewed as something to be avoided. At times, even raising the issue of uncertainty and risk management is interpreted as negativity, noncooperation, or even as opposition to a specific project. In other situations, uncertainty and risks surrounding a strategy or a project are not communicated widely for fear of premature negative reaction by the investor community. Effective communication of threats and response coordination are difficult to achieve at all times, but risk management becomes far more difficult and is a lot less effective when the decision making and the risk management responsibilities reside in the same person or group.

There is no doubt that it is difficult to identify what is needed to have a successful risk management practice because of the great variety in business situations and requirements. This clarifies to some degree why short-term gains and focus on financials are valued more than long-term gains and why a conservative stance is the preferred approach rather than a risk management strategy. The absence of uniform requirements helps to see why there is a preference for handling threats; it is easier than exploiting risks in order to create new opportunities. Also, identifying critical success factors is difficult when the primary objective is getting a balance of risk versus return trade-off acceptable to key project stakeholders. Lastly, the propagation of uncertainty due to uncertainty of information used as input to models and the model errors that result make it difficult to judge to what extent the right moves are followed and the extent of progress made.

14.3 SDF APPROACH AND TECHNIQUES

The SDF team's approach to providing support in risk management is designed to help the uncertainty and risk management team prepare in advance actions and tools to deal with threats when they arise. SDF team members add value to uncertainty and risk management by focusing on the following areas:

A. *External factors*. In this area, the focus of the SDF team is directed to the following external environmental assessment activities:

1. Making sense of a constantly changing world and external PESTLED shocks, industry shifts, and other environmental transformations in a discrete manner without analogs to compare it
2. Understanding how changing user and consumer tastes and preferences, competitor product launches, transformations, extensions, and new competitor entry impact the company's plans and projects
3. Facilitating project team brainstorming about unforeseen, unpredictable, unexpected, and uncontrollable factors and random events and external influence
4. Providing better understanding to little-appreciated megatrends and subtrends impacting business operations and specific projects

B. *Processes*. The activities of the SDF team around developing complete, well-balanced, and effective processes to facilitate creation and execution of risk management plans include the following:

1. Creating appropriate SDF and uncertainty and risk management processes and defining the scope of the SDF team
2. Helping to correct poor decision-making processes and decisions based on conclusions that were supported by flawed information and analysis
3. Assisting the project team to assess rapid technological changes, project complexities, and build internal uncertainty and risk management capabilities
4. Helping to define and clarify key stakeholder roles and responsibilities, outputs, and deliverables, and assume forecasting function management

5. Bringing the broad SDF team project scope to manageable levels to minimize errors in judgment during project definition, evaluation, and implementation stages
6. Communicating clearly project threats to key stakeholders and properly coordinating uncertainty and risk management strategy and instruments

C. *Evaluations.* The analysis and evaluation parts of SDF team role in uncertainty and risk management are the bulk of the support provided in terms of the following activities:

1. Correctly identifying and dealing with business environment uncertainty, information uncertainty, business process uncertainty, and model input uncertainty
2. Eliminating the creation of biased, dubious, and untested assumptions in areas of limited project team experience
3. Making explicit the implicitly assumed associations and identifying model drivers, feedback and links, and the views of key stakeholders on the timing of events
4. Assisting the project team to estimate unknown, inadequately researched, and little-understood or wrongly estimated cost parameters through competitive intelligence and insights
5. Identifying and helping to eliminate corporate business plan budget preparation biases, budgetary constraints, and budgetary untouchables detrimental to successful project implementation
6. Determining the multicourse and action correction multiplier effects, especially in cases of missing change driver information, understanding, and data
7. Subjecting assumptions and scenarios to reality checks and explaining and addressing unexpected strength of relationships, trend reversals, and market shifts
8. Developing comprehensive business operations models, performing a balanced evaluation of results, and simulating "what-if" situations and parameter change impacts
9. Identifying black swans occurring in the duration of the project and developing competitive response plans and forecast realization tactics and moves

D. *Practices.* Concerning uncertainty and risk management practices, the role of the SDF team centers around the following activities:

1. Identifying and dealing with the effects of unsuccessful projects and initiatives not driven by corporate and project strategy or measured by wrong performance indicators
2. Ensuring that project objectives and risks are consistent with those of corporate strategy, risk tolerance, and realities of the marketplace
3. Determining how the divergence of risk-taking and risk-bearing ability relative to corporate risk appetite impacts project execution
4. Reality and sanity checks to manage unrealistic expectations, targets, pledges, goals and objectives, and forced commitments and undertakings
5. Preventing common misreading or misinterpreting market signals and inadequate opportunity screening with extensive project due diligence and validation efforts
6. Correcting unbenchmarked and badly monitored internal, supplier, and partner performance measures by introducing appropriate industry norms, standards, and best-in-class practices
7. Modifying wrongly held views concerning external and internal drivers of company operations, direction of causality, and paths of feedback

The SDF team is in the knowledge creation business and in the uncertainty and risk management area it directs focus first on ways for the future state to materialize according to expectations and predictions and then on damage control. The basic tenets of the SDF approach are that risk is measurable uncertainty and uncertainty can be reduced to residual uncertainty and that project risk is a choice that earns a profit rather than fate and project value created is simply the return for the risk taken. Second, choices are made from risky alternatives with outcomes that cannot be accurately predicted or completely controlled, and accepting greater risks improves opportunities for higher profits and other goals as long as they can be managed properly. Third, risk-reducing moves will increase operational costs and may moderate other goals, but they will improve the chances of higher project value creation.

The interaction among the many factors and sources of uncertainty and risk requires an approach that begins to manage them prior to project definition, during project evaluation, and in the project implementation phase. The outline of the approach used by SDF team members is shown in Figure 14.1, which summarizes the main activities carried out to address effectively threats and risks to strategy and the project. The first module is the uncertainty management component, showing the strategies commonly employed to address residual uncertainty, which is what remains after the interactions and effects of external factors are understood and knowable information becomes available.

The second module of the SDF approach is comprised of the key elements of the project management area, which deals with risks after some of the residual uncertainty is removed as discussed in earlier chapters. The analyses and evaluations performed in this module reduce some more the residual uncertainty, answer questions raised, and create the foundation for future-state forecasts. The third module is the risk management module, which uses the results obtained from the project management module as inputs to its own assessments to determine remaining project risks, their potential impacts, and create actions and instruments to deal with those risks.

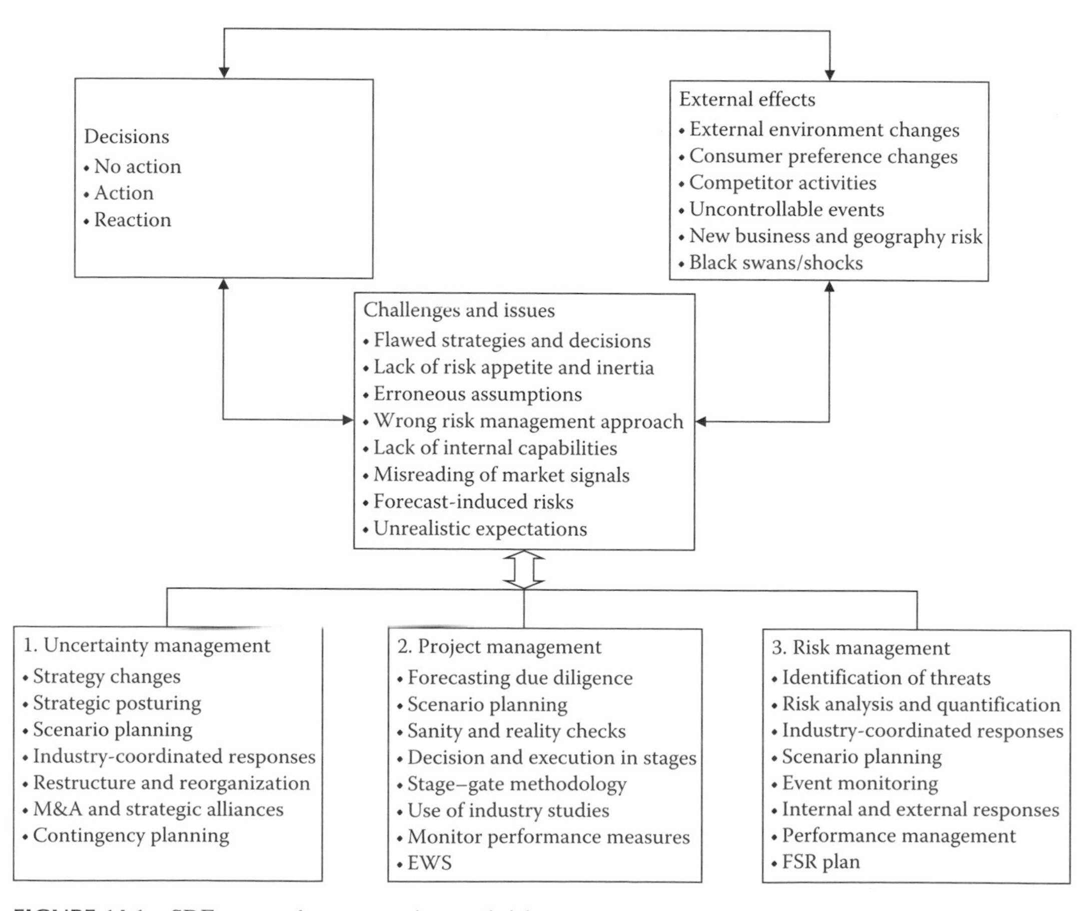

FIGURE 14.1 SDF approach to uncertainty and risk management.

In the SDF paradigm, the essence of risk management is to start with a thorough review of external and internal threats, identify controllable and uncontrollable events and factors, and determine what needs to be done. Once threats are identified, the probability of occurrence of each risk is estimated, the severity of the threat is evaluated, and the implications of exposure of assets and people to threats are assessed. This sets the stage for the risk management team to develop moves and instruments to offset risks. The moves and instruments are of three types: risk controlling, risk transforming, and efficiency enhancing or risk exploiting.

There are several, previously discussed methods and techniques, which are used by SDF team members in supporting uncertainty and risk management, but the ones used specifically to deal with important uncertainty and risk management issues are the following:

A. *Determining the level of uncertainty.* This is a powerful, structured approach used by SDF team members to think about and categorize uncertainty and determine what analytic tools to use in making strategic decisions at different levels of uncertainty. Starting from total uncertainty about a factor affecting the future of the project, the first step in the process is to evaluate the PESTLED environment, identify mega- and subtrends impacting the company, and perform industry and SWOT analyses to make unknown factors recognizable. This helps to shape some parameters that define demand for a product, service, or technology. By the end of this step, a lot more is known and what remains is called residual uncertainty. In the second step, residual uncertainty is placed in four categories:

1. A clear future, where there is little residual uncertainty left and single forecasts suffice for making project decisions and executing strategy
2. Alternative futures, where a few discrete scenarios can describe the future but cannot predict the outcome that will materialize
3. A range of futures, where a range of potential futures is identified and a few key variables define the range, but not the actual outcome in that range
4. True ambiguity, where the future is impossible to predict due to inability to identify variables that define the future and the complex interactions of factors, in which case skilled qualitative analysis is used to develop insights and strategic perspective

B. *Developing strategic posture.* This is a continuation of the previous method to determine and deal with uncertainty, define company strategic intent relative to the present and future state of its industry, and help create new market opportunities. The three main options available for decision makers to implement a strategy or a project under uncertainty conditions are

1. Big bets, which in some instances produce large returns to large capital investments but in others result in big losses. These are moves that shape the future of the industry but need large corporate commitments.
2. Options, which are moves intended to generate large returns in the best case but limit losses in the worst-case scenario. These moves are used when companies reserve the right to play in emerging and uncertain markets or to hedge big bets.
3. No regrets, which are actions that pay off under any future scenario and are usually intended to improve productivity and enhance capabilities and competencies.

C. *Risk identification and prioritization model.* This is a tool used by SDF team members to address known risks with unknown impacts and reduce the number of unknown, but expected risks. The first part of the model deals with the identification and quantification part based on five procedures:

1. Analyzing consistency of project with strategy goals objectives and reality
2. Scrutinizing project budgets, plans, deliverables, and timelines

learns from industry responses to threats and uncertainty, which help distinguish common versus company-specific threats. However, more attention is paid to doing risk management scenario planning using the following sequence of activities:

1. Obtaining senior management support, identifying the major stakeholders, and determining the scope of the scenario planning activity
2. Identifying the questions to be answered by the scenarios and subsequent analysis upfront
3. Determining current and likely future changes in the internal and external environments, industry shifts, and market discontinuities
4. Surveying key stakeholders on their views and expectations of trends and future factors and events that are sources of uncertainty and risks
5. Developing a set of common assumptions in conjunction with the project team and checking for reasonableness and supporting evidence
6. Finding the key uncertainties and risks by clustering responses into patterns and identifying causes, relationships, trends, and driving forces
7. Developing distinct scenarios in a flowcharting fashion, describing the links and their rationale, and incorporating feedback effects
8. Evaluating the consistency of the logic and the conclusions flowing out of the forecast scenarios and assessing their implications
9. Updating computational models of the scenarios, testing again for logical consistency and reasonableness of results, and performing sensitivity analysis of the output to changes in driving factors
10. Conducting hypothesis testing and "what-if" and Monte Carlo simulations to identify the range of outcomes and select the three most likely scenarios, and obtaining project team agreement and support

14.5 SUCCESS FACTORS AND SDF TEAM CONTRIBUTIONS

The success factors in uncertainty and risk management vary by project type and specifics, but there are some common sound practices and success factors that make a difference.

14.5.1 Sound Practices and Success Factors

The four categories in which these factors are broken down to cover uncertainty and risk management are processes; analyses, evaluations, and tools; support and project management; and experiences and capabilities.

A. *Processes.* In the process category, there is unanimous agreement among participants in our discussions that uncertainty and risk management must be an ongoing, never-ending process to deal with strategy and project threats, respectively. The idea that the definition of risk impact must be extended beyond the immediate financial impact to include the impacts on all corporate resources is considered important as long as there is balance between the costs and the benefits and reasonable balance of risk taking versus potential project gains.

Root cause analysis to identify sources and causes of risks and assessment of implications is a key success factor in the process category and a good starting point is the fishbone diagram or the Global 8D approach. Addressing each source and cause of risk when they are controllable and feasible to manage and doing contingency planning when they are not controllable are sound practices. A collaborative and continuous improvement culture and sound internal risk management processes is considered crucial as is independence of risk management from risk-taking decisions. The action–reaction–action brainstorming approach is viewed as central because it creates well-structured and tested risk management processes and forecast realization plans. Also, clearly defined roles and responsibilities and well-communicated risks and planned actions rank among the most important aspects of uncertainty and risk management.

Definition of Global 8D approach: *It is a problem-solving method for product and process improvement based on eight disciples that emphasize teamwork. Those principles are as follows* (Stamatis 2011):

1. *Establishing a team to identify the prerequisites to solve the problem*
2. *Describing the problem in concrete terms: what, who, where, when, why, how, and how many*
3. *Developing interim containment actions to isolate the problem*
4. *Identifying and verifying root causes using the fishbone diagram*
5. *Selecting corrective actions to resolve the problem without negative side effects*
6. *Implementing and validating the permanent corrective actions*
7. *Preventing recurrence by taking preventive measures and making appropriate changes in systems, processes, and practices*
8. *Rewarding the team by recognizing the team and individual member contributions*

B. *Analytics and tools.* The class of analyses, evaluations, and tools encompasses several elements considered essential to successful risk management beginning with identifying what is needed for successful risk management. Analytics to reduce uncertainty levels to residual uncertainty and determine what are controllable and what are uncontrollable factors in the project are essential, but accurate assessment of project uncertainty vis-à-vis corporate risk appetite comes first. Expert modeling of operations, scenario development, evaluation, testing, and scenario planning are crucial in enabling SDF team members to validate assumptions, separate market signal from opinions, and develop well-considered long-term forecasts. Integrated operational and risk management models are key to success, but in-depth market analysis and forecasting due diligence are even more important. Lastly, the use of risk adjusted measures of performance (e.g., risk-adjusted NPV) is essential in comparing performance across time and projects.

C. *Support and project management.* In the support and management class, three items made the top of the list of critical success factors: a centrally managed risk management function driven by corporate strategy and project objectives and which requires a knowledgeable corporate risk management team, sound organizational structure, and close functional alignment. That is, risk analysis should be part of each step of project execution and the risk management plan linked with all planning functions. Also, strong commitment and continuous support from senior management and key stakeholders is another success factor to implement the selected actions and controls successfully and document the risk management system. Ongoing and clear communication of threats identified by a reliable EWS and coordination of responses is crucial and so is coordination to eliminate duplication of controls, costs, and hedging activities. Furthermore, a credible and convincing business case is a critical success factor in projects of high levels of uncertainty, shaping the future strategic postures, and big bets strategic moves.

D. *Experiences and capabilities.* In this category, rigor in reducing uncertainty and risk and ability to assess market signals is highly rated and so is expert knowledge, process, and project management. Good understanding of project objectives to judge reasonableness, determine consistency with market reality and understanding of company operations, and ability to model are critical to risk management success. Experience and training in SDF and risk management are very important and so is flexibility and willingness to make plan changes and adapt to new market realities. Also, aptitude to define meaningful performance measures, create a sound EWS, and monitor and communicate effectively threats as they appear in the horizon is most valuable to ensure success. Finally, ability to conduct independent, critical, and objective evaluations, project reviews, and postmortem analyses are very helpful in judging current risk management performance and ensuring future project success.

Here, the SDF team's focus is directed to preventing risks from materializing in a future state and secondly on damage control. This requires accurate evaluation of corporate risk appetite versus ability to bear versus internal skills and competencies to evaluate risk sources and create moves and instruments. Another sound practice is that uncertainty is evaluated at the strategy stage before projects begin, threats are evaluated during project evaluations and plans are created, and risk management moves take place upon project implementation.

14.5.2 SDF Team Contributions

Following are narratives of discussion participant opinions of SDF team contributions in the uncertainty and risk management aspect of strategic projects:

A. *Framing the problem.* The contributions of the SDF team in this area begin with helping the project team to frame the risk management problem and ensure consistency of strategic with project objectives. Business process mapping and development of better processes is another contribution as is development, testing, and updating of a common assumption set company-wide. Demonstration of risk impacts through the assessment of project goal implications for corporate resources promotes awareness of threats and the need for risk management to be an ongoing process where key stakeholders participate. Also, the process of creating targets and performance indicators, monitoring risk management effectiveness, and explaining deviations from expectations using the strategic management model are significant contributions.

B. *Analyses and evaluations.* In this category, contributions start with determining what really needs to be known, what is knowable and unknowable, and what parts of different factors are controllable and which are uncontrollable. This is followed by construction of operational models of how business drivers are interrelated; identification of feedback effects; assessing forecast implications; and establishing the sources of forecast risks. The creation and testing of common assumptions used throughout the project and updated as required is invaluable help and so is the SDF team's use of industry averages, norms, and benchmarks as sanity checks and reference points. Another SDF team contribution is in the area of megatrend and subtrend evaluations, and PESTLED environment, industry, market, and SWOT analyses.

C. *Evaluation of the business.* SDF team members' independent, critical, and objective evaluations of the business are critical for a sound evaluation of project rationale and economic value creation. Also validating or initiating market research and competitive analysis to create intelligence, competitive insights, and knowledge is important to success. Demand analysis and forecasting due diligence, modeling of project parts, and creating an integrated view of operations are highly valued activities. Just as important though are the assessment of forecast implications, identification of forecast risks, and root cause analysis. Scenario development and planning, testing of hypotheses, understanding what is possible, and quantifying the potential project risk damage are value-adding functions of SDF team members. Further, the definition of performance indicators and creation of an EWS are highly valued as are the creation of response moves to counter risk identified in the forecast realization plan.

D. *Business case.* Providing inputs from analyses and evaluations and assisting in the preparation of a convincing business case is a prime contribution of the SDF team. The documentation of identified risks and description of their nature, the areas they would impact, and the investigation of the sources and root causes through integrated models are of invaluable help. Determining key likely events, their timing, and triggers and identification of black swans are other benefits of SDF team participation as is the creation of a management dashboard obtaining real-time feeds from the EWS and notification of risks before actually happening. The SDF team approach that makes uncertainty, risk, and forecasting function management part of one common project management plan is a substantial contribution. Also, including in the sources of internal threats, capabilities and competencies, processes and policies, organizational and culture issues, production and supply chain challenges, marketing and sales difficulties, and financial and legal problems help immensely the team's ability to manage risks.

E. *Knowledge sharing.* Substantial contributions are noted in the SDF team members' willingness to share experiences and develop project team capabilities in risk management. This is especially helpful in the case of risk management for project financing assignments. The assistance lent in determining and clarifying needs, assessing risk and existing controls, and implementing policies and controls contribute to successful risk management. However, modeling the business operations targeted by the decision in order to demonstrate the damage a risk occurrence can inflict is an eye opener and a critical success factor as is showing the nature of risks arising due to a forecast scenario and inputs to the model. Also, sharing of unique perspectives, insights, and experiences makes the SDF team members valuable business partners. Finally, their critical risk management function review and clear delivery of lessons learned from the project postmortems make their recommendations for future projects most valued contributions.

15 Major Capex and Project Financing

Capital expenditures (capex) are investments or expenses to acquire or upgrade physical assets, such as land, buildings, and equipment, or to increase the scale of operations. They are incurred to create long-term benefits for a company and increase earning capacity or reduce the cost of doing business, but not for items to be sold at a profit. Money spent on production inputs, inventories, personnel hiring, and training are not capex; they are operational expenditures. Hence, planning for capex is part of planning for financing of projects, which is different from financial planning for operations. Capex decisions are mostly driven by strategic planning, which defines the company goals and initiatives to attain them. Consequently, capex impact company operations, investor expectations, what others are investing in, and competitor reactions.

Capex are investments to acquire long life assets from which a stream of benefits is expected. These expenditures represent a commitment to produce and deliver future products and form the basis of future company profitability. Common criteria for capital expenditure decisions are the net present value, the internal rate of return, the discounted cash flow, and the payback period. The economic life of assets, and by extension the required capex level, is determined by their physical and technological life and by the product market life of the goods or services they produce. The physical life of assets is determined by engineering specifications, the technological life being a function of the current state and evolution of technology, and the product market life being a function of the lifecycle management of the products or services they produce. Therefore, a great deal of attention is paid to balancing the components of capex in projecting future costs and benefits.

Financing of capex often takes place through internal funds budgeted for that purpose, through externally on-balance sheet funding, or via off-balance sheet external financing. Financing of the first two types is determined by the expected economic value and the year-to-year funding requirements calculated by a formula based on stable relationships between assets, liabilities, and sales. Although analyses are similar, the focus of the discussion in this chapter is primarily directed to the last type of financing capex projects where (a) large capex occur over a number of years and the market relationships vary and (b) capex requirements are derived from long-term, scenario-based demand models that address the assets' underlying economic life changes. Project financing is used in both the private and the public sectors because companies and government entities can finance projects off their balance sheets. Project financing uses the same approach and type of business operation models as those used to forecast capex requirements for other new strategic projects of long durations.

Long-term financing of large capex requirements based not on the strength of the project sponsor's balance sheet but on the forecasted cash flows of the project is appropriate in cases of infrastructure and industrial projects. For example, infrastructure projects such as water and sewer facilities, power plants, transportation projects, and telecommunications networks often use this type of financing. They are long-term projects with designing, planning, and construction periods anywhere from 10 to 15 or more years.

in project financing are responsible for environmental, industry and market analyses, long-term demand forecasts, and evaluation of project economics. These functions involve engagement in every aspect of generating long-term forecasts and identifying and quantifying forecast risks. The output from the SDF team evaluation is then added to those of the other experts and the financial analysis to produce a complete project assessment.

Because of the focus of the funding sources on coverage ratios, the impacts of megatrends, and industry and political, economical, social, technological, legal, educational, and demographic (PESTLED) environment developments are assessed to justify the purposes of the project company. They are performed at the surface level, and they do not matter very much in the development of the forecasts. Even in cases where project suppliers are also sponsors or owners, current industry and projected future supply capacity are not considered seriously. And if this was not enough, there is a limited understanding of project financing costs: For projects under $100 million, there is a disproportionately high cost to draft and negotiate contracts and other legal agreements, estimated to be around $20 to 25 million. In one project financing assignment, a legal team from Washington, DC, shared that the paperwork supporting the project filled a good size room of boxes containing project-related agreements and documentation. In small projects, the high legal costs and documentation requirements make them difficult to structure so as to be profitable.

15.2 PARTICULAR CHALLENGES AND ISSUES

Project financing is a field that requires a confluence of several areas of expertise, which, in our experience, is never possible outside companies that have established project financing organizations (PFOs). The gaps observed in the requirements for effective project financing are due to factors such as cultural limitations, ineffective processes, and lack of project financing skills, experience, and relationships. That is, expertise is localized to respective areas:

1. Project sponsors know the business in which they operate but not project financing.
2. Funding sources know financing but not the business or the technical aspects of the project.
3. Suppliers know the technical but not the operational aspects of projects.
4. Project company owners sometimes know neither if they are not from the same industry or are suppliers to the SPC.

The absence of commercially available integrated software systems to evaluate project economics and financing proposals, track activities and flows of goods and services and money, and determine project value created is a major issue. This is complicated by ineffective organizational structures and lack of training and relationships with funding sources, credit-enhancing agencies, and project financing advisors. However, there is one issue that presents major challenges in structuring successful project financing, namely the unwillingness of sponsors and other participants to contribute equity to projects outside their core business.

Project participants have different or even conflicting interests and different types of expertise, which result in participants sometimes not obtaining benefits proportional to costs incurred and commitment to the project. Often, off-balance sheet financing is undertaken without all project participants having a good appreciation of how the different types of risks could impact them, how to manage those risks, and ensure value creation. These factors make it difficult to balance the costs and benefits of project financing participants according to their financial and human resource contributions. The end result is that project evaluations are biased in favor of the parties in command of project financing expertise and not based on actual project economics.

Setting up SPCs is a very large and time-consuming effort and involves expensive legal document development. As mentioned earlier, for projects under $100 million, it costs about a quarter of that to create and negotiate the required agreements, which is further complicated by difficulties in obtaining consensus when several participants are involved in the project. Furthermore, there are several types of risks associated with the different aspects and participant impacts in the project classified as follows.

1. *General risks:* These are risks related to agreements and default by counterparties, country risks that have to do with political, economic, and regulatory stability, and business and legal institutions risks connected with enforceability of contracts.
2. *Project-specific risks:* These are the risks that have to do with the quality of sponsors, construction, operations, suppliers, and offtake risks.

There are many myths surrounding the benefits of project financing that lessen the focus on project economics, which is needed to structure effective financing projects. And, the multitude of risks involved in project financing schemes translates into demand forecast uncertainty beyond construction and into the operation phase. Hence, the need for sound risk management pervades all participant contributions and agreements that require commitment and large investment of time and effort. However, project risk management is made more difficult due to varying participant approaches and diverging interests in a project, namely, the following:

1. Suppliers are constrained by conservative financing approaches, limited tools and experience, and unwillingness to provide equity.
2. The sourcing of equipment sometimes impacts financing terms substantially and determines what country equipment is procured from.
3. Suppliers, developers, and sponsors recognize the need to educate customers and the owners of SPC on the role of project financing and evaluation of proposals, but this rarely happens.
4. Differences in the customer and SPC financial evaluation methods provide challenges and opportunities in developing financing proposals.
5. The objectives of other stakeholders in the project finance equation must be considered, but lenders focus on service coverage ratios and sponsors look to maximize return on equity in the project.
6. Each participant has their own demand forecasts and value created based on different assumptions and assessments of the marketplace.

All strategic projects require solutions tailored to fit the situation and address the project needs, and so is the case with project financing. However, there are no off-the-shelf financing arrangements in project finance. Every project is different and each project needs a very carefully tailored deal structure. While the difference in costs between financing proposals is minimal, the risk of a poorly thought-out financing arrangement is colossal. Hence, the assessment of the project and financing structure must be based on an objective, critical, careful, and well-organized and comprehensive project economics evaluation approach. But even then, sponsors may not gain competitive advantage through the use of project financing, and the project perceptions of other participants coincide with vested interests. That is, project financing is not really a source of competitive advantage because one cannot patent project financing techniques; but as a supplier, one needs to offer it to be competitive. Lastly, slow decision making in participant organizations when large projects or when equity contribution requirements are involved, they make project evaluations and risk management difficult.

In summary, the major challenges associated with managing the assessment of project economics and reaching consensus when several participants are involved in the project are due to the following reasons:

1. Each participant having their own SPC revenue, risk, and investment requirement assumptions, and cost–benefit allocations
2. Different interests in the project and the timing and expected benefits for a given equity contribution, which are difficult to reconcile and cause long negotiation periods

3. Stated or implied expectations of participants colored by their interests and experiences and no credible demand analysis and forecasting due diligence performed
4. Project uncertainty and risks viewed from different perspectives and which requires managing them in ways that satisfy those perspectives
5. Differences in the skills and qualifications of the representatives to the project team and their ability to negotiate effectively a scenario, which drives long-term forecasts and to assess their implications objectively

15.3 SDF APPROACH AND PARTICIPATION

Project financing for large infrastructure projects is carried out by a group of highly specialized experts in their fields, including seasoned SDF team members. The broad project financing process is shown in Figure 15.1 and involves several key steps with multiple underlying activities. The process starts with the identification of a potential project, formulation of the business concept, and high-level preliminary analysis and screening to determine whether a project is technically, economically, and politically feasible. Then the assessment and allocation of risk follows. This entails identifying risks associated with start-up costs, operating revenues and costs, technology changes, industry and market shifts, ownership issues, and other risks of doing business in foreign countries.

The project economics assessment part of the process is performed by the SDF team and involves an in-depth due diligence and evaluation of the driving forces, interactions, and feedback effects to establish the assumption set, model the SPC operations, and create long-term forecasts. These forecasts are based on realistic scenarios in order to determine the economic viability of the project. To create these scenarios, a critical assessment of the proposed project elements, participant expectations of returns, and validation and verification of project goals and objectives, market data, and competitor information is required. In the commercial structuring phase, technical designs are completed and permits obtained. The financial institutions involved conduct their own due

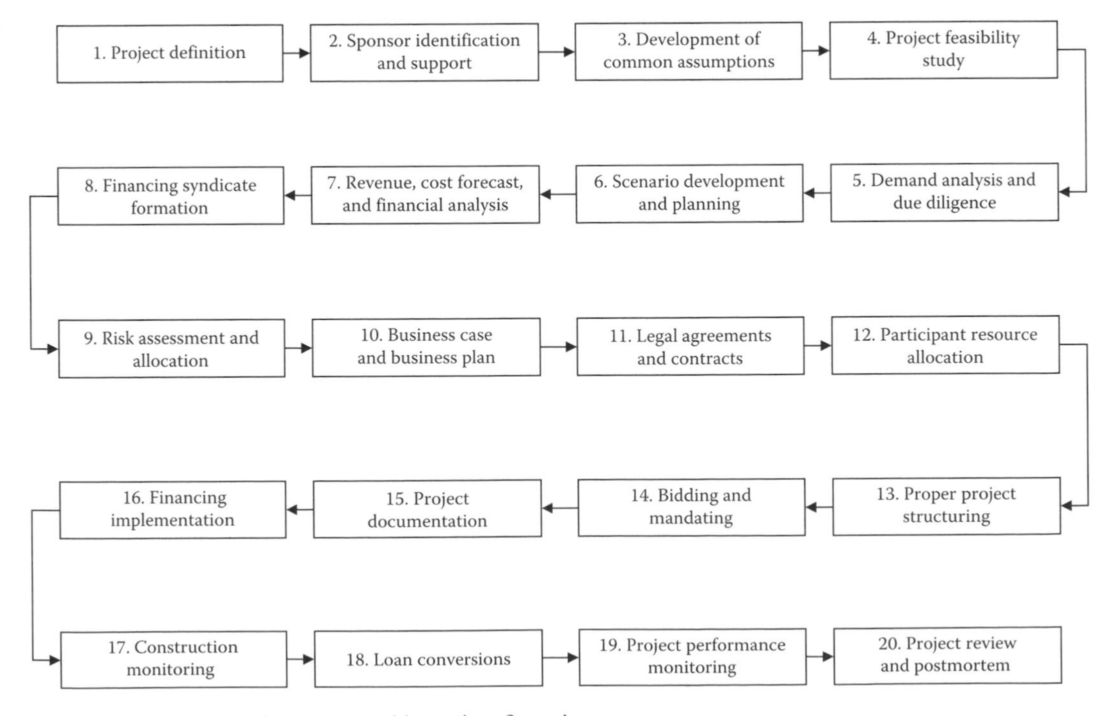

FIGURE 15.1 General process used in project financing.

diligence studies, and major contracts are finalized and agreements put in place so that the financing information memorandum can be prepared and the offering memo issued. In the closing and release of funds step, all contracts are finalized and outstanding issues are resolved so that funds and the project are released for construction.

Leading PFOs engage strategic forecasters in all significant projects, which enable them to be versatile in applying a wide variety of techniques to meet specific customer needs and situations. The responsibility of the SDF team in capex financing projects is to perform the analyses and evaluations it normally undertakes in all strategic decision projects. It also includes project managing the development of consensus assumptions and forecasts used for common decision-making purposes. Therefore, the set of roles the SDF team plays in project financing assignments includes monitoring, investigating, analyzing, and evaluating the aspects of project financing that impacts customers or end users and project economics.

The specifics of SDF team analyses and evaluations performed in support of project financing, and the required skills and qualifications vary by type of project structure and financing techniques. The most common project financing techniques are the following:

1. The build–own–transfer (BOT) structure and its variants, where the sponsors build and own the project for a specified period and then transfer ownership to the ultimate owner
2. Capital market offerings that involve issue of securities such as 144A issues and carry higher potential returns and longer duration
3. Export credit agency (ECA) financing that involves direct loans or loan guarantees against commercial and political risks and is most effective when high supplier content is involved
4. Securitization schemes whereby the owner of the SPC bundles and sells the receivables to pay down the debt incurred by the project
5. Capital lease, Japanese leverage leasing, Islamic financing, and forfeiting export financing, which are other options utilized; countertrade, which was used when trade restrictions were prevalent in the developing world and is now infrequently used

One tool commonly used by external advisors and adopted by the PFOs and the SDF team for their respective areas of responsibility is the enterprise model shown in Figure 15.2. The idea here is to identify gaps in areas crucial to project financibility success and determine approaches to fill those gaps. In project financing, one of the SDF team's attention is participant competency and capability gaps and focus on evaluations related to the following:

1. Assessing megatrends and developments in the external environment and their impacts primarily on the SPC
2. Helping to create, negotiate, and manage a common assumption set to drive project valuations
3. Modeling the project structure, process, and operations and creating scenarios to generate long-term forecasts
4. Assessing project risks, evaluating forecast implications, performing sanity checks, and assessing project economics
5. Coordinating activities and cooperating with internal planning groups and counterparts in the other participant organizations

A major benefit of using the enterprise model in off-balance sheet financing is that it helps to systematically assess each participant's internal organization dimensions from company culture, to competencies and capabilities, to ability to meet project needs and challenges, to deliver on forecasting model input needs. The second advantage of this model is that it helps in the efficient identification of the prerequisites for financing success, namely, it helps to determine the key project and team focus areas, the scope of project participation, the most likely scenarios, and long-term

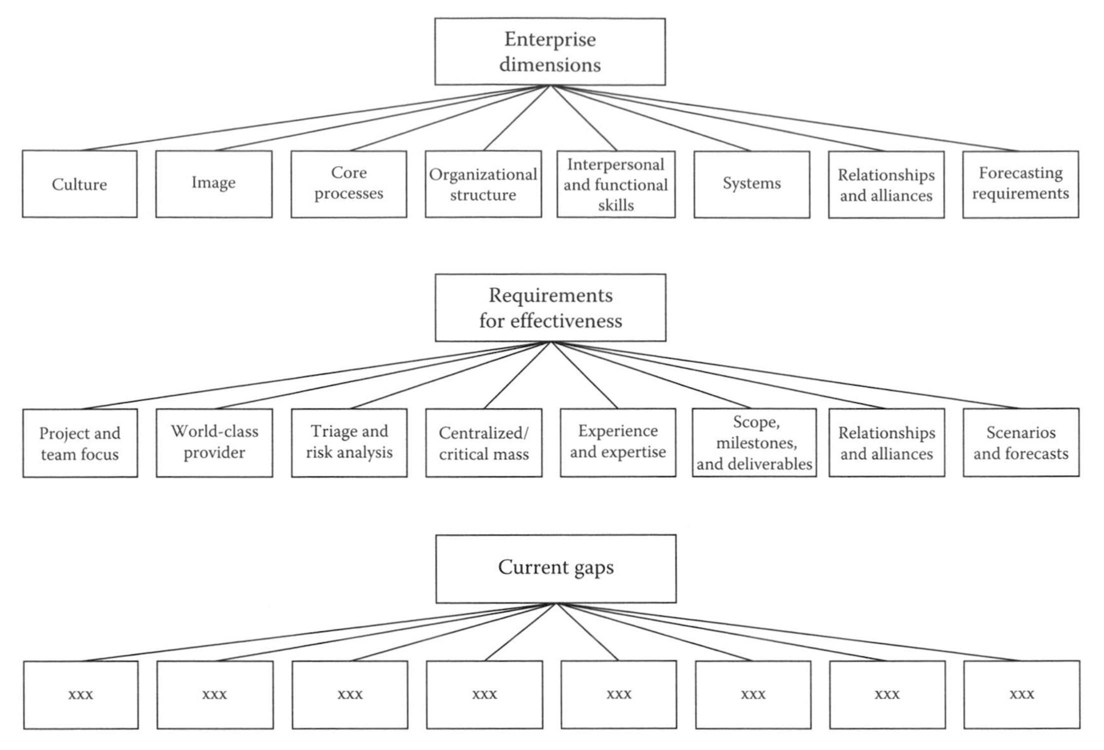

FIGURE 15.2 Enterprise model for off-balance sheet financing. (Adapted from Triantis, J. E., *Project Financing Benchmarking Study,* AT&T Submarine Systems, 1994.)

forecast requirements. The comparison of the requirements for project effectiveness versus the internal organizations capabilities identifies gaps that need to be filled in order to obtain the objectives of each participant and ensure project success. Then, the measures needed to fill key identified gaps are determined and often negotiated between project participants.

15.4 SDF ANALYSES AND EVALUATIONS

The primary role of the SDF team is to support PFOs to understand the external environment and the marketplace, facilitate relationships and knowledge sharing, focus on the process and project economics, and validate proposed financing structures. However, clarity on the SDF team roles and responsibilities comes after a number of decisions are made in the capex financing process. Once the need for future capex arises, the first step and decision in the financing process is to assess the capex requirement rationale and motives and to determine if the contemplated increased capacity, reduced costs, and technology advances warrant serious consideration versus current operational needs. The SDF team is engaged to verify the need for capex in light of demand changes, industry capacity adjustments, new competitor actions, and recent general market conditions. It also validates the consistency of the project generating the new capex requirements vis-à-vis the corporate strategy goals and objectives and alternative growth projects.

The timing and levels of funding needed are determined to coincide with plant and equipment acquisition based on long-term demand forecasts. These forecasts drive the cost–benefit analysis of the proposed capex from an operational, financial, product portfolio improvement, and meeting customer needs perspective for the financing alternatives being considered. Based on these considerations, a decision is made whether internally generated funds or external financing will be used. If external financing is the best option, a decision needs to be made whether to use balance sheet or off-balance sheet financing. If off-balance sheet financing is selected, then an SPC is created to get such funding. Once the SPC is created, the options with their pluses and minuses are evaluated to choose the project ownership structure; that is, build–own–operate–transfer (BOOT), built–operate–transfer (BOT), and so on, which determine the role the SPC will play in the project.

The functions performed by SDF team members in project financing depend not only on the financing alternative selected but also on which participant they support and on the selected project structure. In all cases, however, the concentration of their work is in the project economics and risk identification, sharing, and management areas first. Then, focus shifts to contracts and agreements, the sourcing of equipment, and production inputs and services. The major activities of SDF team members common in all alternative financing options are as follows:

1. Performing megatrend and subtrend, industry, market trend, and key competitor analyses
2. Developing, updating, communicating, obtaining agreement, documenting, and maintaining the project assumption set
3. Evaluating the effects on consumer demand of alternatives to new capex, such as making changes in production and supply chain management processes, acquiring competitor facilities, outsourcing parts of production, or leasing new equipment
4. Modeling the project processes, creating scenarios, and simulating alternative assumptions from the perspective of the client company
5. Generating long-term forecasts, conducting sanity checks, assessing the implications of forecasts, and identifying and quantifying project risks
6. Providing inputs and helping in the capex cost–benefit analysis and considering the alternatives financing structures available
7. Helping the project team in performing needed evaluations, supporting the effort to negotiate agreements, and preparing inputs for a convincing business case and the creation of the SPC's business plan

A. *On-balance sheet, internal capex financing.* Under the "on-balance sheet, internally funded capex" alternative, the primary focus of SDF team members is on generating long-term sales quantity and price forecasts. This necessitates assessing the impact of megatrends, industry developments, and changes in the external environment, and performing market analysis and product and customer preference evaluations. Lending help in establishing discontinuities in plant and equipment needs via scenario planning simulations, in conjunction with technology and production experts, requires SDF team evaluations. Help is also lent to distinguish additions to plant and equipment for existing operations from new project capex requirements for new market or business entry needs. Additional SDF team analyses are performed to identify events, triggers, and risks to the long-term forecasts; conduct root cause analysis, recognize possible black swans, and help in contingency planning and risk management.

B. *On-balance sheet, external capex financing.* In the case of the "on-balance sheet, external financing of future capex requirements" alternative, the attention of SDF team members is directed toward a reoriented set of priorities to meet company and external requirements. Here, the emphasis is on analysis of environmental changes, industry developments, and market trends; competitive analysis; in-depth forecasting due diligence; identification of forecast risks and risk mitigation planning; demand and supply pricing in uncertain environments; and creation of models and long-term projections. Also, scenario development, sensitivity and "what-if" analysis and planning are performed by SDF team members to provide inputs for cash flow projections and help in planning for the SPC's project objective realization. In this case, they also monitor project performance, internal process improvements, and cost trends relative to those of competitors.

C. *Off-balance sheet capex financing.* Under the "off-balance sheet financing" alternative, understanding the project created by the capex requirements and ensuring consistency with corporate policy, strategy, and goals and objectives is paramount. Here, the expertise of the SDF team is brought to bear on a sound project definition and the evaluation of project economics for two reasons: to provide independent, critical, and objective assessment of the SPC operations and to facilitate balance of participant interests and their impacts on project economics. The attention of SDF team members is directed to the following areas:

1. Evaluating the impacts of megatrends, PESTLED environment changes, and the findings of the company strengths, weaknesses, opportunities, and threats (SWOT) analysis
2. Assessing current demand and supply conditions for the products or services of the SPC and the production inputs and making expert judgments about future conditions
3. Providing benchmarks, analogs, and industry practices and standards for analyses and evaluations and ensuring their appropriate usage
4. Identifying key market drivers and modeling the SPC's operations in sufficient detail using agreed-to assumptions
5. Performing scenario planning investigations to test assumptions, do a forecast risk analysis, and identify specific trigger events and threats to the project
6. Actively support the coordination of inputs to financial assessments required to facilitate preparation of contracts and timely funding decisions
7. Understanding the restrictions of financing agreements on the demand and supply aspects when the SPC is a wholly owned subsidiary versus a JV entity
8. Creating a project management dashboard and an early warning system (EWS) to assess progress against milestones and deliver warnings of impending risks
9. Defining appropriate performance measures and the right indicators to monitor, assessing environmental developments, and contributing to the variance analysis
10. Monitoring the quality and consistency of analyses, evaluations, and conclusions for purposes of project rating by external agencies and the preparation of the information memorandum for equity issues

The roles and responsibilities of SDF team members also vary under different scenarios of company participation in project financing and are aligned with the objectives of the client they support as pointed out in Figure 15.3. These analyses and evaluations are geared toward supporting client objectives in a balanced and well-considered independent, critical, and objective manner keeping in mind that interests and objectives may change in the future. The SDF team's role becomes more interesting, however, when the participant client assumes more than one position in the project. For example, a project sponsor may be a supplier as well, or the purchaser of the SPC's output may be the eventual owner.

1. *Client is an owner.* When the SDF team is supporting an owner, a lot of effort is expended toward balancing interests among participants to reduce negotiation time and contract preparation by analyzing current industry capacity and planned changes. Attention is also given to assessing forecasts further out in the future and helping the project team to evaluate the portion of the required equity to assume in order to maximize profit. To evaluate project economics, it is crucial for the SDF team to evaluate objectively the demand for the product or service offered and supply conditions for inputs to the production process of the SPC. But, this is not sufficient; assessing the stability of cash flows and identifying events, intervention factors, and their timing that could upset that stability is needed to help the project team in risk assessment and management and the overall task coordination effort.

2. *Client is a sponsor or developer.* When the SDF team's client is a sponsor or a developer, its effort is directed to supporting a set of client objectives that include identifying the true customer needs. The SDF team provides inputs to estimate the level of enough equity to demonstrate commitment and ensure project financibility success. It also coordinates and assesses the evaluations of other

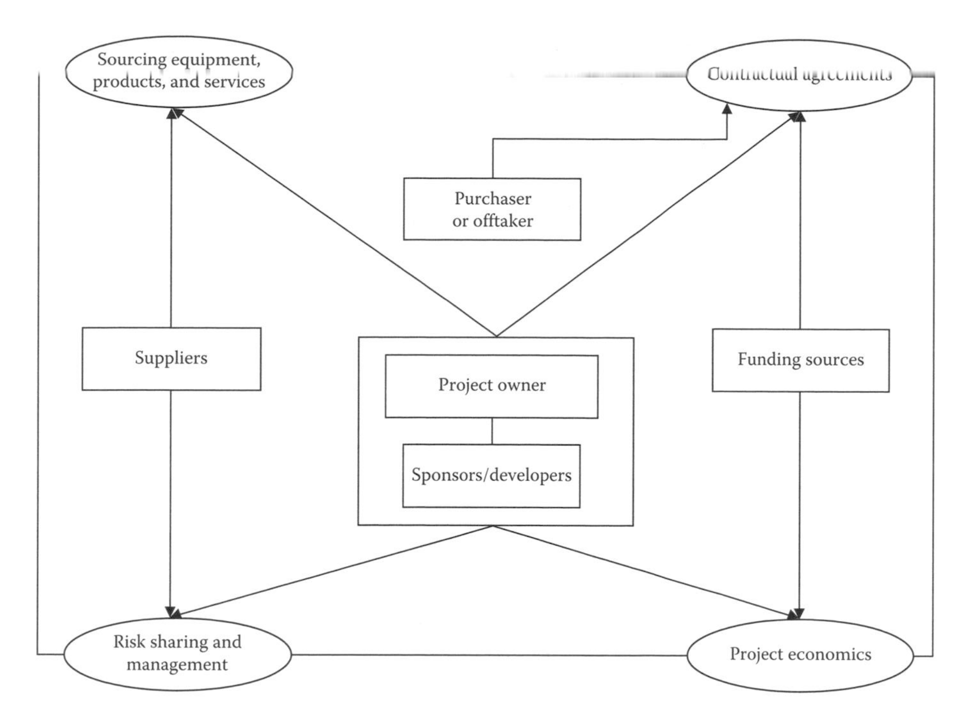

FIGURE 15.3 SDF focus aligned with client interests.

project participants. However, assessing project economics is priority one along with identifying project risks and determining SPC revenue-neutral approaches to spread and manage them effectively. In this case, a sound assessment of project economics naturally entails thorough assessments of the external environment, industry and market conditions, and in-depth competitive analysis.

3. *Client is a supplier.* When the client is a supplier to the SPC, the SDF team support is directed to supporting the client objectives by helping the project team to assess and provide all equipment needs and production inputs to the SPC at competitive pricing. This often means supplier financing for the SPC at competitive rates. Assessing the current state of technology and anticipating and projecting technological changes that would render the equipment supplied to the SPC outdated and alter production input requirements is another role played by SDF team members. Also, assistance provided in the evaluation of competitive bid improvements through choice of sourcing equipment country and credit enhancements comes from competitive intelligence and insights derived from country analysis by the SDF team. Monitoring closely competitor product offers, moves to enhance product quality, and improvements on pricing is done in order to identify forecast risks and ensure an acceptable spread of risks and risk management so that timely payments in full for equipment and production inputs materialize according to agreed-to schedule.

4. *Client is a purchaser.* When the client is a purchaser of the SPC's output, effort is concentrated on assessing correctly its market power vis-à-vis that of the SPC and on agreements successfully balancing their interests. In all cases, SDF team support is needed to create supporting revenue analysis and forecast documentation to obtain permits, licenses, and other government approvals, whether the purchaser is a private or a government entity. Uninterrupted flow of products or services from the SPC is crucial, and it requires objective evaluations of capacity adequacy and supply chain agreements to ensure that. Just as important, however, is the SDF team evaluation of product and service quality and the pricing structure, which require identifying areas of gaps in need of quality and pricing improvements.

15.5 SUCCESS FACTORS AND SDF TEAM CONTRIBUTIONS

The success factors in project financing and the contributions made by SDF team members, when they are active participants in the project team, come from three sources:

1. The independent, critical, and objective assessment of project economics
2. Their extensive, formal benchmarking experiences
3. Their understanding of strategy and other planning functions and relationships with experts in different areas of project financing

Note on benchmarking: *For benchmarking to be successful, participants in benchmarking studies and informal discussions should include a number of large infrastructure project developers and sponsors, investment and commercial banks, unilateral and multilateral agencies, and owners of SPC's and infrastructure projects. Equipment providers, law firms specializing in project finance, and reputable independent consultants and project advisors need to participate as well. Preferably, participants should represent different industries and regions of the world.*

15.5.1 Sound Practices and Success Factors

The success factors identified by project financing experts and experienced project participants are aligned with the participants' perspectives along three categories: project structure and process, significance to the participant, and financing institutions deal terms and conditions. Notice, however, that emphasis is on processes alone and the underlying analyses, evaluations, modeling, and forecasting gets little credit. This is indicative of deficiencies in project financing evaluations, even among companies considered best in class, which results in prolonged project assessments and mediocre results.

A. *Project structure and processes.* Leading sponsors, contractors, suppliers, and developers have centralized units of professionals on call skilled in the key areas of project financing. The success factors shared by project financing experts in this category include the following:

1. Manageable project size and project risks. That is, a project of size suitable for the required financial and human resources that can be dedicated without encumbering the company with risks beyond its willingness and ability to bear.
2. Country and customer credit worthiness. For projects outside the company's home base, the credit worthiness of the country is critical when the government is the eventual owner or when the SPC is to be owned by a private entity.
3. Extent of sponsor support equity and credit enhancements. Substantial equity participation demonstrates commitment to the success of the project and future profit expectations.
4. Strong project economics and projected cash flows. These two conditions are necessary to ensure project financibility because they show ability to repay loans and generate adequate profits for equity investors.
5. Sound implementation plan. This is one where the project is headed by a seasoned project manager supported by a competent project team that knows and is able to produce expected deliverables in a timely and quality manner.
6. Reasonable SPC business plan. This is a plan characterized by realistic assumptions, reasonable performance targets, rational timelines, and sensible risk taking.
7. Close cooperation, coordination, and effective communications within and across each and every project participant. These factors are necessary to ensure timely decisions and project implementation in every project finance engagement.

8. Sponsors getting early advice on what is possible and speedy project financing implementation from established relationships, but financing alliances should be on a case-by-case basis since exclusive alliances are not productive.
9. The need to have a clear vision and "sustain-the-course" philosophy. That is, project goals and objectives are well understood and consistent with strategic goals and objectives. Participants must have a clear strategy and knowledge of how financing supports that strategy, whether they are a sponsor, supplier, utility owner, or a government body.

B. *Project participant factors.* In our discussions, sponsors report five sources of competitive advantage, namely, clear vision of financing goals, meeting customer needs, sufficient equity contribution, project team members experienced in economic assessments and project financing, and an extensive network of relationships. Other participants shared success factors relevant to them according to the role they play in the project:

1. Identifying and meeting customer needs beyond efficient project financing is key to successful project structure and efficient implementation. This success factor is recurring among equipment suppliers attempting to gain a competitive advantage.
2. Owner, sponsor, or developer financing needs are met according to each party's stake in the project and risk is shared in a balanced approach.
3. Favorable market conditions at the time of the project are a precondition to success when the cyclical nature of some industries dominates other elements of project financing.
4. Understanding the competitive environment, changes in customer or user needs, and technology evolution. This is most important for all project participants because it impacts demand and supply conditions facing each participant.
5. Solid and enforceable agreements and contracts are needed not only to protect all involved but also to expedite project structuring and the implementation of financing.
6. Balance of project participant interests through the life of the project is required to negotiate terms and conditions and come to closure effectively.
7. Team orientation, access to quick decisions, and tenacity. Because of the massive multiparty negotiation effort, project teams must have access to and get quick decisions from senior management.
8. Performance evaluation for each participant's team as a unit based on closing projects successfully and evaluations occurring at regular milestones and critical path junctures.
9. Sound project economics determines financibility and its value creation power determines willingness to invest in the project and the cost and speed of financing implementation.
10. Experienced professionals with knowledge of global capital markets and relationships with funding sources, ECAs, and advisors and experts. This is essential not only to structuring and implementing project financing but also to identifying early on those projects that will not make it.

C. *Terms and conditions.* There are several factors necessary for fair terms and conditions, and which are considered crucial by the funding sources participating in our discussions, the most common being the following:

1. Guarantees and other support provided by the sponsor or developer company are crucial in securing loans to the SPC. In projects outside the home country, ECA and other government support is also required.
2. Matching of owner, sponsor, or developer risk tolerance with project risk level is needed to make equity contributions and agreements on risk sharing possible.
3. Reasonable assumptions and financial projections are crucial for developing plausible scenarios and long-term demand forecasts for the SPC's output, development of convincing business cases, and reasonable business plans.

strategy to portfolio is usually done by top-down strategy models, which ensure that the eventual portfolio coincides with the business strategy.

The areas of primary SDF team attention that require broad business knowledge, skills, and experiences are the following activities:

A. *Information needs assessment.* In the first step of involvement, the SDF team helps the project team determine what needs to be known in order to execute the project successfully because different types of PPM projects involve different information needs, analyses, and evaluations. Then, the rationale of and the needs for a PPM project and the uncertainty surrounding it are checked against the strategic plan and corporate risk tolerance levels to decide if it makes sense to proceed. Also, the SDF team determines what is known, unknown, knowable, and unknowable and identifies what factors impacting the project are controllable and what factors are outside of the company's control. Then, it identifies reliable sources of data and intelligence, assembles the information pieces needed, and validates them.

B. *Current portfolio assessment.* In the evaluation of the current portfolio step, SDF team members analyze and evaluate the company's product line in terms of current and expected demand, pricing, marketing, and advertising support. A product-by-product competitive analysis is performed and current productivity profiles are evaluated against established targets, expected capex requirements, and maximum production capacity. That information is used to assess portfolio rebalancing project costs and the resulting project NPV. A reexamination of market drivers takes place and uncontrollable factors are identified and the ways to mitigate potential impacts are examined. By this point, the current portfolio risk profile has been assessed and potential future threats are identified, all of which form the basis of scenarios and FSR plans.

C. *Portfolio validation analyses.* The portfolio validation stage is a parallel step where the proposed new projects and the resulting new portfolio are evaluated by the following SDF analyses and evaluations:

1. A critical, objective, and realistic product lifecycle analysis, validation and verification of results, and reporting of findings
2. Assessment of the impact of megatrends and subtrends relevant to the industry and, more specifically, the company's product line
3. In-depth industry analysis and scrutiny and testing of assumptions that help define some boundaries of the future portfolio uncertainty level
4. Assessment of competitor products, activities, future growth plans, product differentiation, and capacity constraints
5. Well-informed PESTLED analysis which along with the earlier assessments and industry analysis helps define the future operating environment where portfolio adjustments will work out
6. Critical and well-balanced internal and external SWOT analyses whose end product is the identification of the competencies and capabilities needed to ensure success of the proposed projects and the new product portfolio
7. Independent and critical SDF team assessment of project and portfolio uncertainty and risk to establish the company approach to future growth

D. *Scenario development.* Once the current portfolio evaluation and the proposed project and portfolio validation take place, the SDF team engages in the development of plausible scenarios, forecast solutions, and recommendations. It also synthesizes the findings and results and helps the project team in their portfolio representation by incorporating all the insights distilled up to this point into distinct and plausible future-state scenarios, testing different hypotheses, and performing sensitivity analyses. The end product of these activities is a set of long-term forecasts that define the future value of projects and the entire portfolio, which, in turn, are used in the representation of their relative value contributions that allow meaningful comparisons and project selection.

E. *Performance metrics definition.* When projects are beginning to be implemented and portfolio augmentation or rebalancing takes place, yardsticks are needed to measure progress toward accomplishing PPM goals and objectives and overall portfolio performance. In this phase, SDF team members assist in the development of performance metrics and ensure data feeds to populate the business operations dashboard, the EWS, and the balanced scorecard. Some common gauges of portfolio performance include the following metrics:

1. The new product or project incremental NPV to the overall portfolio value
2. The number of portfolio value creating new ideas and proposed projects in a year
3. The percentage of profitable and strategy supporting PPM projects approved
4. The number of new products under development and other approved projects
5. The product and project pipeline throughput during the fiscal year
6. R&D resources and investments devoted to new products relative to the total R&D budget
7. The speed to market of new product launches, uptake rates, and market shares realized
8. The balance of projects and portfolio components by type, market segment, size, and risk profile

F. *Monitoring performance.* In the implementation of projects and portfolio rebalancing phase, monitoring portfolio performance requires a lot of skill and discernment. Here, SDF team members help the PPM project team to interpret results vis-à-vis external environment developments. They provide inputs on how to incorporate performance checks, assumption tests, and risk analyses in planning the implementation of PPM initiatives and projects. They also help to manage the reprioritizing and reordering of portfolio projects in light of recent industry developments. Additionally, SDF team members provide support to allocate resources efficiently across portfolio elements and projects, to ensure realization of expected states based on projected value added, and to rebalance the portfolio when new projects and proposals enter the queue.

G. *Knowledge management.* While all these PPM team activities take place, an Intranet project database is created to house data, information, intelligence, and insights created and evaluations performed as well as assumptions used. The project database, at a minimum, typically contains market research, demand, pricing, marketing, competitive intelligence, investment requirements, operational costs, and past project lessons learned. In the analytic information section of portfolio management database, the development of the GE/McKinsey matrix by SDF team members is shown, which enables assessment and ranking of the likelihood of each product, service, or technology profitability and investment success.

16.5 SDF TOOLS AND TECHNIQUES

Many of the SDF processes, analyses, evaluations, and additional special tools discussed in Chapters 6 and 7 are used to support assessment, ranking, and prioritizing PPM projects. There are, however, some additional special tools used in these projects, such as the alignment timeline balance chart, roadmapping, bubble charts, and GE/McKinsey matrix.

A. *The GE/McKinsey matrix.* In the PPM context, market attractiveness for the GE/McKinsey matrix is assessed by the SDF team by defining attributes such as market size and market growth rate for a product or product line, industry rivalry and intensity of competition, and barriers to entry. Also, the ability to differentiate products, services, or technologies, pricing trends and market profitability, and industry, geography, and other types of risks of investment returns are factors defining market attractiveness. On the other hand, competitive strength is assessed by characterizing factors that impact it most such as strength of underlying physical and human resource assets, management competencies and capabilities, and brand recognition and strength of marketing and sales support. The other three determining factors of competitive strength are customer loyalty, market share and share growth, cost structure and profit margin relative to competitors, and ability to reinvest profits to strengthen sources of profit.

This matrix is constructed through the following steps: first, specifying the key elements of market attractiveness and the company's competitive strength relative to each portfolio component and determining the significance of every element in achieving corporate goals and objectives. Then, assigning weights to each element of market attractiveness and competitive strength through consensus; scoring each portfolio product, service, or technology, or asset under study; and ranking the scores by plotting them and examining and interpreting the results take place. Based on the results of these steps, SDF team members then construct different scenarios, conduct hypothesis testing, and "what-if" simulations, and perform sensitivity analyses. The results of hypothesis testing and simulations are used to determine the responses needed, if there are competitor reactions to the project and their effect, the costs and benefits, and the underlying support for each element of the portfolio.

B. *Alignment timeline balance chart.* The essence of the PPM alignment timeline balance chart is to show how current projects and initiatives, customer and user requests, and new product or project ideas and proposals are aligned with corporate strategic goals and objectives, financial and human resources and competencies and capabilities, and project uncertainty and risks. This is illustrated in Figure 16.3. But, how does one achieve portfolio balance? A desired balance of projects is achieved via a number of specific parameters such as risk versus return and short term versus long term, and across various markets, business arenas, and technologies. Typical methods used to reveal balance include bubble diagrams, histograms, and pie charts. Here, the SDF team support is focused on assessing the competitive landscape; developing processes, scenarios, and forecasts;

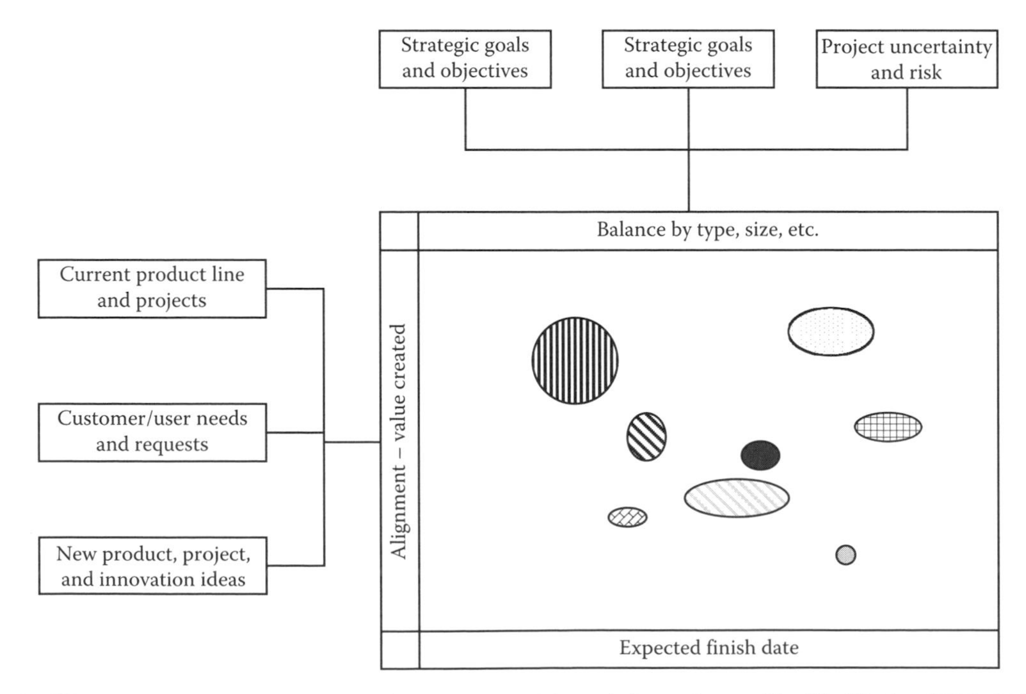

FIGURE 16.3 PPM alignment timeline balance. (Adapted from Contec-X, CA Clarity PPM, 2007. http://www.ca-clarity.com/cms/upload/EN/pdf/CA_Clarity_Functional_Overview.pdf.)

conducting evaluations and sanity checks; identifying forecast risks and triggers; and providing data and analyses to populate the bubble charts.

C. *Prioritizing rebalancing proposals chart.* The pictorial tool shown in Figure 16.4 is used to show the stage of different portfolio rebalancing projects and the impacts they have on the various corporate assets and components of the portfolio. Here, a distinction is made between project and product portfolio management. The SDF team's participation here is in activities of the latter even though they affect each other:

1. Prioritizing projects by helping build the centralized repository for tracking project information
2. Developing data inputs and insights for a holistic representation of the efforts to determine overall project commitment
3. Modeling and performing "what-if" analyses to determine and show the impact of adopting a new project or canceling a project
4. Representing the diversification of projects within the company and showing what projects are in different critical areas

The prioritization of portfolio management projects makes extensive use of the performance metrics scorecard, which shows the component or the performance metric being measured, its importance based upon the severity of the operational priority or value added, and whether portfolio management is in control, that is, whether the value of the component is in the expected range. The value of the component being measured in NPV dollar amounts is also shown along with the process owner or the project team member responsible for a component or process and the functional metrics in terms of measurable items.

D. *Bubble diagrams.* They are plots of the different project values, shown as bubbles, along the NPV and probability of success axes. The size of the bubbles shows the value created by different projects and their position on the chart shows the likelihood of realization. These diagrams are used in conjunction with the alignment timeline balance chart shown in Figure 16.3 and other tools in

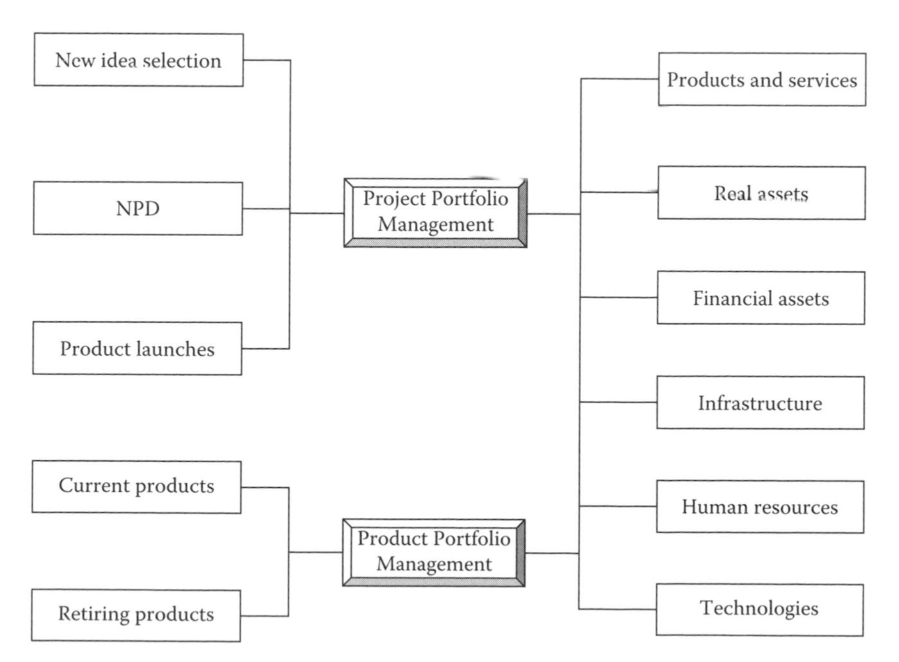

FIGURE 16.4 Prioritizing portfolio rebalancing proposals. (Adapted from notes taken in a 2001 Oracle Corporation portfolio management forum presentation.)

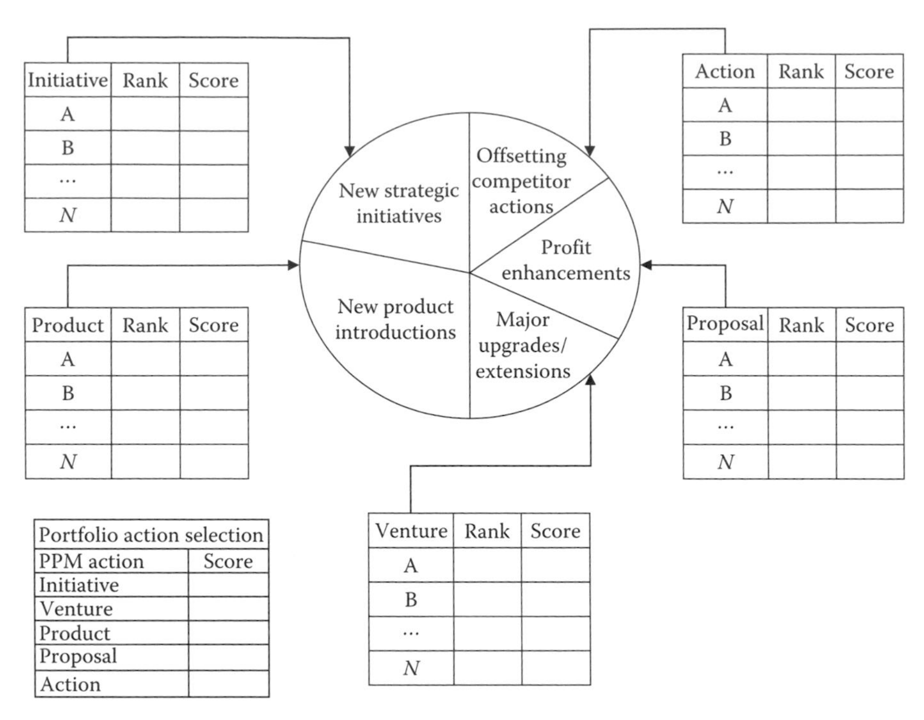

FIGURE 16.5 Example of strategic buckets evaluation chart.

making portfolio management decisions. Other pictorial tools and charts, such as Figure 16.5, show the proportion of projects supporting strategic objectives by size of project, type of the project, or ranked projects by some portfolio balance criteria.

E. *Roadmapping.* It is the creation of summary time-based matrix of information to support PPM objectives and decision processes. Its rows list the key areas impacting a project, such as product development, technology, industry structure, and internal capabilities and competencies. The columns are project definition and strategy (know why), direction (know what), tools and means needed (know how), and action plan (how to do). Roadmapping is developed using input from SDF team members who provide information and intelligence, sanity checks, and insights in this process, and the project team outlines the migration path that brings the portfolio from where it is today to the desired future state. When used as part of a strategic planning operation, it can foster innovation by elements used to address future technological needs or market demands. It also makes it possible to communicate long-term strategic plans throughout the organization in a consistent format, resulting in a unified and synergistic portfolio rebalancing vision (Garcia and Bray 1997).

F. *SDF portfolio validation matrix.* Validation is needed to help the project team prioritize and schedule, account for costs, and apply resources across the portfolio efficiently. This tool is used to track changes in a portfolio as changes in the operating environment take place. The columns of this matrix are the long-term strategic objectives, the alternative projects and proposals, the evaluation results, the prioritization of projects, the rebalancing when new projects enter the queue, and the implementation requirements. The rows of the matrix may include the following elements:

1. Portfolio management goals and objectives
2. Project risk structure, profile, and preference
3. Expected project and portfolio NPVs driven by market demand, pricing, sales support, and competitive actions

They may also include non-risk-adjusted costs and expected peak sales, expected and maximum use of available capacities and portfolio productivity, realization of risk threats and uncontrollable events, changes in market conditions, technology, introduction of new projects, and availability of critical resources impacting project prioritization.

G. *Project scoring table.* This is a PPM technique sometimes used in conjunction with the prioritizing rebalancing proposals chart. It is considered an effective tool to reach decision consensus, and many of the inputs and sanity checks required to populate it come from SDF team analyses. The rows of this table are scoring criteria for projects such as strategic fit and alignment, product advantages, market attractiveness, technical feasibility, uncertainty and risk, expected returns, and other relevant criteria. Column-wise team members rate each rebalancing alternative project along each of the criteria on a 1–10 scale, average scores are computed, weights are assigned to each criterion, and weighted scores are computed and summed up. The scores of projects are then compared and ranked in a descending order.

H. *Strategic buckets chart.* This chart illustrated in Figure 16.5 is a technique used to determine how each type of portfolio management project supports the strategic objectives. It is also used to demonstrate the balance of portfolio components according to major corporate or departmental objectives. The different project buckets show projects of various types, such as offsetting competitor actions, profit enhancement initiatives, new product introductions, and major product extensions and upgrades.

16.6 SOUND PRACTICES AND SDF TEAM CONTRIBUTIONS

Top performing companies emphasize the link between project selection and business strategy. When conducted and implemented properly on a regular basis, portfolio management is a high-impact, high-value activity because it communicates priorities and brings focus, links project selection and business strategy, and enables objective project selection. Portfolio management relies heavily on external environment and competitive analysis, market research findings, and market and individual product long-term forecasts. Thus, a key success factor is for SDF teams to maintain functional links and personal relationships with counterparts in Strategic Planning, R&D, Product Development, Business Development, Marketing and Sales, Product Management, and Finance.

16.6.1 Sound Practices and Success Factors

The themes of skills and discernment, sound processes, analyses and evaluations, and communication and relationships are universally thought to be key success elements. The factors and best practices considered essential to success by PPM experts and project participants in our discussions are shown by their respective areas of concern.

A. *Value maximization and efficiency.* The accepted principles of PPM are stated in terms of value maximization, balance across projects, and linking the portfolio to business strategy. They are achieved through application of best practices, and project teams commonly look for projects to be aligned with company objectives and the portfolio to contain high-value projects, that is, high profit-generating projects or high customer satisfaction products, investments to reflect strategy, projects done on time to capture market opportunities, and the portfolio being balanced and having the right number of projects. Some practices related to efficiency are stated in the following requirements: have a strong senior PPM champion, use appropriate tools and techniques beyond NPV, and iterate, validate, and conduct sanity checks, but being efficient in building processes and defining criteria and scoring guidelines. In all projects, include strategic forecasting experts, keep the project team small, select a few criteria, and always challenge assumptions, models, scenarios, and resulting forecasts.

B. *Improving the PPM function.* The success factors that lead to improving portfolio management are captured in the following practices:

1. Maintaining a sharp focus on defining a PPM problem and having clear objectives
2. Involving senior management and assigning the most knowledgeable and best qualified people to the portfolio management team
3. Tailoring the analysis and evaluations to the current portfolio management problem and creating sound end-to-end processes to deal with it

Besides these practices, data integrity and reasonable and tested assumptions, use of a consistent gate process, and insertion of sanity checks throughout the process are essential. But, adopting an incremental commitment or options approach is also an important element to PPM project success. Also, knowing when to walk away or look at alternative projects, using a scenario planning and forecasting approach, and triangulating forecasts are vital since there is no single best way to pick successful projects.

Other success factors in improving the PPM function include creating scorecards using unambiguous criteria to judge success, defining reliable indicators, and using the right financial valuations with skill and discernment. And, in addition to ranking tables, learning from competitors and the best in business and building in portfolio reviews to rank projects are crucial factors. It is also important to use the most qualified strategic decision forecaster and an expert implementation manager in every project and have clear review procedures. Also, utilizing objective sounding boards, objective third-party opinions, and competitor and industry experiences to facilitate communication and training are essential to enhance the company's portfolio rebalancing capabilities.

C. *Portfolio risk analysis.* Portfolio management is a dynamic, ongoing process, and to be successful, the SDF team conducts ongoing uncertainty and risk and impact analyses for each project. This is the determining success factor for some senior managers involved in PPM projects, and in their views, decisions should be based on the projected impact on the portfolio and whether the decisions are in line with strategy. Because project decisions influence portfolio decisions and vice versa, the project team should be aware of the questions regarding project decisions: how a decision impacts the existing portfolio, how it affects the growth strategy, and what it means for the future portfolio.

Factors expressed as the most contributing to sound risk management are also considered overall success factors in PPM projects: must have a thorough financial analysis and forecasts of a product's NPV created. But, testing of assumptions, demand analysis, forecasting due diligence, product quality assessment, speed to market, and revenue are more important in driving portfolio performance. In certain projects, efficiency and speed in development to meet expected customer demand are crucial. Also, the product's market opportunity or market viability must have a critical assessment of risks and the company's ability to implement its growth strategy through portfolio rebalancing. In other words, decisions need to be made on product cost and profitability adjusted for risk.

D. *Right projects and implementation.* On the issue of how to choose the right projects, the opinion of PPM experts of what makes a difference includes the following factors: an integrated strategic planning and portfolio management process, transparency in the project team and decision-making process, and good forecasting, risk analysis, and project evaluation processes. Rigor in developing, questioning, and getting agreement on the assumptions is essential and involving the right people on the project. A different set of success factors determines project success according to the same experts and includes the following:

1. Portfolio management adjustments must meet corporate strategic objectives.
2. The evaluation and rebalancing process should be standardized to allow meaningful comparison of assessments performed in competing projects.
3. Budgetary and human resource availability constraints are removed in order to achieve results efficiently.

E. *Best practices*. In this category, best portfolio management practices are captured in the following statements:

1. The purpose and goals of portfolio augmenting or rebalancing projects need to be clearly defined.
2. The team should evaluate evidence from different sources and the perspectives of different experts.
3. The background, context, goals and objectives, and evidence need to be linked to allow for assessment of portfolio effectiveness and performance of individual projects.
4. The analytics, evaluations, and long-term forecasts must capture the dynamics of market growth and environmental changes.
5. The evidence should be objective to establish a correspondence between portfolio activities and real-life experiences through reality checks.
6. The project stakeholders determine the goals to be met and what evidence and forecasts to use because they are responsible for portfolio success.

There are five more factors considered best portfolio management practices and are expressed as follows:

1. Agreeing with project stakeholders on the strategy for growth and on criteria and milestones early in its development
2. Establishing clear, quantifiable, go/no-go criteria agreed by the project team and management
3. Assessing results critically and objectively and conducting surveys covering satisfaction, quality, value added, and transparency of decisions
4. Encouraging team members to give objective feedback, specific examples, and ideas on how things could be done better
5. Getting feedback from stakeholders, evaluating it, and implementing it in the next portfolio rebalancing project

F. *Winning portfolios*. In the discussion on successful portfolios, the attributes of winning portfolios include the following sound practices:

1. Internal processes allow more projects to make it from concept to market faster, but to increase success likelihood, projects need to match and leverage company competencies.
2. The number, nature, and timing of projects are balanced with the company's existing resource capacity, and product launches are spaced to allow for attention to quality and have support requirements identified by scenario planning and strategic forecasting.
3. PPM projects are well balanced by type and projected value created for major upgrade and extension investments and profitable new products.
4. Company resources are strategically allocated to new products and important innovation projects.
5. Project process, analysis, and evaluation duplication are minimized and cost reduction synergies are well coordinated among key stakeholders.

G. *Operational distinction*. The success factors for achieving operational excellence are captured in the following requirements of industry experts:

1. Clearly defined criteria, well-defined strategic forecasting inputs, and realistic financial goals are a must to evaluate and implement projects.
2. Well-communicated growth strategy, vision, and strategic intent drive all portfolio management activities.
3. Scenario-tested market attack plans and product and technology roadmaps incorporate environmental forecasts.

However, these factors must be accompanied by a strong team approach to managing the product portfolio with fully engaged executives and supportive decision teams. They also require use of the strategic buckets technique to drive the evaluation and prioritization criteria to rank projects, and timely data and analysis, alternative scenario-based forecasts, identification of sources of risk, and risk management.

H. *Regional government lessons.* The portfolio management performance lessons shared by regional governments are indicative of several factors to be mindful in making successful physical asset portfolio adjustments. First, portfolio management in public entities is not much different than in private companies; they all have similar problems and similar solutions. Second, senior management objectives should cascade down the organization creating champions and a team approach to achieve the project mission. But, it is crucial to create reasonable assumptions, to test them and assess their implications through scenario analysis, and to get project team and senior management buy-in. It is equally important to listen attentively to the views of customers and portfolio management stakeholders, and to what line managers have to say. Also, the project team should adapt, not adopt; that is, make a best practice work for the specific portfolio rebalancing situation and keep in mind that project team partnership and collaboration with customers, stakeholders, and subject matter experts lead to quick portfolio adjustments and success.

16.6.2 SDF Team Contributions

Experienced portfolio managers are not happy with the current state of forecasting support for their projects and actively seek support from specialists experienced in forecasting for strategic decisions. They express the view that from experienced strategic forecasting groups they get rigorous, balanced, and objective analyses and evaluations needed to generate sound forecasts and make rebalancing decisions. The contributions of SDF teams are embodied in the opinions of portfolio managers and subject matter experts in the following categories:

A. *New product strategy and objectives.* In helping the project team to develop a new product strategy to augment the current portfolio, SDF team members add value first by ensuring that the link between the new product goals and objectives and the company's growth strategy is supported by demand data, competitive intelligence, and external environment assessment. Their help in articulating the assumptions, conditions, and requirements to make the new product goals achievable is essential. Defining the critical areas to concentrate on efforts and helping to develop product attributes and marketing support that would make a product superior to the competitors are significant contributions. Also, clearly communicating the new product forecasting goals, processes, methods, and results and how to use forecasts appropriately help the project team in using strategic buckets, creating product roadmaps, and assisting the team avoid the NPV trap. Another significant contribution is in helping to obtain project support and long-term senior management commitment through effective communication of forecasts and their implications.

Value added by SDF teams that contribute to the project team's success is stated in the following ways:

1. Determining the information needs and performing critical studies for all priority projects
2. Performing the analysis and evaluations and developing scenarios and forecasts for business cases to fund projects that create the most value
3. Identifying objectively projects that have to be abandoned or licensed out

These contributions make it possible to create realistic objectives. But, what is valued particularly well is the SDF team's determining the sources of project risks, creating an EWS, and helping create a plan with strategy and tactics to manage portfolio risk and respond to competitive threats efficiently.

B. *Rigor of analysis.* A common statement showing the value of SDF team contributions is that "...their participation in our projects enhanced our project selection capabilities by bringing rigor in the process." How did they do it? They did it by

1. Having structured methods to evaluate and forecast for different PPM projects and using reason, prior project experiences, and best practices.
2. Challenging assumptions, data, criteria, scenarios, forecasts, and their implications and getting project team support.
3. Performing sanity checks in every stage and grounding for decisions in each gate.
4. Explaining the results of analyses and findings of the forecasting due diligence throughout the process.
5. Engaging the project team in the scenario development, sensitivity analyses, and simulations which result in reaching forecast consensus effectively.

In addition, their input, tools, and guidance in defining indicators, monitoring performance, and conducting thorough variance analysis helped the project team assess the success of portfolio rebalancing plans. But, leading the project team in the competitive response and forecast realization planning, helping in assessing project uncertainty, and creating an EWS for effective risk management are most valued. Lastly, assisting the portfolio manager to conduct project postmortems and distill lessons learned are contributions above expectations.

C. *Project selection.* The contributions of SDF team members to the portfolio project selection area are due to their specialized expertise and support they provide portfolio managers in

1. Assessing project fit with corporate business strategy through collaborative, critical, and objective analyses, data, and evaluations.
2. Performing objective SWOT analyses to determine the firm's capabilities and competencies vis-à-vis project needs.
3. Generating and managing assumptions, providing analytic support, and producing scenario-based long-term demand forecasts.
4. Helping Finance produce accurate return-on-investment (ROI) evaluations for each project.

Also, identifying sources of risk and assessing the likelihood of occurrence for each project using a structured approach, and helping manage risks as they appear are very helpful. Furthermore, demonstrating the use of scenario analysis and project prioritization techniques for meaningful project ordering makes a big difference. Yet, evaluating technology forecasts and competitive intelligence for project time and cost estimates helps to generate more profitable solutions.

D. *Best practices.* Bringing best practices to PPM projects in a professional and effective manner is a highly valued SDF team involvement because they lead to higher success rates by

1. Helping the team to understand the sources of potential value through portfolio management forecasting best practices.
2. Identifying discrepancies of project objectives with those of corporate strategic planning based on evaluation of performance targets, assumptions, and forecasts.
3. Enhancing the current portfolio assessment and rebalancing processes by introducing practices used effectively by other companies in the industry.
4. Creating an Intranet database and made available models to the project team to do scenario planning for PPM implementation.
5. Introducing system dynamics modeling and forecasting methods, techniques, and sanity check points in all phases of the PPM process.
6. Defining key success factors and support requirements for successful product upgrades, extensions, and introductions requiring substantial capex infusions.

7. Freeing clients of the responsibility of documenting, updating, and managing the assumptions, the process and tools, and timelines and deliverables.
8. Providing the framework and measures of performance and conducting variance analysis and quality monitoring audits.
9. Helping to perform project postmortem analysis, sharing lessons learned, and incorporating them in the PPM process for future rebalancing cycles.
10. Increasing team collaboration in the competitive response and forecast realization phase and providing actionable recommendations.

E. *Contributions across projects*. The SDF team contributions when brought in early in PPM projects benefited project teams by

1. Proving and improving the value of the product line through their analytics, especially the evaluation of the impacts of megatrends and environmental changes.
2. Increasing the transparency and accuracy of unit forecasts and total product costs.
3. Helping to manage unit costs to industry benchmarks and enhance overall productivity through insights from competitor experiences.

The SDF team recommendation of common performance metrics across all products, increasing alignment of product development functions and their identifying conditions, and supporting requirements for faster development cycles are instrumental in having more profitable new product launches. Finally, their conducting "what-if" scenario analysis and providing sanity checks help the bottom line by leading to improved capacity utilization through better long-term forecasts.

Important lesson learned in PPM projects:

- *Different projects have different decision context and objectives that forecasters need to understand well.*
- *Different projects are supported by dissimilar project team competencies and capabilities.*
- *Different project stakeholders have different needs and interests and assign different weights to success factors.*
- *Senior management and client priorities do not always coincide with those of analytical results.*
- *Project participants come from different cultures and have diverse mindsets.*
- *There is no single best way to do things and rank success factors and contributions.*
- *Models work well in evolutionary, but not that well in revolutionary environments.*

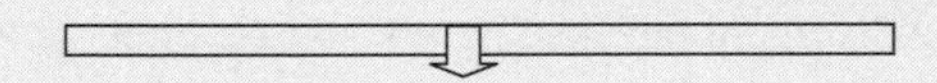

- *Take participant views of success factors for what they are worth; what works for some cases may or may not work in others.*
- *Success factors, like democracy, cannot always be transplanted successfully; they need to be adapted.*
- *Don't make another company's success factors your own before testing their impacts and understanding their implications.*
- *You must create your own objective success factors in each and every strategic project because you will be judged by them.*

17 Negotiations and Investor Relations Support

Two key elements of strategic project success are successful negotiations to meet stated project goals and objectives and effective communications with the investor community. They are important because they ensure maximum value obtained and because they help communicate the significance of the project to the outside world. In the current paradigm, they do not receive attention from forecasting groups and are given no support by them. Under the strategic decision forecasting (SDF) paradigm, however, SDF teams play an important role supporting negotiations behind the scenes, making valuable contributions to the development of the project communication plan for investors, and shaping the project message to the outside world.

Negotiation is a process intended to compromise through exchanging positions and reach desired outcomes. It takes place in order to obtain understanding, resolve issues and differences, create trade-offs and bargain for gains, satisfy interests of the parties involved, and reach agreement at the end. To that extent, negotiations first require sound assessment of where the two parties are today and of the internal and external environments they face. They also require cognizance of where they are headed, how they can get there, and what each company needs to get out of the negotiation process.

Once negotiations are well under way, the Investor Relations group is informed of negotiation particulars and agreed to terms and conditions in order to understand what the project means to investors and create an effective communication message. Investor Relations is a corporate function responsible for handling investor community inquiries, providing guidance about expected future company performance, and briefing industry analysts about important company developments, strategic initiatives, and projects. This function is especially important when projects of strategic importance are undertaken that need to have important points properly explained to shareholders, industry analysts, and the broader investor community.

A key requirement for effective negotiations and investor communication plans support is ongoing, careful screening, testing, validating, and verifying of data, information, models, scenarios, and forecasts. Another requirement is a synthesis and integration of all the analyses and evaluations performed by the SDF team. In fact, effective negotiations and investor communications support are not possible without ongoing confirmation and integration of the pieces of work the SDF team touches. This takes the form of an independent verification and validation (IVV) step in the forecast and other strategic project processes. In Chapters 8 through 16, we discussed the evaluations and analytics performed by SDF teams in different kinds of strategic projects. In this chapter, the focus is on ensuring the validity of the work performed and the resulting forecasts in support of negotiations and the Investor Relations group.

How is IVV different from the forecasting due diligence and how are IVV, negotiations support, and Investor Relations group support related?

- *The forecasting due diligence uncovers information that impacts the development of SDF solutions with the objective of identifying, quantifying, and minimizing uncertainty and project risks.*
- *IVV confirms if the right approach, data, models, and scenarios are used, whether processes are followed, and the SDF team activities were performed the right way.*
- *In the support of the SDF team provided in these areas, the forecasting due diligence is used along with the IVV effort to generate supporting evidence for business case development and corporate approvals. But, part of the business case, supporting evidence comes from the outcome of project negotiations and key elements of the business case support are, in turn, inputs to the Investor Relations group's project communication plan.*

17.1 INDEPENDENT VALIDATION AND VERIFICATION

The validity of long-term forecasts and recommendations is partly based on verification and validation of each part of the SDF process, methods, models, scenarios and tools used, and their outputs. Verification is the process of determining if the informational inputs and judgments are correct and complete; the models, analyses, and evaluations comply with professional standards and client needs; and the SDF solutions and recommendations are consistent with the SDF team's values and standards. The output of the verification process is to provide evidence of correctness for each element. On the other hand, validation in strategic forecasting is the process of evaluating conceptual and computational models and plausible scenarios during their creation or at the end of the SDF process to determine whether they satisfy the specific requirements and meet the project goals and objectives.

IVV of forecasts for large impact decisions is an objective, sound, and trusted approach to ensure that long-term forecasts are developed that meet the needs of the strategic decision and are created with the right forecasting processes, methods, assumptions, models, scenarios, and data. It is a required total quality management (TQM) approach, which involves an in-depth scrutiny conducted by the SDF team to make sure that all the components of analyses, evaluations, and projections lead to well-grounded decisions. The purpose of the SDF team's IVV is to answer three questions: Was the forecast solution developed right, was the right forecast developed, and do we trust the forecast enough to use it to make decisions based on it? It is a multifaceted quest that entails several broad reaching considerations:

1. Confirming the validity of inputs to the development of long-term forecasts, which includes understanding of company operations, processes, data, assumptions, methodology, models, and scenarios
2. Substantiating the validity of the outputs from forecasting processes, which include the results of analyses as well as evaluations of their implications
3. Identifying and correcting defects in processes, methods, and models and errors in data, assumptions, and results in a timely fashion to prevent wrong conclusions and decisions
4. Determining the root causes of defects and errors and initiating an effort to address them effectively and improve the SDF function
5. Ensuring that the best decision options are developed through the use of sound methods, tools, scenarios, and future-state forecasts
6. Establishing forecast credibility, increasing client trust in the achievability of the forecast, when the appropriate support is provided, and reducing time to decision

7. Increasing client and senior management support by delivering forecast solutions with a high degree of confidence
8. Providing accurate assessments and judicious advice on the quality and readiness of a project to move to the next decision gate

Strategic forecast IVV is an ongoing review of the SDF team's work, it is a time-consuming endeavor, and the degree to which is applied is determined by the importance of the decision to be made. Thus, SDF team's IVV activities derive their importance from the intended use of forecasts in strategic project decisions. The basic principle followed by SDF teams conducting an IVV is the saying from the software development area: "Too little testing is a crime, too much testing is a sin." Thus, careful consideration is given to the IVV timing, focus, process, and reports created. This carries over to a well-considered definition of the role of the SDF team in project negotiations support and investor relations guidance formulation.

The SDF team's approach to the project forecast IVV parallels that of creating strategic decision forecasts and its elements are outlined in Figure 17.1. It starts with external assessments based on 5Rs (recognition, relevance, receptivity, response, and relationships). It proceeds to 4P (product, price, place, and promotion) evaluations and then addresses questions about (a) systems, processes, techniques, and tools used; (b) organizational capabilities and project support; (c) project management competencies; (d) uncertainty and risk mitigation plans; and (e) performance measures to judge how well the analyses and assessments are performed. The IVV process consists of the following elements:

1. Outlining an efficient and well-balanced IVV plan to address questions and concerns of decision makers
2. Validating the real business needs and client wants, identifying the decisions that need to be made, and determining the SDF solution needed

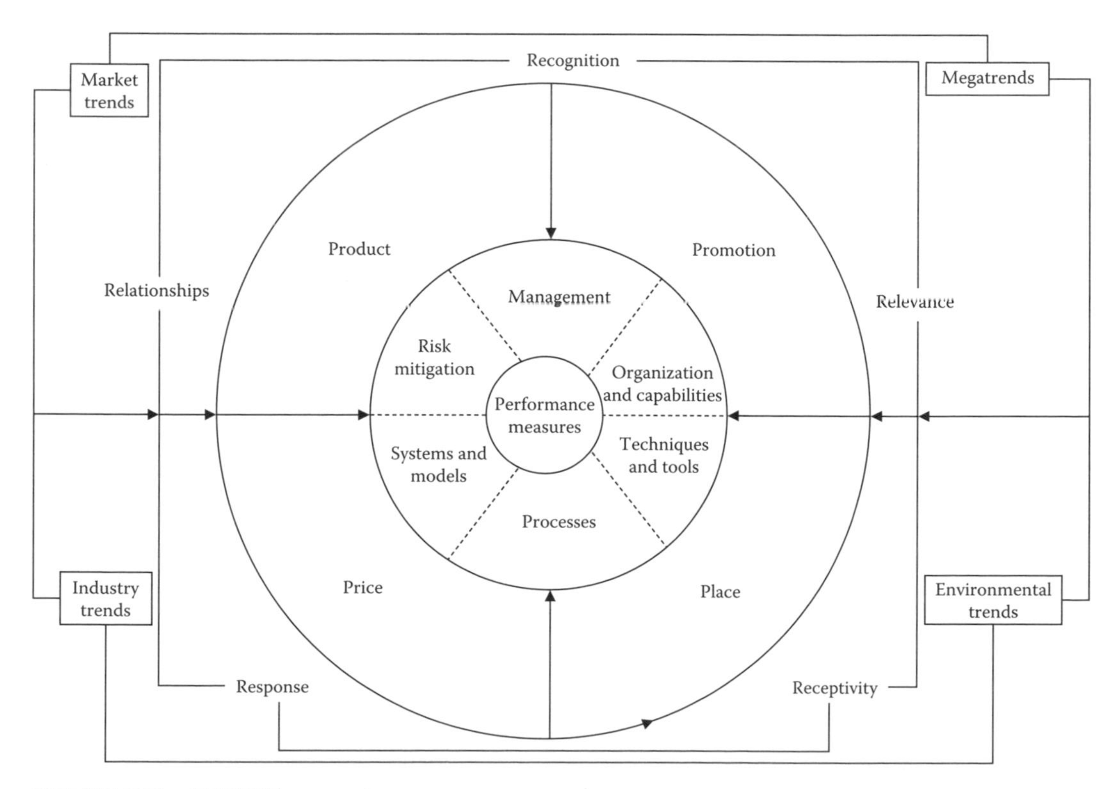

FIGURE 17.1 SDF IVV approach.

3. Establishing that the right project participants, expectations, objectives, and mindset are present in the project in a critical manner
4. Verifying the project rationale and structure vis-à-vis corporate strategic goals and objectives as the project moves along the evaluation process
5. Validating the methods and inputs to forecasts, reevaluating their use in decisions made in different phases of the project, and reexamining their implications
6. Establishing that processes used are complete and consistent and verifying and validating models, data, assumptions, analyses, findings, scenarios, and forecasts
7. Determining that assumption or forecast changes made in handoffs from one project stage to another are appropriate and properly recorded
8. Verifying the quality of project management, information and decision record keeping, and the accuracy of communication of findings
9. Producing an interim report, sharing it with key stakeholders, and obtaining their input, feedback, and comments
10. Recalibrating IVV findings and producing a final report, which contains specific recommendations for project decisions and improvement of the entire strategic forecasting function

In some instances, external experts may perform the forecast IVV function, inform senior management about potential problems, and provide recommendations on how to deal with forecast risks. In other instances, the IVV function is performed by SDF teams on long-term forecasts generated by external advisors. Here, their focus is on detecting inconsistencies in assumptions, biases in methodology, and overly optimistic scenarios. Well-established SDF teams use IVV for the purposes of ensuring the following:

1. Reasonableness of assumptions, methods, and models used and quality of company and environmental assessments, industry data, and market intelligence
2. Adequacy and completeness of processes followed from observation to the creation of SDF solutions and recommendations
3. Application of stringent criteria in performing sanity checks and the use of benchmarks and standards in the development, testing, and evaluation of reasonable and feasible scenarios
4. Identification of forecast risks, their evaluation, and recommendations to offset their impacts that is completed and the recommendations derived from SDF solutions that are implementable
5. Consistency of reasoning and transparency of decisions and validation of the future-state realization (FSR) plan

Areas where the SDF team's IVV is applied to, other than projects, negotiations, and investor relations guidance, include the following:

1. Studies conducted by third parties such as industry groups or external advisors
2. Results of evaluations performed by other internal organizations
3. Forecasts underlying different options or earlier decisions
4. External experts' recommendations to senior management for strategic projects
5. Internal decision stalemates due to differences in future-state outlooks
6. Project stakeholder conflicting strategies, goals, and objectives
7. High uncertainty and risk decisions based on market research studies

Special issues and challenges associated with the SDF team's IVV center around how to evaluate changes in uncertainty and risk levels with key assumption changes, focusing on identification of key project issues and prioritizing their importance, and defining IVV analytic requirements and corresponding performance criteria. Validating project participant's market familiarity and the company's way of doing business is a sensitive issue. Also, dealing with qualitative assessments of resources allocated to the project, availability of qualified employees, and senior management support levels are also a challenge. Lastly, judging performance relative to past projects and industry benchmarks, standards, and common practices requires extensive project experience not readily available.

The SDF team's IVV techniques used contain the critical thinking tools mentioned in Chapters 6 and 7 including root cause analysis, application of questioning and fact finding, and other analytical techniques. Retesting of hypotheses, assumptions, and model structure and asking "so what" and "what-if" questions at each juncture are commonly used to confirm reasonableness and validity of results as well as using industry standards, benchmarks, and earlier experiences and results. However, in some cases, other more effective but difficult to administer tools are used, such as the following:

1. Duplication of models, scenarios, assumptions, and inputs to obtain results identical to those obtained by other sources
2. Simulation of early warning systems (EWSs) and forecast realization models to ensure that company reactions to competitive threats work as intended

Some of the issues and challenges SDF teams face conducting project IVVs include absence of clear standards and guidelines on how to do it for a strategic forecast and strong interest on the part of the project client. Determining if and how much blind spots and missing things in the SDF process really matter is not easy due to difficulty in identifying them. Also, there are no methods to demonstrate the value and effectiveness of IVV for strategic decision forecasts because it is difficult to determine the range of uncertainty in forecast IVVs themselves. This is true particularly in large residual uncertainty, scenario-based projections of revenues and costs. Furthermore, the high cost of IVV relative to the perceived benefits coming out of it and the timing of IVVs often changed by other project priorities make it difficult to justify it except in large projects. Another challenge is that the cooperation of project participants is inhibited by fears of negative consequences because cultural resistance can obstruct and marginalize the importance of forecast IVV. Criticism for prioritizing the forecast IVV tasks to be performed by plotting their importance and easiness of performing them, and mistrust of ascertaining that absence of problems means that there are no issues to be concerned with and that there are no faults in the IVV process are formidable issues to overcome.

17.2 NEGOTIATIONS SUPPORT

SDF models and solutions are decision-making support tools, not just forecasting models that can be applied in many business areas including negotiations. The main assumption of the SDF team is that negotiations are governed by reasonably fair processes, understandable explanations of positions, and clarity of expectations, and that ownership and responsibility reside with the negotiations team. The four pillars of the SDF team's negotiations support approach shown in Figure 17.2 are as follows:

1. Knowing well the project goals and objectives and what is needed to have successful negotiations
2. Hearing the positions of the other side and understanding the underlying motives, needs, and desires
3. Providing analysis and evidence for positions taken or against them
4. Exploring and developing trade-offs and evaluating the costs and benefits of options and helping to assess offers and counteroffers by the other party

The other elements of this model include factors common to SDF team involvement in other strategic projects and the culture and mindset elements required to develop SDF solutions. Notice the central role of identifying correctly what needs to be known in order to provide excellent support to the negotiations team.

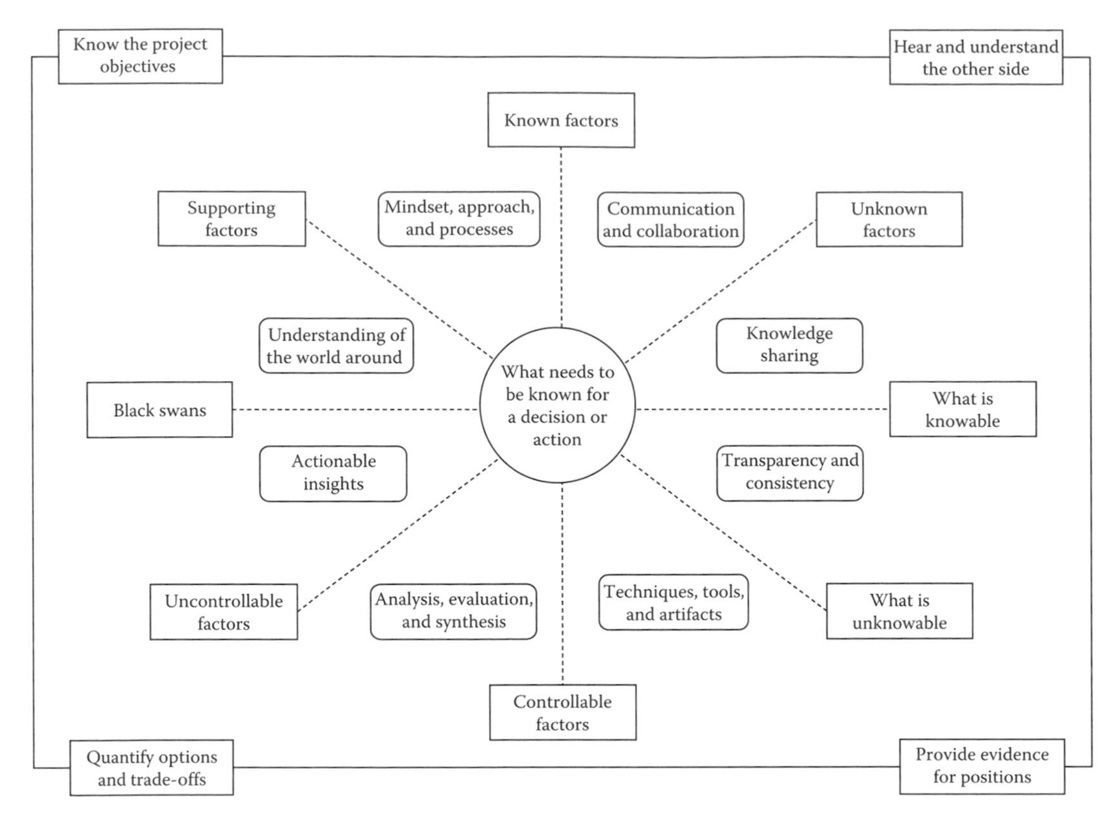

FIGURE 17.2 SDF negotiations support model.

In preparation for supporting the negotiations team, SDF team members develop a good appreciation of what is really needed and desired to be obtained in the project. Then, they evaluate the current state of business and identify gaps present, which provide the basis to build its negotiations support in terms of the following:

1. Clarifying the strategic decision, determining the consistency of project objectives, preparing a list of questions to be answered, and recommending ways to address them through the negotiations process
2. Addressing competing priorities of internal stakeholders and balancing competing interests of the negotiation parties in the support they provide
3. Answering the questions: What does it take to realize the desired state, what results should come out of negotiations, what is needed to obtain the desired outcomes, and what is the company willing to give up or trade in the process

The SDF team's way of delivering negotiations support is based on understanding the context in which the company and the other party to the negotiations operate in. However, this support entails several factors and challenges in the following areas:

A. *Knowledge identification.* Actionable SDF team insights are developed systematically following a structured procedure of determining first what needs to be known, what is known and unknown, what is knowable and at what cost, and what is unknowable. Then, factors driving the project value creation, interrelationships, and feedback loops are identified. Effort is concentrated on reducing the unknown factors and on the process evaluating what is controllable in the project and what is not and how to leverage the controllable factors. This enables the creation of some preliminary options for initial negotiation positions. However, while negotiations support is based on analytical models and forecast scenarios, the possibility of black swans is also considered and attention is directed to the task of finding offsetting factors, which would be acceptable to both parties.

B. *Prenegotiations support.* In this phase, SDF team members help the negotiations team understand how negotiation positions can be linked with the scenario models that generate forecasts and get an appreciation of the impact of changes in the drivers on project value. They bring into this process, leverage internal and external relationships and knowledge and insights from them, research companies and competitor precedence, and assess the motivations behind proposals of the other party. They also investigate the impact of negotiation options on product quality, customer acceptance, demand, pricing, costs, and potential risks associated with each option. This enables them to develop alternative negotiation positions by creating and evaluating hypothetical scenarios, testing different assumptions, and providing answers to "what-if" questions. And to demonstrate the likely outcomes of negotiations, they simulate and assess conjectural proposals and counterproposals in order to create reasonable and profitable trade-offs.

C. *Development of support.* The SDF team determines the best approach and processes to follow in a highly collaborative and communicative fashion and with a problem-solving mindset. It exchanges information and shares intelligence about the external environment and of competitor positions and activities, and the results of its own analyses and evaluations with the negotiations team in a transparent manner. Practical techniques and tools are also shared and the implications of the analyses and evaluations are fully explained. Then, the SDF team synthesizes its findings and translates them into actionable insights and recommendations regarding the sensitivity of the company's product, service or technology demand, pricing, and cost to factors that can be affected by negotiations.

The SDF team helps to clarify company positions based on its project assessments, competitor experiences, and industry standards and norms. It also identifies items that can be exchanged or traded, their value, and what concessions of equal or greater value the company can get in return. And, it assists in the evaluation of how the other party's proposals and counterproposals impact the expected value of the project. In some cases, the SDF team creates high-level models of the other party and simulates different positions to develop a better understanding of the impact of changes on the status quo of negotiations.

D. *Elements of support.* Business negotiations involve give and take over a number of areas and issues and successful negotiation plans involve the creation of a negotiation agenda and a range of outcomes for each item on it. In most strategic projects, negotiations are conducted by senior managers responsible for the outcome with no involvement of the forecasting organization. However, in companies where SDF teams are well established, their support applications are used in the following manner:

1. Creation and evaluation of forecast scenarios representing the opening position, which is a desired (often being the best case) and defensible position
2. Identification of elements in negotiations that can be altered and assessment of the impact of deviations from the opening position on project value
3. Description of events and development and evaluation of the walkaway position (worst-case) scenarios on key issues
4. Creation of backup alternatives performed jointly with the negotiations team and ranking their value and importance
5. Development and evaluation of trade-offs based on forecasted value changes from some anchor scenario or position
6. Evaluation of proposals from the other party or changes in the current terms and conditions

E. *Skills and qualifications.* The role the SDF team plays in negotiations and the support it provides are determined by the organizational position and governance structure of the SDF team, its functional links with other corporate-wide functions, and the personal relationships with clients and counterparts in those organizations. However, other considerations decide the acceptance of SDF team members as partners in project negotiations. They include considerations such as knowledge of the company business, understanding of the decision-driving forces, aptitude to identify key issues impacting negotiations, and ability to negotiate with project team members and win their support. The SDF team members must demonstrate negotiations experience and qualifications to participate, facility to listen to client positions with an open and constructively critical mind, and skill to take note of the other party's points. Other SDF team participation criteria are good business judgment, willingness to probe and verify information and interpret findings, and self-confidence, insightfulness, and ability to uncover underlying factors.

Wisdom in order to be constructively involved with and understand the negotiating partners' issues, needs, and objectives and tolerance for uncertainty, changing targets, and ambiguity are essential as are commitment, integrity, and unwavering team support. Negotiations support is a difficult task, but nowhere is the tangible return on SDF team support as high as in this effort. In every case, it requires a combination of diverse traits and associate skills and the process of negotiating demands through knowledge of the business. This type of SDF team support also requires good business judgment, a keen understanding of human nature, and ability to see motives, which are skills not evenly present across all SDF team members. And, it has been said that besides negotiations, there is no other business area that is characterized by lack of experience in the art of persuasion, finance, motivation, and dealing with organizational pressures that come together in a concentrated fashion in a short time frame.

F. *Support challenges.* Negotiations teams sometimes negotiate on behalf of several organizations, yet little attention is given to overcoming the challenges posed by diverse team dynamics by using the insights and analytical support of the SDF team. Ordinarily, negotiations support is delegated exclusively to Finance associates, and in these cases, there is a need to justify SDF team involvement. This practice leads to wrong positions and trade-offs because Finance groups and project negotiators have partial, second-hand information: They do not fully understand all the assumptions, models, and scenarios underlying the forecasts, how they can be impacted by negotiated terms and conditions, and how forecast models can be used to fully assess the impacts of changes in a future-state scenario due to negotiation concessions.

Another challenge in SDF team negotiations support is that credible information and a good understanding of the other side's needs are hard to obtain as is assessing the differences in the company's and the other side's position on every objective they wish to accomplish. Hence, sometimes the SDF contribution is limited to helping define concessions and absolutes, determine justification and supportive reasoning for them, and develop trade-off approaches to solve key negotiation problems. This problem is accentuated when individual negotiators are not ready or able to understand the other party at all due to time constraints, and as a result, they go unprepared into negotiations with a "just wing it" attitude.

Other issues in negotiations support include that company negotiators are usually in a reactive mode and responding to the other party's positions, lack of patience, criticism, and derogatory remarks coming from them to test the negotiations team. Negotiators being too rigid and avoiding last-minute changes to prepared positions are an obstacle to helping them obtain gains in negotiations as is lack of confidence to press for positions outlined in the negotiations plan and strategy. Lastly, when many parties are involved in the negotiations, the complexity of assessing motives and trade-offs is increased exponentially. In that case, and especially when the negotiations team faces a coalition of two or more negotiator parties, it is very difficult for the SDF team to create trade-offs that are acceptable to all parties at the same time.

17.3 INVESTOR RELATIONS SUPPORT

Investor Relations is a corporate support group usually reporting to the CFO or the Treasurer organizations and is often part of and managed by the corporate communications department. It is a strategic function responsible for integrating pieces from forecasting, finance, communication, marketing, and securities law compliance. This integration is essential to achieve a fair valuation of the company's securities through effective, two-way communication between the company and its shareholders, the financial community, and other constituencies. The Investor Relations group's main functions are to handle inquiries from investors and shareholders and provide guidance about expected future performance.

In large companies, Investor Relations groups provide consistent and credible information about corporate strategy, financial performance data, and future strategies and guidance to industry analysts and the investment community. They are proficient in piquing long-term investor interest in the company and they have the ability and competence to develop and retain good relationships with them. It is an important function that, if not maintained properly, can become a liability. For these reasons, the involvement of the SDF team is needed to help them in terms of the following:

1. Understanding the context of strategic projects or decisions and the key elements involved in creating long-term forecasts that support them
2. Ensuring understanding of key assumptions and the rationale behind them, data and their sources, methods and models, and the processes and scenarios used to derive forecasts that drive decisions
3. Getting a good appreciation of the project uncertainty, the risks involved in the forecasts and their sources, and the company's risk mitigation plans to address threats
4. Familiarizing them with the scenario used to describe how the company moves from the current to the future state, what it takes to achieve the desired state, and the validations and sanity checks conducted
5. Identifying possible forecast risk-related questions that may be raised in their presentations and coaching them on how to explain key forecasts and answer questions about them
6. Anticipating likely competitor reactions and the impact of announcements of strategic decisions and projects and providing satisfactory explanations for regulatory compliance requirements and forward-looking statements
7. Facilitating the shaping of a message from a senior management perspective, which can easily be communicated across the entire organization and externally
8. Making them aware of current and upcoming issues, events, and uncontrollable factors that may impact the strategic decision and black swans that would affect the company's operations
9. Instilling a good appreciation of the FSR plans and how they are to be implemented in the event they are needed
10. Providing fact-based, well-thought-out answers to inquiries that touch on forecast issues in order to maintain industry analyst and investor community confidence

In the past, the investor relations function focused on just the release of financial information, but the trend is that almost everything issued for external communication is now the responsibility of the Investor Relations department. The investor relations function is concerned with any company information that may affect its earnings and stock price, and companies maintain web sites dedicated to investor relations. Earnings releases, earnings forecasts, annual and quarterly reports, and most press releases are all part of investor relations and the responsibility of that department. Consequently, it is very important that information released to investors is correct and timely to ensure that the market place remains a level playing field for all participants. This also explains the need for the SDF team's thorough IVV effort in strategic projects.

The responsibilities of the SDF team in helping the Investor Relations group are determined by the organizational positioning of the SDF team and relationships with clients responsible for strategic projects. Other determinants of the SDF team involvement include the following:

1. The importance of the strategic decision or project, the financial and human resource commitments required, and the expected value created by the project
2. The qualifications, experience in addressing investor guidance issues, and reputation of SDF team members based on performance in past projects
3. The nature of Investor Relations group's needs, whether reactive to address specific or crisis situations or a proactive stance to manage industry analyst and investor community expectations
4. SDF team member ability to explain in layman terms how strategic decision forecasts are created, provide knowledge about market developments and competitor activities, and share knowledge and insights derived from complex analyses and evaluations
5. The need of the Investor Relations group to be coached on the information and insights to acquire in meetings with investor groups and industry analysts, which helps gauge the SDF team's work to provide satisfactory answers that meet the needs of the Investors Relations group

In most cases, the approach of the SDF team support and knowledge sharing with the Investor Relations group is outlined in Figure 17.3. The process begins by explaining the situation, circumstances, and context that created the need for the decision to be made and the project rationale. Then, the extent of strategic and other fits and the technical and operational feasibility are

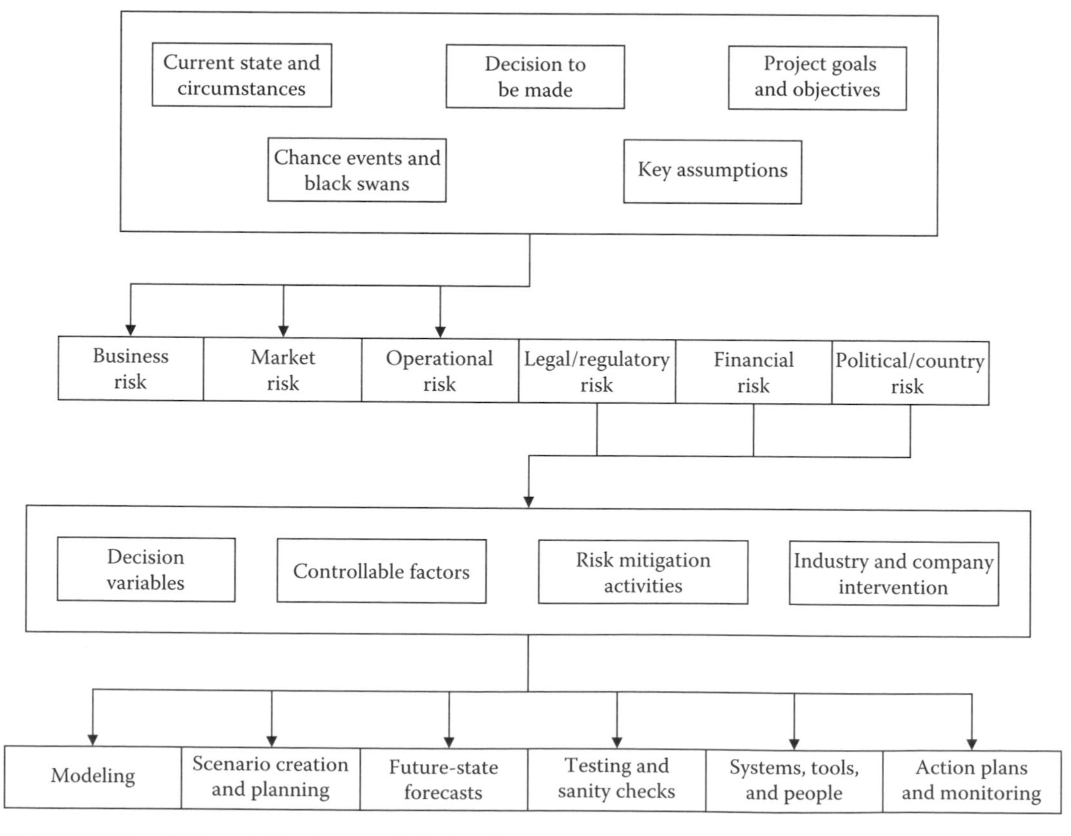

FIGURE 17.3 SDF model for Investor Relations Support.

articulated. Specific project goals and objectives, how and what will help the company realize them, and the expected costs and benefits are explained. However, no explanation of project goals and objectives and benefits is credible without a good explanation of the assumptions made and key factors underlying the strategic decision forecast. In addition, understanding processes, methods, techniques, tools, data, models, scenarios, and internal and external environment analyses and evaluations performed is essential. Also, a discussion of chance events and black swans needs to take place to enhance the Investor Relations group's appreciation of how such events might affect the strategic decision and project success, and the group's ability to handle questions about such factors.

Extensive dialogue between SDF team members and Investor Relations associates takes place about the uncertainty and risks involved in the strategic decision, how total project uncertainty is reduced to residual uncertainty, and how it is to be managed. In this dialogue, questions are raised and answers are provided concerning the presence and potential impact of business, market, operational, legal and regulatory, financial, and external environment risks. Naturally, this dialogue brings in analyses and evaluations related to

1. Megatrend and subtrend impacts on the industry and more specifically on the company's operations and product demand.
2. Political, economical, social, technological, legal, educational, and demographic (PESTLED) conditions, developments, trends, and changes likely to impact the predicted future state.
3. Industry, market trend, and competitive analysis to determine how they may affect the demand for the company's products, services, and technologies.
4. Strengths, weaknesses, opportunities, and threats (SWOT) assessment to determine internal competencies and capabilities to execute the project successfully and ensure that stated objectives are realized.

To further demonstrate that the SDF forecasts are well founded, thorough explanations are given to the Investor Relations group about what the decision variables are, their importance and ranking of impact, and how they may affect each other. A discussion takes place on options of appropriate company and industry intervention to preclude the occurrence of potential risks or, in the event they materialize, their mitigation with specific actions, methods, and activities planned. Once again, uncontrollable factors and those which are controllable and how they will be used to ensure that expected results are discussed. And, to make sure that the coaching by the SDF team is effective in helping the Investor Relations group prepare sound guidance to the outside world, a review of draft releases to ascertain a good understanding takes place. This review involves, once again, the following:

1. The methodology, models, assumptions, and data used to create the long-term forecasts
2. The rationale behind the plausible scenarios created, the selection of the model used to describe them, and the outcome of model calculations
3. The description of future-state forecasts that come out of each plausible scenario and how the most likely scenario is selected
4. The testing of assumptions, models, and scenarios and the sanity checks and evaluations they were subjected to
5. The systems, tools, dedicated people, and financial resources in place along with the senior management support commitments obtained to ensure project success
6. The controls in place to ensure high levels of communication, collaboration, and coordination needed to execute the project successfully
7. The project performance measures and monitoring that needs to take place, the future realization plans prepared, and when and how they will be activated

SDF team support to the Investor Relations group, like any other area of its involvement, has its own issues and special challenges. The most important issue is that SDF team involvement is often late in the process even when clients suggest that SDF team support is needed to make the Investor Relations group more effective. Challenges also come from Finance organizations and other stakeholder groups with widely differing perspectives and expertise. The reason that the latter are a challenge is that investor education entails sharing the company's view of the future, future plans, company projections, analyses, evaluations, and results, which is a major concern due to fears of sensitive information being disclosed. At the same time, management of investor expectations requires producing convincing evidence about strategic project decisions and actions, which is difficult to develop under tight timelines. Additionally, diffusing and dispelling investor fears about company performance, due to market changes and competitor actions, are real worries for the SDF team if the revised guidance is based on ad hoc forecasts that create unfounded expectations.

Guidance to industry analysts requires reasonableness, consistency with past performance, grounding of results on evidence, validity and sanity checks, and well-thought-out arguments. But Investor Relations groups sometimes do with incomplete or dated information. Also, reporting on progress in a strategic project, decision, or initiative requires a good understanding of the nature of the results, the monitoring process, and reliable performance measures. This understanding is not always present. This is a problem because using market relationships, causality, drivers, and controllable factors expectations in investor guidance may involve comparison with industry norms and common practices, which are not always or entirely appropriate. Also, the common practice of guidance erring on the conservative side does not consider the implications in both long-term costs and benefits associated with such guidance. At times, the divergence of SDF team objectives from those of the Investor Relations group is a source of discord, which may carry over to other projects and future guidance support. Lastly, explaining complex technical concepts and the multitude of processes, models, analyses, and evaluations so that they can be understood by layman is a time-consuming task for both the SDF team and Investor Relations groups.

A note on strategic project analyses and evaluations: *Project participants are overwhelmed by talk about processes, methods, analyses, and evaluations needed to assess and implement a strategic project successfully. To put them at ease, we offer some thoughts to keep in mind:*

- *Things do not always have to be complex; only when necessary.*
- *Processes can be as simple as drawing an outline and checking the adequacy and consistency of activities in it.*
- *Methods can be techniques and tools used under certain conditions or a sequence of steps to deal with a problem.*
- *Models are ways to calculate the impacts of relationships and can be as simple as Excel spreadsheets.*
- *Analyses and evaluations may be as simple as comparisons with benchmarks or calculations to examine an issue.*
- *The tools of the SDF trade are applied by SDF team members with ease and effectiveness.*
- *Some strategic project assessments may be fine with simple versions and complex solutions reserved for certain large projects.*
- *Bring out the big guns only when necessary.*

17.4 SOUND PRACTICES AND SDF TEAM CONTRIBUTIONS

Negotiations and Investor Relations support provided by SDF teams in strategic projects are different than that provided in the types of applications discussed earlier. This is due to differences in the kind of help needed and its importance, the nature of goals and objectives in these functions, the company's negotiating position, and the personalities of negotiators and the Investor

Relations associates involved. And, negotiations success comes not from creating a perfect plan, but from sound modeling and forecasts and the ability to adapt to negotiation dynamics. This, in turn, requires feedback assessment mechanisms and structured processes for analyses and evaluations in order to learn, understand, and interpret the situation correctly.

17.4.1 Sound Practices and Success Factors

Over time, SDF team managers, clients, and negotiation experts have found that a number of factors are necessary to successful negotiations support. The most important factors shared in our discussions are as follows:

1. Appropriate SDF team's organizational positioning, close functional links with Finance and Corporate Planning groups, and personal relationships with clients
2. Early involvement of the SDF team in negotiation preparations to get a first-hand understanding of the participants' needs and the interpersonal dynamics, and to demonstrate how it intends to add value to the process
3. Prenegotiations preparation, modeling of positions, flexible negotiation plan and strategy, and a common understanding of project objectives by all participants in negotiations
4. SDF team member ability to transfer knowledge and understanding of key operational performance drivers, how they interact with one another, and how they can be impacted by negotiations
5. Thorough understanding of assumptions underlying expected project value forecasts, how they may differ from those of the other party, and what it means if both parties adopt common assumptions
6. Understanding the other side's needs, goals and objectives, what they must get out of negotiations, and what they may be willing to trade in exchange for some concession from the company
7. Preparing a set of trade-offs that include what is to be given, their value to the party parting with it in exchange for what is to be obtained, their value to the party getting it, and possible uses of trade-offs in stalemate cases
8. SDF team flexibility and capability to assess the implications of proposals and counterproposals for the company's products, services, and technology; its customers base; and demand and pricing
9. Appropriate SDF team member temperament, previous negotiations experience, business acumen, and ability to provide convincing arguments
10. Tested and well-functioning linkage of SDF models, analyses, evaluations, and results with Finance models and joint assessment of implications

The SDF team objective in Investor Relations support is to help that organization do their job better. To provide support of excellent quality, a number of required factors need to be satisfied, such as participation of SDF team members in meetings with industry analysts to gather competitive intelligence and obtain analyst reports and industry and company releases. SDF team member experience and qualifications, close functional links with the Investor Relations group, and skillful interactions are considered essential. They facilitate ongoing dialog and support continuity to understand each organization's processes and how to maximize SDF contributions to investor guidance. As in the case of negotiations support, the SDF team members' ability to explain in simple terms their processes, assumptions, models, techniques, and scenarios used to generate forecasts and demonstrate how project uncertainty is reduced to residual uncertainty, what risks remain, and how those project risks are managed is paramount.

Completeness of explanations and advice through Q&A sessions and testing of the Investor Relations group's understanding of key elements surrounding forecasts is a beneficial practice as is consistent and reliable SDF coaching and advice that translates to effective investor guidance.

Thorough and critical SDF team checking of the reasonableness of the intended investor guidance and explanations given to eliminate gaps and points of contention require intensive preparations of the Investor Relations group. This is needed to be able to describe the scenarios considered and black swans identified so that there would be no surprises in meetings with investor groups and industry analysts. Another helpful factor is ensuring that key talking points developed for presentations and releases are fact based, accurate, relevant, and timely.

Other sound SDF team practices and beneficial activities in negotiations and Investor Relations support articulated by experts in those functions include the following:

1. Understanding of customer needs, product attributes, customer satisfaction levels, price sensitivity, and sales and promotional support needed
2. Recognition of driving factors, measuring their relevance in the value created by the project, and bringing together analyses from other planning groups
3. Expertise, intelligence, and insights made available through SDF team's functional links and relationships with other planning groups and external entities
4. Inferring the impact of megatrends, environmental factor changes, and market and industry trends and developments on the company's future
5. Sharing market research intelligence, competitive analysis, and insights used to support conclusions flowing out of analytical models
6. Determining current and upcoming events, controllable and uncontrollable factors impacting the project, identifying black swans, and assessing their potential impacts
7. Analyses, evaluations, and tests performed in order to arrive at the most likely scenario selection and generate long-term forecasts
8. Providing convincing evidence for the reasonableness of long-term forecasts, recommendations, and the negotiation positions developed
9. Communicating the need for and helping create an FSR management plan
10. Unique ability to perceive and exploit options and possibilities and being a source of trusted advice in negotiations

17.4.2 SDF Team Contributions

A major challenge in supporting negotiators and Investor Relations groups in strategic projects is ensuring effective communication and coordination across different organizations. Due to its organizational positioning, its charter, functional links, and personal relationships with internal and external entities and member qualifications, SDF teams are the logical central point capable of developing a common understanding and model the company's operations and a common set of assumptions used in planning. SDF teams do deliver on these two important project success factors in a significant measure consistently.

SDF team contributions common to negotiations and Investor Relations group support are recognized by experts in these functions in the following areas:

1. Ascertaining where the company is headed by internalizing the corporate strategy, determining strengths and weaknesses, and creating measurable performance goals and targets, which establishes reasonable project goals and objectives
2. Identifying, from a demand perspective, the gaps the project needs to fill by assessing where the company is today and analyzing the internal, external, and competitive environments
3. Using the identified gaps to develop negotiation positions and trade-offs and demonstrating how closing those gaps affects the value created by the project
4. Describing how the company gets to the future state and ensuring that corporate strategy, business plans, and project goals and objectives that are logically consistent, which helps determine scenarios to move the negotiations forward

A number of negotiations support clients stated in our discussions that the value of the SDF contribution in strategic project negotiations lies primarily in its ability to bring together thinking from different disciplines, to model, and to simulate the operations of the business. Another important value-adding activity is the evaluation of positions of the other party's reasonableness of the forecasts they are working with and their foundations and implications. Yet, a third area of SDF team contribution is helping to create an alternative negotiations plan that can achieve strategic objectives outside the range of the negotiations agenda by creating positions to offset surprises by the other party. Naturally, the creation of trade-offs and the evaluation of proposals and counterproposals are of critical importance to all negotiators.

The benefits of the SDF team support to Investor Relations associates who participated in our discussions include the following contributions:

1. Sharing intimate knowledge of competitor operations and performance, external environment intelligence, and evaluations; developing company operations models; and assessing project uncertainty and forecast risks
2. Incorporating investor expectations and feelings related to strategic decisions in the synthesis of their analyses and evaluations
3. Evaluating research obtained by the Investor Relations group, industry analyst opinions, and client newsletters and providing well-reasoned responses
4. Providing a comprehensive assessment of the global economic, political, cultural, and technical landscape that is likely to prevail over the next few years and how it would impact the project
5. Describing as an independent, critical, and objective observer the probable company's future and how megatrends will affect the company for the next 15–20 years
6. Explaining how the SDF team looks at industry and market trends transforming customer needs and how it incorporates those changes in forecasting the company's future business
7. Giving details on how the forecasting due diligence and the verification and validation of forecasting and project valuation are used to create well-founded and convincing scenarios and forecasts
8. Bringing together the perspectives of all planning groups and providing feedback and input to the Investor Relations group on draft releases and talking points for presentations
9. Preparing the Investor Relations group to explain with confidence the key business drivers, conceptual models of operations, major assumptions underlying the project valuation, residual project uncertainty and risks, and how they are addressed in FSR management plans
10. Ensuring that statements made by the Investor Relations group are indeed supported by facts, evidence, and the best possible research, analyses, and evaluations at all times

18 Future-State Realization Planning

Future-state realization (FSR) planning is a corporate-wide project responsibility to identify, prepare for, and implement all actions necessary to ensure that the expected strategic project outcome becomes a reality. It is a broader and more inclusive approach than risk management, and it requires first-, second-, and third-order company reaction planning to competitor responses to be impactful. Forecast realization planning occurs prior to project implementation, and strategic decision forecasting (SDF) team involvement in FSR planning takes place before and after project implementation. SDF teams play a big role in forecast realization planning by determining and communicating to the project team what is needed to ensure that the forecast on which the project decision was based materializes.

The main motive for creating an FSR plan is to understand what external and competitive threats have the greatest impact on the project and focus the project team on responding more quickly if need be. Another reason motivating the creation of a comprehensive FSR plan is understanding how all the pieces of risk management and the company responses to environmental changes and competitor reactions fit together to avoid confusion at the time of execution and finger pointing when things go wrong. Also, FSR planning helps to minimize the level of silo mentality, assign unambiguous ownership of actions to the right project stakeholders, ensure that the internal processes and communication, coordination, cooperation, and collaboration (4Cs) work as expected, and execute the project as planned with all the required resources and support. Additionally, FSR planning verifies that conclusions reached prior to implementation, which led to the decision to proceed with the project, are well founded and it prepares a ready-to-deploy plan to offset project setbacks coming from competitor and environmental sources.

Identifying the triggers, timing, and actions and assigning responsibility to appropriate stakeholders to execute the FSR plan is essential to create effective response moves and reduce reaction time to competitor and external environment threats while increasing the likelihood of project expectations being met. Another reason for creating an FSR plan is to maintain or even improve project performance and value creation by controlling or offsetting advantages that competitors may obtain from their reaction to the company's strategic project. Again, this is achieved by anticipating possible threats to materialize and preparing actions to offset them. These FSR planning activities enhance the effectiveness of project risk management, help to avoid knee-jerk reactions, address threats effectively, and develop and communicate a common approach to deal with competitor reactions and the occurrence of uncontrollable events. The SDF model for FSR planning purposes is shown in Figure 18.1.

Forecast realization planning is the foundation of FSR plans and the synthesis of analyses and evaluations required to build successful plans to competitive and environmental threats. The first set of forecast realization planning parts include megatrend, subtrend, PESTLED (political, economical, social, technological, legal, educational, and demographic), industry, and SWOT (strengths, weaknesses, opportunities, and threats) analyses; reduction to residual uncertainty; and development of strategic posture. The second set includes the forecasting due diligence; validation and verification of forecasting processes, data, and information; and testing of assumptions, models, and scenarios. The third set encompasses evaluation of the forecast implications and identification of risks and their sources. The fourth set of FSR planning encompasses definition of project performance measures, creation of a management dashboard; development of an early warning system (EWS) and risk management and contingency plans; and creation of company response options and actions to

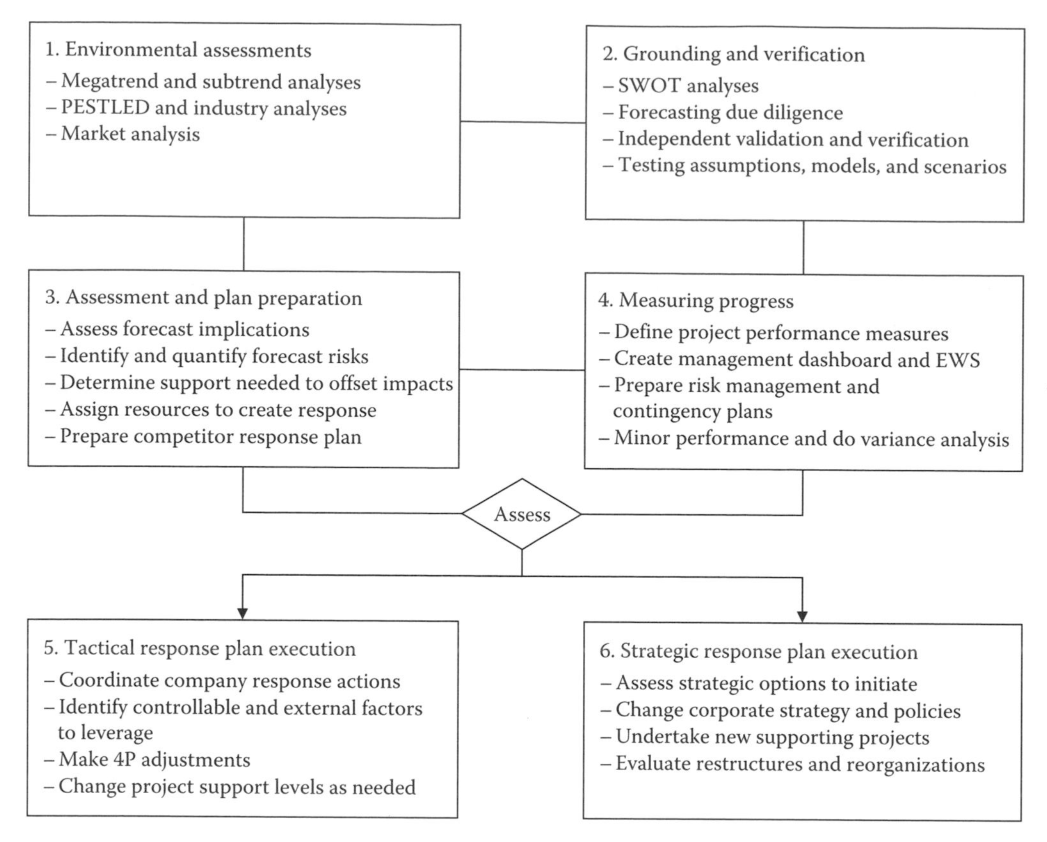

FIGURE 18.1 The SDF model for FSR planning purposes.

meet challenges from specific sources. The fifth set of the FSR plan is the development of a response plan to competitor reactions or environmental changes and the determination of the resources and support required to ensure that the project forecast does materialize. These tactical responses are market-based moves to enhance project implementation and success and require some resources, but are relatively easy to implement or reverse.

Additional FSR plan components include identification of controllable factors to leverage company, industry, and third-party intervention; product changes, pricing, and advertising and promotional adjustments; and changes in support levels. Yet another group of FSR plan components deals with corporate strategy and policy changes, reorganizations, and restructures to address threats from competitor reactions. The set of strategic responses involves significant commitments of organizational resources, is difficult to implement and reverse, and is used selectively in those instances where large potential losses can occur.

The company responses to competitive and external environment factors need to be highly rehearsed and well coordinated actions intended to counter first-, second-, and third-order competitor decisions and reactions to the company strategic project and to manage the threats arising from them. Proactive FSR planning is evidence of the company's willingness and ability to respond to external changes and competitive reactions in order to achieve expected project results. However, proactive FSR planning requires a highly coordinated and well-executed set of responses that add substantially to the projects' cost. The type, form, and strength of planned competitor responses are determined by the following:

1. The impact that the strategic project is expected to trigger the competitor reaction
2. The potential impact of the competitors' reaction on the project and the risks associated with it

3. The estimated costs and benefits to the company and the competitor of implementing the project
4. The company's ability and preparedness to minimize the impact of changes in the external environment, the occurrence of uncontrollable events and factors, and the occasion of black swans

18.1 CURRENT STATE, ISSUES, AND CHALLENGES

In most companies, FSR planning is nothing more than risk management and contingency planning, and it is usually undertaken either by the Marketing and Sales organizations to stop revenue losses or by Finance departments to reduce expenses. It is a reactive, ad hoc approach to responding to external threats, with no serious planning involved and characterized by dismal results. The reasons often cited for not making FSR planning a systematic effort and an integral part of preparations prior to project implementations include internal politics, the presence of silo mentality, ignorance of how all the company response pieces fit together, and inability to centralize the FRP function. Other reasons for the absence of FSR plans in strategic projects include frequent stakeholder reassignments to positions unrelated to the project, lack of experienced managers to lead or lend support to the FSR plans, and lack of management dashboards and EWSs to detect competitor reactions early on.

Often, difficulties in coordinating competitive responses among different business units and functional organizations are causing the absence of FSR plans. In other cases, trivializing performance monitoring and giving low priority to it beyond the first year of project implementation is the culprit. Also, potentially high costs associated with creating and deploying FSR plans and the uncertainty surrounding the success of FSR plans are responsible for this important function's absence. Furthermore, conflicts between FSR planned responses with existing programs, current strategic initiatives, and established company policies are real impediments to effective FSR planning as is the fear of provoking further competitor retaliation and possible industry-wide war that could lead to deterioration of project profitability.

These are valid reasons that require careful examination of company policies and practices on competitive responses, which are further complicated by some commonly encountered issues in the development of FSR plans. Issues such as the scope and complexity of the work in strategic projects with multiple parties involved increase to unexpected and often unmanaged levels. Also, the absence of effective project performance monitoring, accountability management, and lack of project team focus on understanding subpar performance are responsible factors. Furthermore, inability to resolve quickly conflicts between needed responses and corporate strategy and policies waste time and valuable corporate resources. Other challenges to effective FSR planning include erroneous understanding of customer needs and assessment of reaction to project effects on customers.

Mistaken assessments of corporate risk tolerance and senior management risk appetite, project uncertainty, and wrong strategic posture are common occurrences, but they are detected after the fact by which time it is too late to address threats effectively. Also, deficient understanding of driving factors, interactions, and feedbacks resulting in misspecifications of operational models, wrong predictions, and inconclusive tests of assumptions, models, and scenarios are challenges that are difficult to address effectively. And even though conflicting signals from different evaluations are handled by combining forecasts, they are a source of SDF team anxiety and ongoing investigations.

Lack of established and well-understood processes by key project stakeholders is a big gap in FSR planning. Also, there are challenges, such as competitive intelligence limitations, misinterpretation of competitor intents, timely understanding of their reactions, and difficulty in identifying variables needed to drive EWSs. Also, because FSR plan management is a highly collaborative corporate-wide effort involving stakeholders with different and potentially conflicting interests, it presents a problem in getting to agreements efficiently. FSR planning is the crux of decision execution effectiveness, beyond project implementation, and it involves flawless coordination of actions. But it is made more difficult by the following:

1. The complexity of identifying the current levels and assessing the necessary levels of 4Cs
2. Inability to ensure adherence to plans, competence of personnel assigned, and availability of financial resources
3. The meager quality of project management and handoffs across multiple project stages

In addition to the above factors and because project expectations change as corporate strategy and the operating environment change, it is a challenge to reach consensus on what course to pursue, which levers to pull first, and which controllable factors to influence and to what extent. Another challenge to effective FSR planning is the lack of experienced SDF team members able to communicate the need for and convince senior management of the benefits of effective FSR plans. And, when there is no SDF team involvement in monitoring project performance, this responsibility is assigned to the client or other stakeholder groups. This practice always leads to poor results.

When the objective of creating effective FSR plans is to shape the company's future by increasing the likelihood of expected and sustained value creation of strategic projects, four questions must be answered: What needs to be done to do FSR planning right? What should be the right mindset of project stakeholders? What is the correct approach to follow? What tested processes and tools should be used? The framework adopted by SDF managers to guide the work of SDF team members is shown in Figure 18.2, where the components of the right mindset,

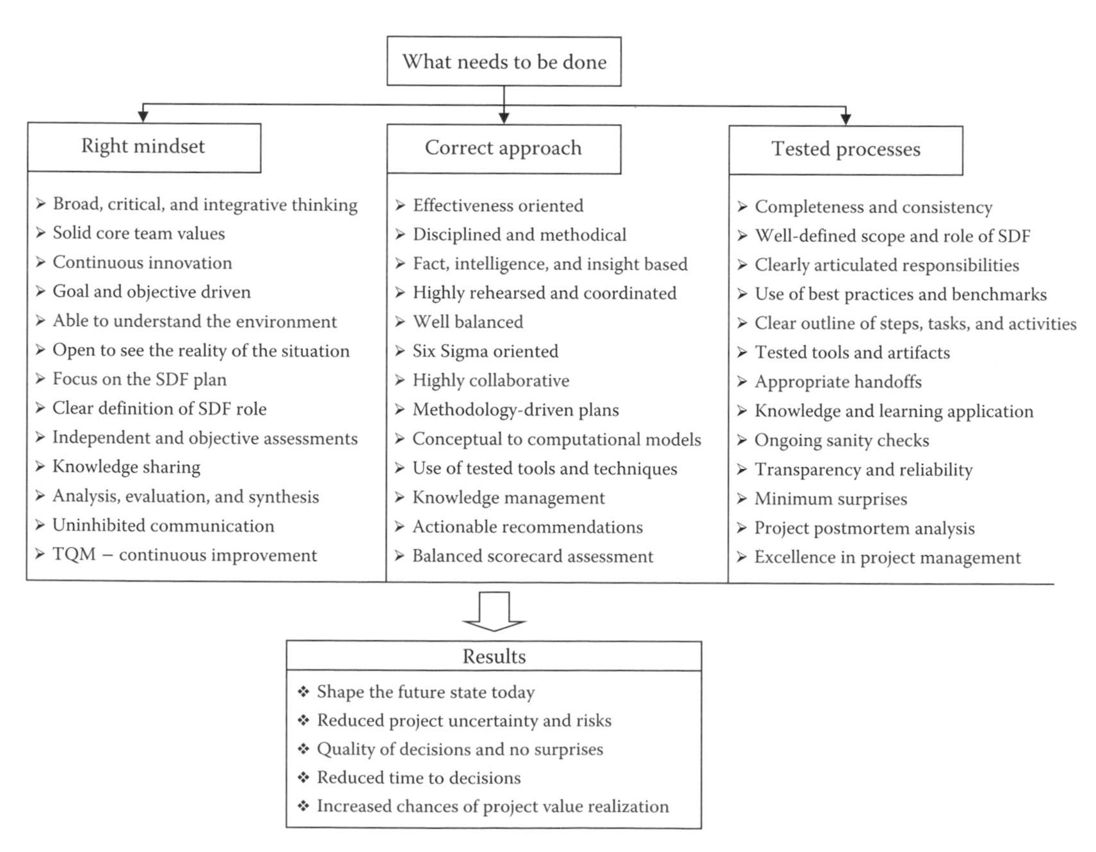

FIGURE 18.2 SDF framework for FSR plan development.

the correct approach, the tested processes that define what needs to be done, and the results of doing what needs to be done are outlined.

Notice, however, that while this framework is developed to address many of the issues and challenges in FSR planning, it is a very helpful framework to use in all other types of strategic projects.

18.2 SDF APPROACH, PROCESSES, AND TOOLS

Different types of strategic decision projects require different kinds of FSR plans and processes, but the elements common to SDF team involvement in all projects are illustrated in Figure 18.3. Here, the process outlined is used to maximize the SDF team's contributions in creating effective plans. The key elements of this process are as follows:

1. Understanding the context and the needs that created the project, the project rationale, and the goals and objectives to be achieved
2. Conducting all the necessary analytics and evaluations in an independent, critical, and objective manner
3. Identifying the effects of the project on competitors prior to the decision to proceed and assessing the impact of the competitors' reactions quickly
4. Identifying the nature and sources of forecast risks and determining their implications and those of project uncertainty and risks
5. Providing inputs, evaluations, and constructive feedback to the adaptive remedial response planning effort

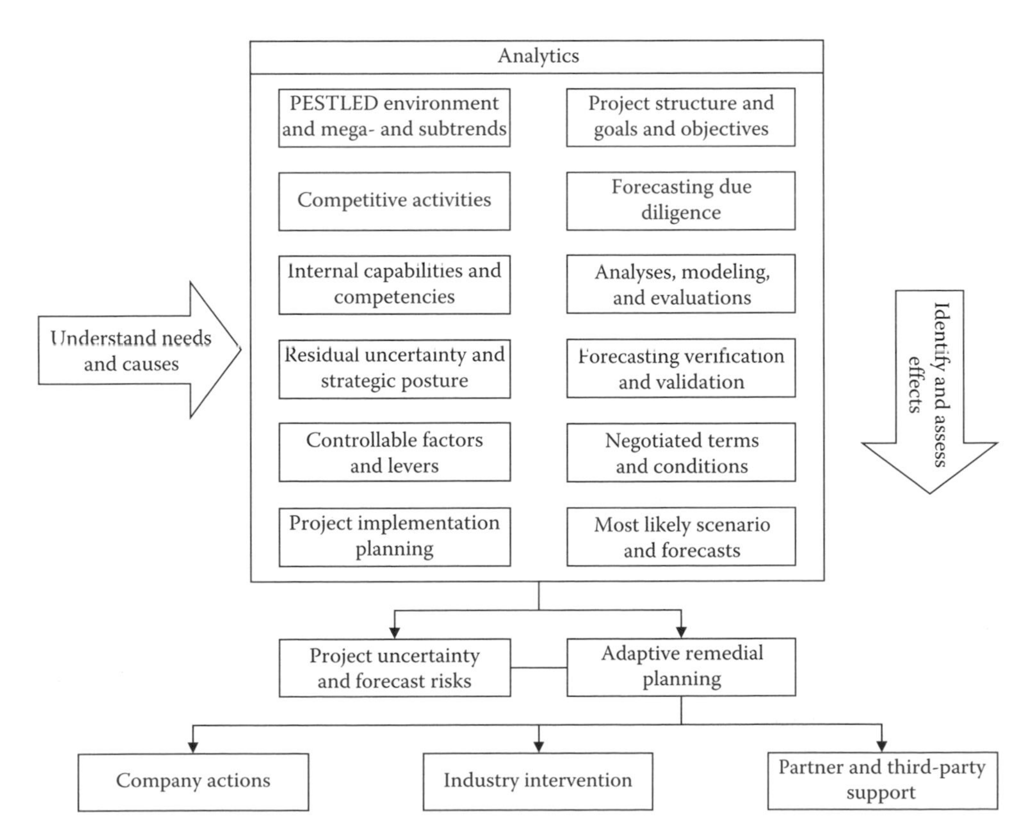

FIGURE 18.3 SDF future-state realization process activities.

Elements (1) through (4) are done well prior to project implementation and are revisited in the postimplementation period. The last element is mostly a postimplementation involvement for the SDF team, but it is as important as any of the other elements.

The preproject implementation SDF process in support of developing FSR plans begins with a description of actual versus desired state through situational analysis to identify corporate and client needs. Then come clarifying project rationale and consistency with corporate strategic goals and objectives and verifying the decisions that need to be made, reasonableness of time frames, and the right project scope. Determining what needs to be known to make sound contributions to FSR planning is a crucial component guiding the SDF team's activities and its project goals and objectives. This is done concurrently with the assessment of project uncertainty and the development of strategic posture. Reducing overall project uncertainty to residual uncertainty through analysis, analogs, and past experiences and developing strategic posture are key elements of the SDF process.

Identifying what is known and what is knowable at what cost is a concern always present in the minds of SDF team members as is determining company controllable factors and events, those outside its control, and probable black swan eventualities. Next, the SDF team focuses on megatrend and subtrend assessments, PESTLED environment, and industry analyses; evaluation of current competitive activities; assessment of internal competencies and capabilities; and the reevaluation of controllable factors and levers. The part of the SDF team analytics and evaluations that follows deals with validating the common assumptions, the project structure, its goals and objectives, and the modeling of operations, and conducting a forecasting due diligence. Then, SDF team members create and test a few distinct and plausible scenarios, select the most likely scenario, generate forecasts, and evaluate their implications. At this point, forecast risks and their sources and overall project risk are determined through scenario development, hypothesis testing, sensitivity analysis, and assumption validation. In the meantime, the consistency of corporate risk appetite versus the identified project risks is reascertained before any recommendations are developed.

Notice: *Many activities of the SDF team in FSR plans are the same as in other types of strategic projects, but are repeated here. So, why the repetition? It helps to show the importance of these activities to project success and help in learning their common applications.*

Another important element of the SDF process for FSR planning is assessing offers and counteroffers and evaluating the effects of negotiated terms and conditions on the likelihood of forecast realization. The SDF team then proceeds to determine, along with the project team, the likely first-, second-, and third-order competitor reactions corresponding to the project. Also, the management of the FSR plan and the company's comeback are included in forecasts, and their implications evaluated. At this point, the SDF team helps to create a management dashboard for the project, define appropriate project performance measures and indicators to be monitored, and create a performance monitoring and variance analysis process. These are prerequisites to the creation of an EWSs to identify threats as they are about to appear on the horizon and communicate them and their impacts effectively. SDF team members also assist the project team to prepare and communicate a common set of response actions and moves to offset competitor reactions, external events, and uncontrollable factors.

When the SDF team is involved in the postproject implementation period, its focus shifts to evaluating actual current performance versus expectations and the projected future-state value creation. For this, it reviews earlier verification and validation findings and updates data and market intelligence, assumptions, models, and scenarios. It also evaluates the incidence of project implementation crucial success factors such as, 4Cs, adherence to plan, and the quality of project management.

At this junction, the SDF team reexamines the controllable factors to leverage, such as changes in product introduction speed, product quality, attributes, and pricing and rebranding. A determination is also made of the need to enhance project support through increased senior management involvement in the project, additional resources shifted over to the project, and adoption of better project management practices.

In some cases, it is prudent to consider external intervention as an instrument of FSR plan management, and the project team requests the support of partners or third-party entities to be enlisted in order to address unexpected problems and correct practices detrimental to the FSR. On other occasions, it makes the case for the company to call for industry association intervention to offset the impact of unexpected barriers to the company's FSR. When all planning activities are completed, the SDF team assesses options and their possible impacts and recommends execution of the actions prepared when and to the degree needed. In other cases, it may have to escalate the need for strategic interventions to address FSR problems more effectively. Finally, when conditions warrant, a new forecast of the future state is generated and evaluated, which reflects market realities, the new strategy, and revised project goals and objectives.

Adaptive response planning: *It is the art of managing the competitors' market behavior and an SDF team critical thinking approach that balances competitive response needs with the company's capabilities and resources. It is based on the learning model of perceiving, planning, acting, adjusting and entails the following key elements:*

1. *Assessing uncertainty and risk and learning through experimenting, and sometimes, by trial and error*
2. *Creating organizational adaptation skills and capabilities to make sense of events as they occur*
3. *Targeting company responses to specific competitor or environmental conditions*
4. *Identifying competitive information quickly, interpreting it correctly, and applying it appropriately*
5. *Measuring how fast and to what degree the competitive response resulted in the desired effects*
6. *Initiating process changes and project team motivation and feelings toward 4Cs*

18.3 SDF ANALYSES AND EVALUATIONS

Faced with the problem of how to address FSR planning issues and challenges, SDF managers follow a model framework developed to put them in context and deal with them in a more efficient manner. This systematic method, illustrated in Figure 18.4, guides along the SDF team's thinking in creating plans and is bound by four elements: the corporate strategy; the project scope, goals and objectives, the competitive intelligence limitations; and the resources allocated to the project and the organizational competencies and capabilities.

Once these four key elements are addressed, attention is directed to scenario planning, the management dashboard creation, the creation and workings of the EWS, competitor response modeling, and business war gaming. The pre- and postimplementation analysis and planning are performed and tactical and strategic response moves are developed. Notice that in the center of these activities is the core of the FSR plan, namely, assessing the impact of the project on the key competitors' business and environmental changes impacting the project. These activities begin the adaptive response planning in earnest. The two main instruments of adaptive response planning are corporate actions, moves, and controllable levers and partner, industry, or third-party interventions. When all these parts come together, the FSR plan emerges.

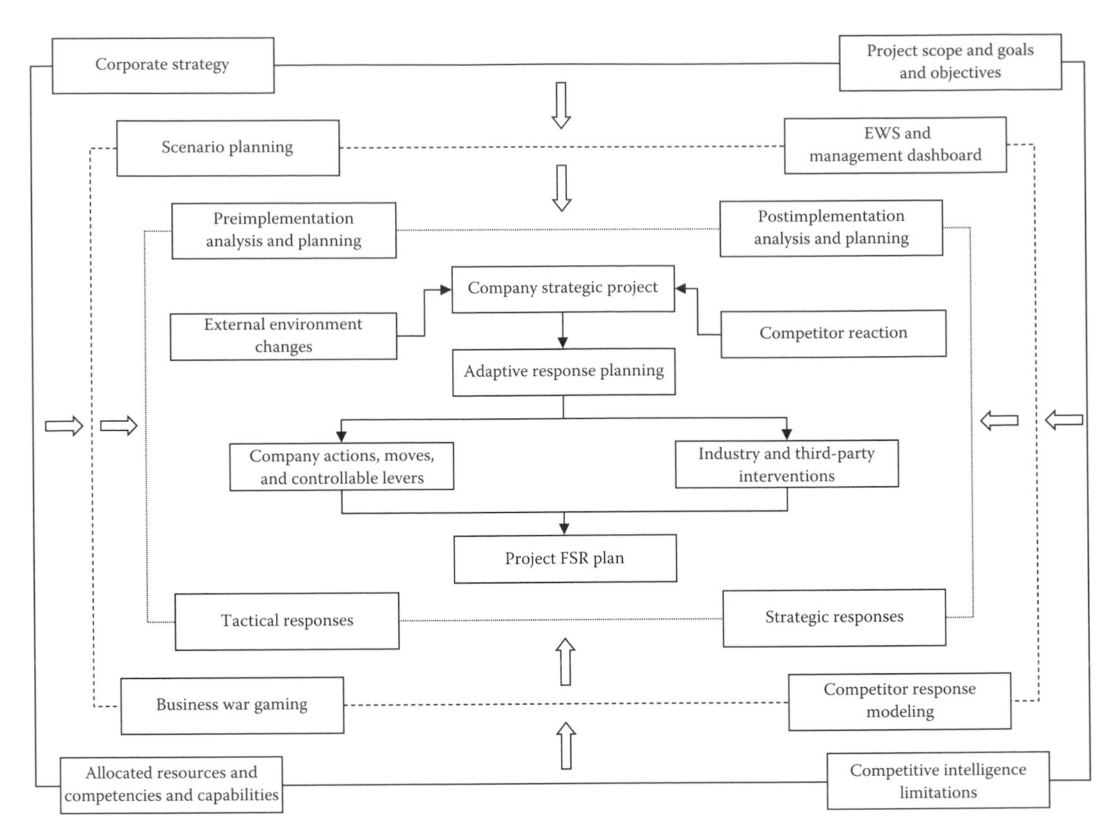

FIGURE 18.4 SDF model to address FSR planning issues and challenges.

The SDF model for support in this area is a flexible framework, which allows the roles and responsibilities of the SDF team and the specific analyses and evaluations they perform to vary according to the nature of the project. In FSR planning, the SDF team's environmental assessment answers the question of what has changed since the last assessment and how it impacts the company's business. Depending on the nature of the project, an appropriate review of the current industry structure and analysis starts the process along with the evaluation of current technological platform adoptions by the company. Megatrends, subtrends, industry, and particular market trends are studied and their implications for the company's future are evaluated. This is followed by an assessment of recent PESTLED changes to assess their impact on the business. Also, changes in customer and consumer preferences, tastes, and choices are analyzed to determine the implied needs in order to deliver the right products, services, and technologies.

The competitor assessment part of the FSR plan requires attention of SDF team members focused on key issues in the pre- and the post-strategic project implementation phases. More specifically, on an update of earlier SWOT analyses with the main objective of determining competitor competencies and capabilities vis-à-vis those of the company. Key competitor actions independent of their first responses to the company's project are evaluated to determine other possible future actions on their part and what the company can do to nullify their impact. Also, a reevaluation of competitor new product introductions, extensions, updates, and enhancements and how they impact the expected future state takes place. This entails an evaluation of changes in their pricing, marketing, sales and customer support, promotion, and advertising levels. Naturally, industry reactions to the company's project are monitored closely so that they can be addressed and channeled appropriately.

While the competitor assessment takes place, recently announced technological advances and future competitor adoption of new technologies are noted versus those of the company to determine how they may affect the competitiveness of both sides. Also, changes in the company's and competitors' policies and strategies related to project responses are analyzed to determine the direction and kind of likely responses to the company's project. Key competitor internal organization, technology, and management team changes are examined to appraise changes in their competencies and capabilities. Additionally, the possible effects of new R&D, product development, and business development projects and initiatives are evaluated objectively.

The SDF team's analytic and support involvement in the development of the FSR plan includes ownership of the following responsibilities in order to provide superior support to the project team:

1. Preproject implementation creation of an EWS, which entails defining the variables to be tracked, how to determine changes significant enough to warrant action, and how to communicate the need for corporate-wide action
2. Postproject implementation reassessment of the assumption set, scenarios used to ensure current validity, and changes needed to enhance their relevance
3. Reevaluating the risks identified and their estimated impact and revisiting the earlier risk analysis and risk management plans
4. Reexamining the decision scenarios, models, and parameters used and testing and simulating "what-if" occurrences to obtain insights from recent developments
5. Monitoring project performance and obtaining key stakeholder assessment, opinion on the performance of the forecast, and the causes of deviations of actuals
6. Developing a convincing case for and creating an updated strategic forecast, which reflects the changes that have impacted it since it was first created
7. Separating the impact of competitor reaction risks from those of external factors and interpreting and communicating the significance of early warning signals and the need for action when warranted
8. Outlining steps to take, identifying actions to offset the impacts of risks, and identifying parties responsible for these actions

The definition of the first-, second-, and third-order company actions to competitor responses is a team effort, which necessitates involvement of the SDF team in several areas. What is really needed is a critical evaluation of costs and benefits of responding to competitor reactions to shape the FSR plan in the following areas:

1. Revisiting the strategic project goals and objectives to determine whether to stay the course or whether revisions may be in order in the face of market realities
2. Basing company responses on a recent verification and validation of its competencies and capabilities to determine if and what changes may be in order
3. Performing an objective review of senior management actual commitment levels versus those pledged at project start and identifying modifications needed
4. Adjusting company policies to facilitate efficient execution of the FSR plan after evaluating the effects of changes in product attributes, quality enhancements, extensions, as well as modifications on pricing, customer preferences, and service levels
5. Evaluating new technology introductions to offset the impact of competitor reaction and realization of other risks
6. Researching industry association practices to determine the efficacy of industry intervention or a collective industry-wide response to changes in the external environment
7. Determining if and what legal or regulatory interventions are appropriate to address certain threats to expected FSR, which are other considerations in shaping the FSR plan

SDF team support in the FSR planning is directed toward gauging company actions in response to competitor reactions to the strategic project, to uncontrollable factors occurring, and to unexpected environmental changes. To be effective in that, the SDF team must evaluate and answer some basic questions, such as the following:

1. What is needed, namely, what company response actions are appropriate for competitor reactions, for the occurrence of certain uncontrollable factors, or for any unexpected environmental changes
2. When should each of those responses be initiated, in what manner, and whether they should be staggered or simultaneous
3. How much of each response is needed to offset the impacts coming from different sources and for how long they should be sustained
4. What organization and specific individuals are responsible for taking what actions and seeing that they are executed properly
5. How long it will take and how much it will cost to implement each of the planned responses and what the company will get in return to make it worth pursuing
6. What reasonable assurances can be given that these response actions will indeed achieve expected results and how would one know how successful they have been
7. What is likely to transpire if the SDF team's recommended actions are not taken or if the FSR planning effort is not fully funded and deployed

18.4 SOUND PRACTICES AND SDF TEAM CONTRIBUTIONS

The source of the success factors and SDF team contributions is a set of lessons learned in strategic projects where FSR plans were generated with degrees of success ranging from modest to outstanding. In this area, there is general agreement among industry experts concerning beneficial practices in creating FSR plans.

18.4.1 Sound Practices and Success Factors

The first set of sound practices and success factors in FSR plan development includes sound competitive assessments, intelligence, and insights to understand what drives key competitors; what they are capable of doing; and their resources, strategies, and capabilities. These practices are important to outthink competitors, for accurate assessment of potential project impacts on key competitors, and for determining possible individual competitor reactions. Success also depends on identifying and developing effective company actions to counter the effects of competitor reactions. Effective actions require open, 360-degree communication, flawless coordination of steps and activities corporate wide (i.e., of all parties involved), and a high degree of coordination and collaboration to make the company actions work together to reach a common objective.

Obtaining and maintaining senior management support throughout the project is crucial, which is secured by demonstrating ability to separate the impact of competitive reactions from external factor effects and by having a well-defined and clearly articulated FSR planning processes and ownership assignment. Scenario planning to envision different future states, how external shocks could impact them and developing strategies to achieve the desired states in the presence of external shocks is a must although, at times, they may require internal adjustments, reorganizing and restructuring of operations. In some instances, making preemptive strikes or escalating from tactical to strategic responses to counter or prevent certain types of competitor reactions may be necessary to achieve future-state realization.

Preemptive actions can take the form of increased product and process innovation and considering possible cooperation with competitors not usually inclined to react to the company strategic project. In all projects, however, it is critical to have an effective project implementation team, a good project performance dashboard, and an effective EWS. Also, scripting and practicing the responses, escalating when further action or strategy changes are needed, and using performance scorecards to assess the effectiveness of a particular FSR plan are best practices.

18.4.2 SDF Team Contributions

The contributions of SDF teams recognized in the creation of FSR plans span a wide spectrum of solutions, but the ones most valued by clients who participated in our discussions are support activities, which include the following:

1. Evaluating correctly the impact of competitor reactions through competitive response modeling, which allows SDF teams to investigate how competitors are likely to behave under different pressures
2. Creating reaction functions and carrying out the analysis and evaluations to the second- and third-order competitor reactions
3. Ensuring that there are no conflicts of project or company response objectives with current strategic goals and objectives
4. Using external experts to determine independently the effectiveness of responses and validating response assumptions and project team expectations
5. Assessing correctly the causes of deviation of actuals from forecasts through a good variance analysis, which is necessary because it lays the foundation for creating good actions and moves
6. Producing convincing analyses, evaluations, presentations, and recommendations of response strategy and actions
7. Developing an EWS that combines analysis of threat scenarios with a good performance monitoring process and effective variance analysis

At the other end of the spectrum of valuable SDF team contributions, clients are excited by the game theoretic and competitive response models to test strategic project impacts against likely competitor reactions and develop contingency plans to handle future uncertainty. Others find that simulating business war games when the industry and the competitive business environment are undergoing significant changes create a lot of value for their decision-making processes. Yet, another set of clients state that the SDF team's introducing competitive response best practices in FSR plans, such as having a central intelligence repository, an EWS based on pattern recognition, and providing timely alerts are their most useful contributions.

19 Business Cases, Plans, and Operational Targets

The focus of this chapter is on the support strategic decision forecasting (SDF) teams assigned to strategic projects provide in the development of business cases, business plans, performance targets, and dashboards and scorecards. A feasibility study precedes the business case, and it is a preliminary evaluation of a proposed project to determine if it has a good chance to create value. It is the first gate in the decision process that determines whether expanding corporate resources to further assess the project is justified. A good part of the work performed in feasibility studies and business cases is performed by SDF teams in the course of creating long-term forecasts because they can best perform the required analyses.

SDF teams are responsible for providing assessments of the future and many of the inputs that go into making project decisions. And, although they do not have project ownership, they are the most qualified in conducting these types of analyses, modeling, due diligence, and other investigations. They have the experience to perform a good portion of the business case analyses and evaluations in the course of developing forecasts and in their validations and verifications. Their analyses and evaluations along with the creation of future-state realization (FSR) plans enable them to assess the likelihood of project success and produce convincing evidence for business cases, when the evidence is there.

A. *The business case.* A business case means different things to different people but, generally, it is a document used in projects requiring large capital and human resource commitments to obtain approvals to move forward and funding. It captures the knowledge the project team develops about current company operations and how it will operate when the project is implemented. It also verifies that the selected project option substantiates and fills the gaps and the needs identified. The business case validates the rationale for undertaking the project, and it helps senior management to prioritize it against other business opportunities and projects. It describes in summary form the results of analyses and evaluations performed and it highlights key facts and knowledge gained. Furthermore, it determines the remaining project uncertainty, risks and issues, provides a strategic and financial justification for the project, and communicates them company-wide objectively and effectively. Thus, it provides a common and consistent message across the company in a structured and effective manner.

The quality of business cases and processes varies by type of project and corporate resource requirements to be committed to it, but in all instances they create the foundation to base the decision for initiating a project. The business case format usually has the following components:

1. The background, context, and needs that lead to considering the project or the opportunity presented and the objectives to be achieved by it
2. The particular project's rationale and justification and the strategic, cultural, operational, and technology fit assessment
3. The evaluation of the operational, organizational, and technical integration feasibility of project implementation
4. An objective assessment of costs and benefits associated with the project and an estimate of value expected to be created over a period of 10–15 years out
5. Alternatives considered, the ranking of those alternatives, and the reasons for selecting the project and not another option

6. An implementation plan and the final state of the business, which includes milestones, deliverables, and resources required to execute it well
7. The FSR plan, the remaining project uncertainty and risks, and the plans to enhance the chances of expected value creation
8. The opportunity cost of doing the project and an estimate of the opportunity costs associated with not doing the project or some other strategic initiative
9. A set of actionable recommendations on whether and how to pursue the project, or restructure it to make it more acceptable, or reject it
10. Appendices that document the data and information, the models, analyses and evaluations performed, and the results and findings, and a summary of the business plan resulting from the implementation of the project

Feasibility study clarification: *SDF teams provide support to feasibility studies, but that support is an abbreviated version of the business case support and for that reason, it is not treated separately.*

B. *The business plan.* A business plan is one of the five elements required for achieving the objectives of the project, the other elements being the project feasibility study, the business case, the implementation plan, and the FSR plan. It is another decision-making tool that formalizes, depending on the project, the company's or a newly created entity's goals and objectives; presents convincing evidence of their achievability; and describes how they will be realized. The business plan draws heavily from the knowledge gained in developing the business case and it captures the results of the entire business planning process. Its main objective is to develop the blueprint and convincing evidence of successful operations. A business plan discusses the business of the entity, its goals and objectives, the markets it will operate in, and the financial forecasts. It also address additional human resources needed to execute plans, explains the actions and plans that will be used to achieve the entity's goals and objectives, and makes the business of the entity easy to understand and credible, especially for external audiences.

A good business plan creates a blueprint for building the entity's operation, it defines its mission and vision, its goals and objectives, the opportunities it will pursue, the strategy and action plans, and the monitoring, evaluation, and control activities. Like the business case, it is a communication tool for internal and external audiences and contains sufficient evidence for making management and prospective investor decisions. And, as with the business case, the type and quality of the business plan varies according to the need to convince internal and external audiences, the size of the corporate resource commitments, and the expected value creation. But, most importantly, it varies by the collective experience of the project team.

A well-structured business plan for a newly created entity is characterized by descriptions of elements that determine its achievability beginning with an assessment of the company's current state of business, the new entity's goals and objectives, and how they support achieving the corporate goals and objectives. It also has a clear definition of the entity's business, its governance structure, and the management team to lead its operations. The new entity's strategic plan, goals and objectives, strategic intent, and key initiatives are created and an evaluation of the entity's ability to attain them is performed. Business plan information is derived from the SDF team's review of the megatrend and subtrend, environmental, industry, market, and SWOT (strengths, weaknesses, opportunities, and threat) analyses and a synthesis of their implications for the entity's operations.

The operational and supply chain management plans are prepared with clearly defined roles and responsibilities, information flows, and feedback loops. Updated, important market research findings, customer satisfaction surveys, competitive intelligence, and distilled intelligence and insights are evaluated along with earlier long-term forecasts. Then, a complete financial plan based on the consensus forecast, which includes an income statement, a balance sheet, a cash flow statement, and a financing plan when needed. The business plan describes the new entity's deliverables at different milestones and performance targets, which are developed in sufficient detail and validated to make them credible.

C. *Operational targets.* Operational targets are part of the business plan and are simply planned or expected values of performance indicators, which may express quantity, quality, and efficiency values to be achieved in a given period. Companies establish a target for each performance indicator to assess progress toward strategic objectives and operational results. As such, they represent commitments on the part of project stakeholder organizations, provide a forward-looking perspective on the range of an indicator, and show whether earlier stated objectives are being realized.

Targets are sometimes used in conjunction with management dashboards and performance scorecards to show performance across several areas of company operations. Some targets are stretch objectives, and in many cases they are based on competitor or best-in-class benchmarks. But their development is based on a good understanding of the company's operations, its strategic goals and objectives, its competencies and capabilities, the project objectives, and the support committed versus the support needed. Targets bring the purpose of pursuing a project into sharp focus and they serve to orient stakeholders to the activities to be performed and motivate them to ensure that those targets are met. Once established, operational targets serve as reference points to judge progress made on schedule at the levels expected in the business plan. It is important to note that operational targets are part of the business plan and the business plan is part of the business case. Therefore, gaps, errors and omissions, issues, and challenges in one area are transmitted to the others, and in the process they get amplified.

In the sections that follow, we review the current state and issues and challenges associated with business cases, business plans, and operational targets; but the main focus is on the role of the SDF teams, the approach and processes used, and the support provided in each of these areas.

19.1 CURRENT STATE, ISSUES, AND CHALLENGES

The current state of developing business plans, business cases, and operational targets is based on some long-established processes and principles, but they are still characterized by a number of shortcomings. The most notable inefficiencies are incomplete environmental assessments, ad hoc industry and SWOT analyses, and heavy reliance on financial modeling to the exclusion of understanding of market dynamics, current competitor activities, project uncertainty, and forecast risk implications. The result is that they tend to present an optimistic or unrealistic view of the project prospects and its chances of success. The major challenges and issues are discussed by area in which they are encountered.

A. *Business cases.* Due to limited resources and internal organizational issues, late engagement and premature disengagement of SDF team members from projects is a major issue. Even in cases where they are involved, their participation is often limited to providing long-term forecasts alone and no further involvement is welcome. As a result, unreliable sources of information and conflicting signals from competitive and market research analyses lead to incorrect conclusions and decision delays. Unrealistic levels of savings, synergies, and partner support expected by project sponsors are very common, while competitor response assessments and FSR plans are missing all together.

Implementation plans included in business cases routinely understate the support levels needed in senior management commitment, financial resources, and qualified personnel to implement the project. Business cases are only snapshots of the business situation and evaluations at one point in time, and they are not updated enough to reflect the most current state of information and assessment. Under usual circumstances, they are done without a forecast due diligence and an independent project validation and verification, especially when they are driven by the sponsor organization whose intent is to prove the value of the proposed project. In those instances, there is less emphasis on the findings and more on client gut feelings. This is also true when internal politics are conflicting with the project team's conclusions and recommendations of the business case evaluations. Lastly, frequent changes of key business case stakeholders responsible for different aspects of the project result in discontinuities in knowledge and commitment to the project's original goals and objectives.

B. *Business plans.* The major issue in creating and assessing business plans has always been the exclusive reliance on financial modeling and evaluations with little understanding of the operating environment, underlying assumptions, and the implications of financial forecasts. Some business plans wrongly assume a short-term focus on net present value (NPV) and pay little attention to strategic returns on investment, which includes financial performance, market position, technological leadership, customer satisfaction, and progress toward development of competitive advantage. Another challenge is that initial operational plans for new entities created by a project are drafted at the conceptual level with no actual experiences to go by and with limited cooperation and coordination of effort between existing and newly formed entities.

Due to pressures to implement the business plan of a strategic project right away, insufficient attention is paid to new external environment developments and recent competitor moves. This leads to inability to adjust plans to deal with threats from these events when they materialize. Business plans are known to routinely overstate revenue potential and understate start-up costs, cultural differences, system incompatibilities, technical difficulties, and other unresolved issues. At times, there are new entity business plan differences with corporate objectives and policies and conflicting interests of goals and objectives of different stakeholder organizations. In the absence of SDF team participation, business plans suffer from the ever-present overly optimistic marketing plans. Such plans are based on overstated revenue estimates and understated costs of advertising and promotions and cause tough adjustments by the new entity's management. It is not uncommon for business plans to ignore costs of building and carrying inventories, which are left out of financial evaluations and supply chain management plans. And in practically all cases, new entity business plans do not consider business risks or exit plans from operations and shutdown plans and associated costs.

C. *Operational targets.* When operational targets are created in the absence of SDF team participation from testing of the underlying assumptions and developing forecasts, the result is performance targets based on management wishes. Also, most of the time, targets often bear no relationship with market realities and forecasted future performance. Most of the time, however, targets are set without reasonable reality checks, or reference points and evaluations often lack the strong support across all organizations, which is necessary to reach them. In those instances, operational targets are considered nothing more than senior management dreams and lack the support of line managers.

Ad hoc evaluations of the implications of setting operational targets result in targets being missed and unjust blame assigned to certain stakeholders. This is always true when targets are set outside ranges indicated by long-term forecasts and appropriate competitor benchmarks. Also, stakeholder organization commitment is absent when targets are set by the new entity's senior management team with little or no input from those responsible for making them happen. Another issue is that targets are fixed for each reporting period and the need for understanding variance of actuals is delayed until the next period, by which time the opportunity for adjustment is also postponed.

On the other end of the spectrum, targets that are based on best-in-class benchmarks often ignore the type of human and financial resource inputs required to achieve them. Further, due to focus on achieving results in one area, targets are often used and assessed selectively with undue focus in only that area and not in other areas of operations where unusual results are obtained. Even when used in management dashboards, operational targets are sometimes fraught with performance definition problems. In some business plans, we find that very granular or high-level performance indicators and extreme value targets are created and monitored.

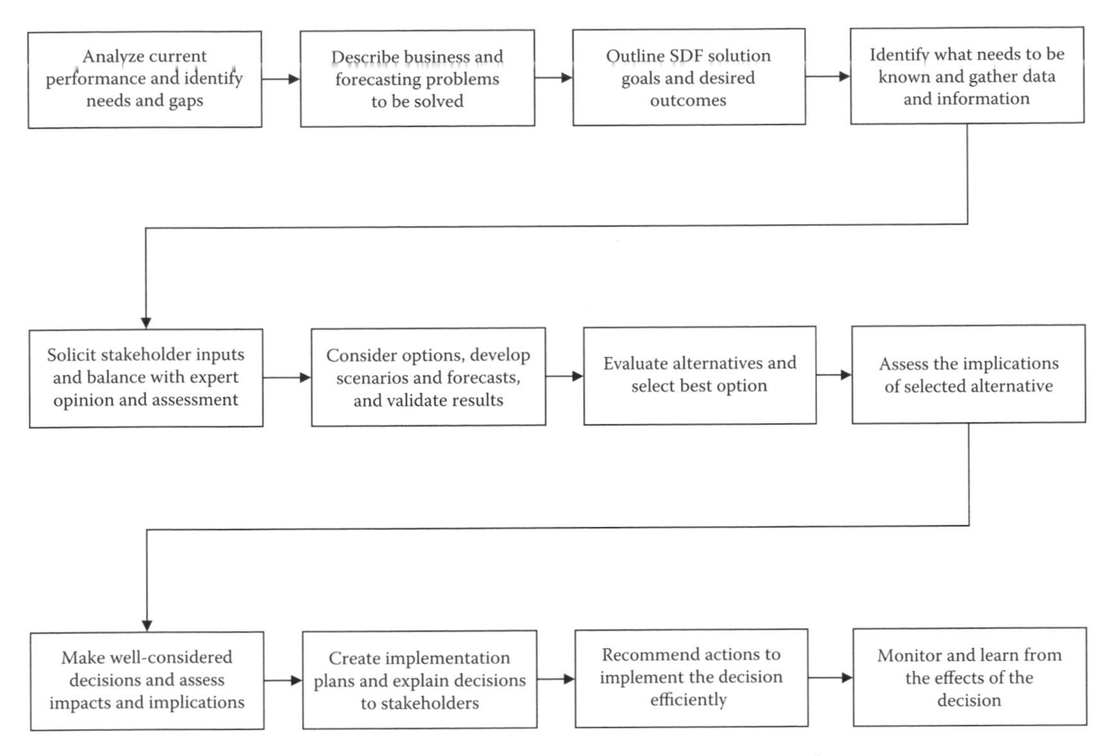

FIGURE 19.1 SDF business case, plan, and operational target creation approach.

In order to deal with the many issues and challenges in business case and business plan support in a systematic and effective way, SDF teams have adopted a method illustrated in Figure 19.1. The key feature of this method is that SDF team members proceed systematically to identify and address issues and problems impacting forecasting in each step of the SDF process before moving on to the next activity.

It is a variant of the validation and verification process, except that the need for problem and issue resolution now is not only noted but also escalated so that appropriate action is taken early in the process. It is the approach used by SDF team members to deliver support efficiently and thereby reduce time to agreements, decisions, and project approvals.

19.2 SDF APPROACH AND PROCESSES

The SDF team's charter is to lend support to the strategic project's business case and decision, the development and evaluation of business plans, creation of FSR plans, and setting operational targets. This support draws from and leverages the SDF team's strengths in environmental scans, competitive assessments, market research, modeling of operations, and analytic and forecasting experience. The approach used to deliver quality support in these areas is outlined in Figure 19.2.

The main idea in this approach is the creation of a solid business case support network, which is characterized by strong internal organizational functional links and personal relationships and access to external sources of intelligence, opinions, assessments, benchmarks, and advice. It is the creation of this network that makes for effective SDF team participation and substantial contributions in the development of business cases, business plans, and operational targets. The analytic and forecasting parts of the support are shown on the left-hand side of the chart, while the outputs of the SDF team analyses and evaluations are shown on the right-hand side of the chart, and they form the inputs to other investigations and planning functions. One key element of the SDF approach

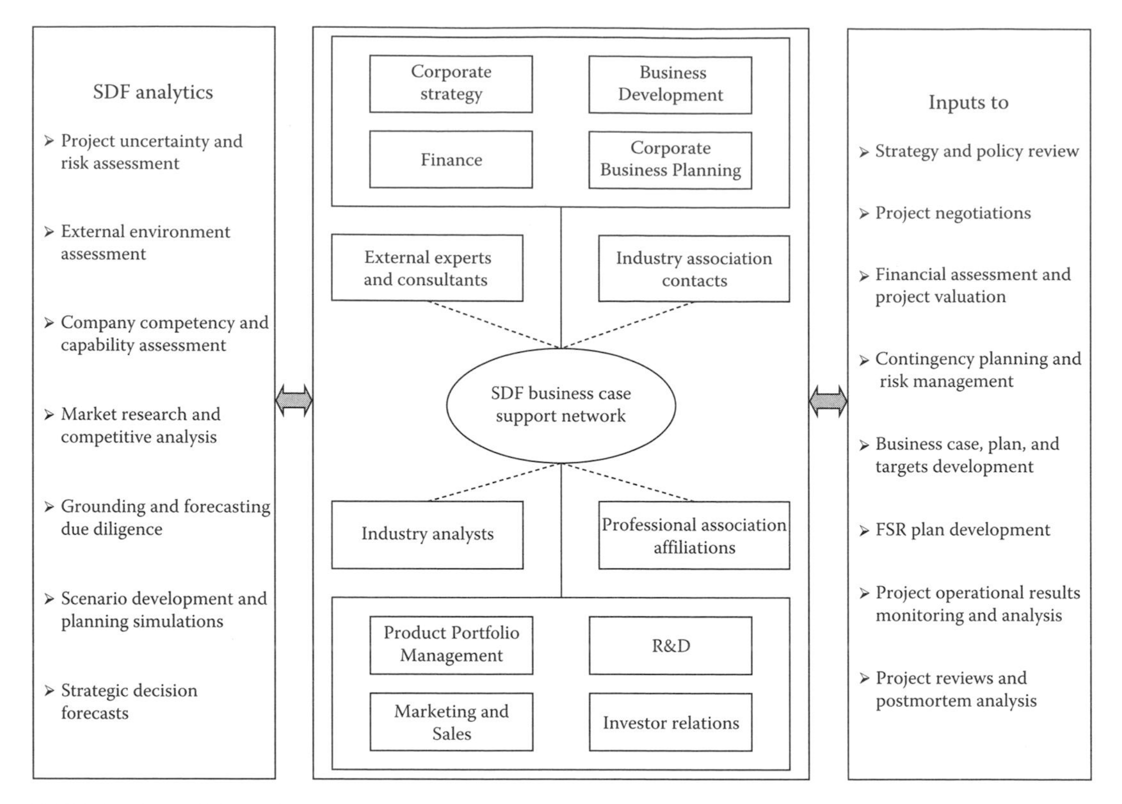

FIGURE 19.2 SDF model for business case and business plan development.

is that all analytic and evaluation elements benefit from this team's functional links and personal relationships with internal planning groups and external alliances and contacts. And since their evaluations and conclusions have the support of that network, they are usually accepted as reasonable and valid inputs to business cases, business plans, and operational targets. Furthermore, the same network, analyses, and evaluations are used in project negotiations, risk management and contingency planning, FSR planning, and project postmortem reviews.

A general process implied by the SDF approach used to develop the business case, the business plan, and the performance targets of a new entity is one sketched out in Figure 19.3. That process is a general framework, which is adapted to project specifics and places with different emphases in business cases, versus business plans, versus targets. Yet, it is a systematic and well-structured process that guides the project team, but attention is required because certain parts of the process need to be well understood and put in the right context the negotiated terms and conditions. For example, in the case of operational targets it means that the levels of performance targets need to be negotiated in advance between the head of the new entity and the organization responsible for attaining the target.

19.3 BUSINESS CASE SUPPORT

The SDF team's process shown in Figure 19.3 is tailored to provide support in the development of a strategic project business case and a new entity business plan. Notice that the steps in this process mirror those in developing strategic forecasts used to make a decision. However, the role of SDF team members in business cases is mostly of a collaborative nature and varies in each phase of a project. The evaluations and activities are listed as sequential steps, but in actual projects, business

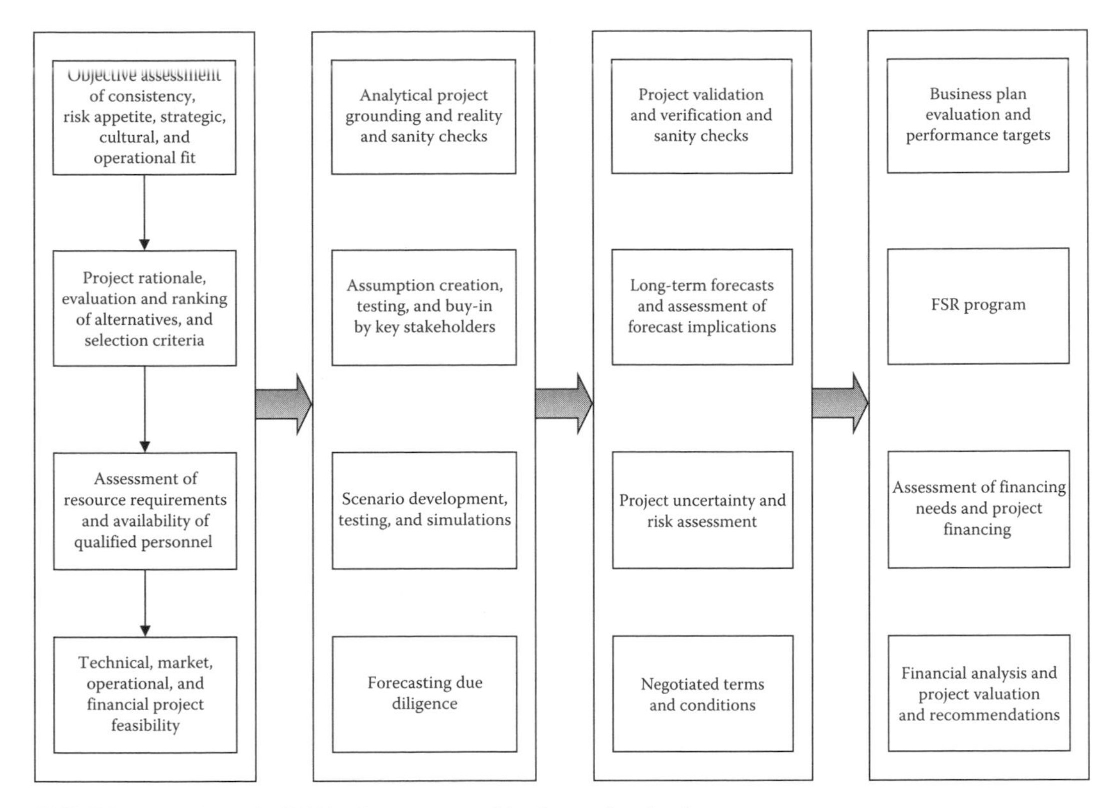

FIGURE 19.3 Generic SDF business case and business plan development process.

case, business plan, and target development are iterative processes and some activities may be done more than one time or in more than one step, before or after the order in which they are shown.

The support of the SDF team is required, but involvement in the creation or evaluation of business plans varies depending on whether a new entity is created and on that entity's management team's perceived value of the SDF team's contributions. The business case segments where SDF team support is needed include the following:

A. *Consistency of goals and objectives*. The first step and the starting point of the SDF process is a situational analysis to verify the consistency of project with strategic goals and objectives and with corporate policies. Checks are performed to ensure the viability of project goals and objectives vis-à-vis marketplace realities. A coarse gap analysis is performed to identify the major needs to be filled in the human resources, technology, market presence, and operations areas. Also, congruence of residual project uncertainty and risks with corporate and client risk tolerance levels is needed to make sure that senior management and client support will be there as the project unfolds.

B. *Project rationale and evaluation of alternatives*. This step is about the evaluation of the project's strategic, cultural, operational, and technical platform fits, which is essential prior to moving to other evaluations. A review of the initial or preliminary assumption set follows, and recent information and evaluations are used to update those assumptions. In the area of project selection criteria, the role of SDF team members is to provide their assessments and advice on the strength of each alternative's project economics. In the ranking of projects and the selection of the best alternative, their role is to ensure that the selection is consistent with environmental and competitive intelligence beyond that of financial assessments that abstract from project uncertainties and risks.

C. *Assessment of resource requirements.* The first SDF team activity in this step is to determine independently the financial and human resource needs to evaluate and implement the project and establish the availability of such talent internally or externally. Assessing the availability of internal resources is difficult, and identified gaps are always challenged under the pretext that the company has qualified employees to implement the project. Lastly, the costs and benefits of external resources to supplement company resources are evaluated along with estimates of management agreements and costs of secondments to the new entity created by the project.

Q: Why would a strategic forecasting team be assessing resource requirements?
A: Because it has performed SWOT and capabilities evaluations, knows how competitors implement strategic projects, and can validate that the right level and type of needed resources are available, can thereby assess the chances of achieving the forecast future state.

D. *Technical, market, and financial feasibility.* This is a step where other project team members perform most of what is required, but SDF team members identify and assess market changes and determine dependencies in the project implementation that would impact the realization of future-state predictions. The same holds true for the different fit assessments and the competency and capability evaluation because they have implications for future-state forecasts being realized. Again, this is because SDF team members understand the implications of residual project uncertainty and risks and play an active role in risk analysis and management. Also, based on their review of earlier company experiences they satisfy themselves of the project's technical, distribution and sales, and market acceptance feasibility. Lastly, they provide findings and insights from their analyses and evaluations and create the basis for the decision to move forward.

E. *Project grounding analyses and evaluations.* In this step, megatrend and subtrend analysis and evaluation of the implications for the company are the major SDF team activities followed by thorough industry and comprehensive political, economical, social, technological, legal, educational, and demographic (PESTLED) analyses. Then, changing customer needs, tastes, and preferences and market trends are examined to determine the impact on demand and product changes needed to remain competitive. And, depending on the nature of the project, other activities may include assessments of market research studies, economic conditions in a different country, and analysis of a different industry than the company's. Other important elements in this step are critical and objective assessments of not only the opportunities and threats facing the company, the company's competencies and capabilities vis-à-vis those of current key and potential competitors but also the implications of not doing the project.

F. *Development and testing of assumptions.* Assumptions are one of the main drivers of forecasts and financial evaluations, the others being actual operational data and postulated relationships. The first SDF team action in this step is to examine the sources and rationale of prevailing assumptions and scrutinize them by subjecting them to sanity checks. Then a review of industry standards, practices, and benchmarks and competitor and industry common assumptions takes place whose purpose is to understand the differences and determine which assumptions to adopt, which ones to revise, and which to drop. SDF team members review analog experiences and pick those that are reasonably close to the project's requirements. This, however, is not sufficient in large-impact projects, and the SDF team obtains independent opinion and input from subject matter experts on the validity of the assumptions. After these activities, the assumption set is updated, support from the project team is secured, and a universal set of assumptions is created for use throughout the project evaluation, planning, and implementation processes.

How does the SDF team test and validate assumptions?
This important and commonly asked question deserves an answer. In addition to the activities related to developing and testing assumptions discussed earlier, the SDF team performs the following checks:

- *Validating the consistency of assumptions with past experiences and observed company and competitor performance*
- *Confirming the basis of assumptions by validating the reliability of sources and information with logical inferences and empirical studies*
- *Assessing the sensitivity of results when some market drivers are omitted and their stability of models when values of drivers are changed*
- *Evaluating the reasonableness of results through assessing their implications for resource requirements*
- *Simulating the impacts of external changes and random, uncontrollable events to examine how they impact the major assumptions*

G. *Scenario development and planning.* This important step follows the testing of assumptions and model alternatives and begins by describing the events, activities, and their timing that lead to the future state. The scenario development activity is based on models of company operations, determining the sequence of events and possible timing, describing how the future unfolds, and picking a few realistic scenarios that depict the project team's collective view of how the company's performance will change with the implementation of the project. Hypothesis testing and sensitivity analysis are performed to confirm or reject beliefs and expectations and to see the impacts changes in key parameters make on the value of the project. "What-if" evaluations are performed by simulating the effects of shocks to the models and establishing the stability of model parameters and scenarios. Then, scenarios are ranked by some value creation measure, ease of implementation, and likelihood of success. After that, the most likely scenario is selected and a consensus forecast is presented to the project team.

H. *Forecasting due diligence.* This step involves the independent, critical, and objective assessment of elements that go into producing a strategic forecast. Sometimes, the forecasting due diligence takes place after assumptions are created and before scenarios are developed, but in all cases the following five major activities are common:

1. Customer survey evaluation and assessment of market research findings and their implications
2. In-depth assessment of recent competitive intelligence and creation of insights
3. Confirmation of the findings of megatrend and subtrend assessment, environmental scans, and industry and SWOT analyses
4. Validation of data sources and reliability, methods and models, and results of analyses
5. Evaluation of forecast implications and consistency with industry benchmarks and market realities
6. Synthesis of the findings to increase confidence in the forecasted value created by the future state
7. Comparison of SDF forecasts with forecasts from alternative sources, models, and assumptions

I. *Project validation and verification.* This is a broader activity than the forecasting due diligence effort that begins with verifying once again the consistency of corporate with client risk appetite and strategic with project goals and objectives. Then, a reexamination of project rationale, strategic fit, and consistency of goals and objectives takes place. A second-order validation and verification of the universal assumption set is made as well as revisions according to facts uncovered in the forecasting due diligence step. SDF team members take another look at the validated and verified forecasting processes, data, methods, models, and scenarios to eliminate inefficiencies and increase confidence in future-state projections. And to preclude future surprises, they verify senior management support and availability of appropriate resources to implement the project.

J. *Forecast development and assessment of implications.* This is a continuation of the scenario development and planning step where long-term forecasts are developed or fine-tuned through selected scenarios and adjusted by stakeholder feedback and senior management input. Reassessment of the forecast implications for corporate resources and support needed is another key activity in this step. Then, the identification of potential project impacts on key competitors and their likely responses is a crucial element followed by second- and third-order competitor response evaluations. Once these activities are completed, a final determination of the forecasts to drive subsequent planning and implementation activities is made. This determines the project support levels needed to implement successfully, namely, the financial and human resources and senior management involvement, commitment, and support required.

K. *Project uncertainty and risk assessment.* In this step, SDF team members revisit the uncertainty and risk issues. Project uncertainty and risk assessment is part of the forecast implications evaluation and begins with a review of the company history and earlier experiences in implementing projects like the one under consideration. Then, the SDF team determines which are controllable and which are uncontrollable factors, how the latter may impact the project, and how the former can be used to prevent or offset risks being realized. They also identify the sources of forecast risks, the timing of events and triggers, and the likelihood of threats occurring. In the meantime, an evaluation of market acceptance and technical feasibility of the project are performed and project uncertainty is determined. Overall project uncertainty is reduced to residual uncertainty through SDF team analyses and evaluations and the remaining risk and the company strategic posture are assessed.

To minimize the risks creeping into the forecast future state, SDF team members reevaluate the assumptions and forecast scenarios through "what-if" types of analyses. This brings SDF team members to the point where they are able to determine more firmly the project residual uncertainty, forecast realization risks, and risk impacts and how to deal with such threats. After that, they help the project team to create the risk management strategy and plans to avoid, shift, share, mitigate, or accept selected risks. By now, discussions about black swans have taken place and contingency plans are developed with input from the SDF team.

L. *Negotiated terms and conditions.* In most strategic projects, value created also depends on the terms and conditions negotiated with customers, partners, suppliers, or with sellers or buyers of a company. Hence, developing negotiation positions, identifying, quantifying, preparing, and evaluating trade-offs is probably the most important SDF team contribution in this support function followed by assessing proposals and counterproposals jointly with the Finance organization. With help from the SDF team, when the negotiations team develops negotiation positions it is fully aware of the costs and benefits of different concessions. A more involved SDF team activity is assessing the impact of negotiated terms and conditions on customers and the lagged effects on revenues, which determine the ultimate value created by the project. Thus, SDF team support enables negotiators to reach a balance of costs and benefits to each party and create win–win arrangements.

M. *Business plan creation and evaluation.* SDF team participation in a business plan creation and evaluation first occurs when a new entity is created by the project. Then, the support it provides is similar to that in other strategic projects. However, if the SDF team is not involved in its creation, its involvement in a business plan assessment starts with an evaluation of entity goals and objectives for consistency with corporate strategy, market realities, and project goals and objectives. Then, an assessment of marketing, sales, and distribution plans follows in light of recent competitor activities and announcements.

An operational plan review and evaluation also takes place to validate its achievability and verify its cost components. Also, the new entity's competitive standing is evaluated to determine how likely it will be able to deal with future competitive pressures. Business plan performance evaluation necessitates development of indicators, a management dashboard, an early warning system (EWS), and performance targets. SDF team members are actively involved in their development and assessment and provide insights into monitoring and evaluating performance. These evaluations are inputs to and supplement the financial plan evaluation. The last SDF team activity is to evaluate the implications of the new entity's operational and financial plans and develop inputs to the project team's recommendations.

N. *Project financing.* This is a highly collaborative effort between the SDF team and the Finance organization with a major component being the evaluation of project economics. More specifically, this effort involves the assessment of market potential, project acceptance, pricing options, customer impacts, and revenue generation and stability. The relationship of financial requirements to achieve the expected value is examined to ensure the adequacy of allocated resources. Help is provided by SDF team members to evaluate the possible impacts of internal versus external financing options on customers, product quality, pricing, and revenues and costs. The determination of on- and off-budget financing options is next and last is the assessment of the changes in the company's financial standing as a result of securing a certain type of financing. Here, the role of the SDF team is to determine consistency of financing requirements and terms and conditions with the realities of the external environment and expected project economics.

O. *FSR plan.* The FSR plan is a logical extension of the forecasting due diligence, project validation and verification, and risk management. It begins with a review of the preproject implementation analyses and postimplementation evaluations. In this step, the SDF team identifies and assesses potential project impacts on key competitors and possible reactions to the company's project. It provides inputs and assistance in the adaptive response planning effort to develop first-, second-, and third-order responses to competitor reactions. It also helps the project team to develop and assess tactical actions and moves to counter the effect of competitor reaction, and it escalates the need to strategic moves when it determines that tactical actions are insufficient to offset external threats and strategic initiatives are needed. Furthermore, the SDF team helps responsible stakeholders to create processes to monitor and assess project performance and communicate need for timely risk management actions.

P. *Financial analysis and recommendations.* Project financial analysis is clearly the responsibility of the Finance organization and the role of the SDF team is to provide long-term forecasts, perform sanity and reality checks, and to validate market information and results. In that capacity, the SDF team creates intelligence, models, and forecasts and provides inputs to financial models, such as market drivers, scenarios that generated the forecast, the baseline, and alternative views to be evaluated by financial and other planning models. It then validates the financial model assumptions and assesses financial model outputs against earlier company history and competitor experiences to ensure reasonableness and consistency. Key assumptions validated are those driving future capital requirements to accommodate expected output, cost of capital vis-à-vis industry averages, competitor practices, and demand forecasts. Helping in the development of joint recommendations is an important activity because a strong SDF team–Finance organization consensus conveys confidence in future-state projections and provides additional comfort to clients and senior management.

19.4 BUSINESS PLAN SUPPORT

The contributions of the SDF team in the development and evaluation of the business plan vary by type of project and whether support is provided to the company's or a new entity's business plan. In the case of a new entity, support is given across all its plans from the strategic plan to the monitoring and reporting plan. A key consideration in this type of business plan support is to ensure that linkage to the company's strategy and continuity of purpose are preserved in the new organization along with unbroken realism of expectations and objectives throughout the project. Other key considerations are an effective management of both the parent and the new entity expectations and an efficient integration of entity with parent company operations. The major areas of SDF team support in business planning are in the subplan components.

A. *Strategic plan.* This is the foundation of the business plan and SDF team support starts with a review of the entity's business definition, its mission and vision, and its goals and objectives. The purpose of the review is to make sure that a strategic intent drives operations, that the entity's mission and vision are in accord with the thrust of the parent company's business plan, and that there are no gaps that may affect FSR plans. Then a minienvironmental assessment and industry and market analysis are performed to make sure that the entity's objectives are consistent with market realities and key stakeholder expectations. Also, reality checks are conducted on entity goals and objectives to ensure consistency of entity strategy and goals and objectives with those of the company, determine their achievability, and identify issues to be addressed.

B. *Operational plan.* The operational plan is the second component of the business plan, which tells how the new entity will make things happen in accordance with business plan projections. In order to validate that and provide context, the SDF team investigates the external forces shaping the future of the business. It goes on to validate key assumptions underlying the operational plan and assess the validity of forecasts created by the new entity in which SDF team members have not been involved. An assessment of technology issues takes place along with an assessment of the entity's technological platform and systems compatibility with those of the parent company. The SDF team also helps to develop reasonable entity targets consistent with the projected future state of the company and ascertains that parent company support established in the business plan is indeed present. This helps to establish consistency of business plan expectations with financial and human resources availability. Lastly, the SDF team may assist the new entity's management team to develop incentives comparable to industry practices and create the entity's FSR plan.

C. *Human resources plan.* The SDF team's involvement in this area is primarily to share information, results of analyses and evaluations, and insights to identify required key positions to fill for business plan success. The SDF team shares the findings of its SWOT analysis concerning company and entity competencies and capabilities, and the availability of talent to determine the extent of business plan feasibility. It may also present alternative forecasting organization structures when creating new entities and identifies the skills and types of employees that successful competitors have used to implement business plans. Another SDF team role here is to validate the costs of human resources employed by the new entity, share its knowledge from prior project experiences, and provide inputs to the new entity's management incentive plans.

D. *Marketing plan.* Strategic project clients like to see the SDF team play an active role in this part of the business plan to temper the optimistic views of the Marketing organization with earlier company project results, industry benchmarks, analogs, averages, standards, and competitor experiences. To do that, the SDF team evaluates the competitive landscape and then reviews market research and customer survey results to determine the level of market acceptance of the new entity's product line offering. It conducts product cycle reviews to identify the timing of competitor events and threat actions. This way, the SDF team is in a position to help the project team identify new market opportunities and challenges.

When called upon to develop forecasts for new product development (NPD) and product enhancements and extensions that require large capital expenditures, the SDF team identifies gaps and issues in all aspects of new product launches. More specifically, the SDF team evaluates product attributes and advertising and promotional support levels needed to achieve established targets.

It also assesses the effects of price changes on demand and revenue using analytical models and prior knowledge and simulates the effects of changes in competitor advertising and promotional activities. Furthermore, the SDF team helps to identify support needed to maintain performance expected once competitor reactions materialize and the FSR plan is activated.

E. *Financial plan.* The financial plan of a new entity is a summary view of its needs, goals, and objectives, its revenue and cost-generating activities, and investment plans in light of external environment realities. The participation of the SDF team enhances the financial plan in the following ways:

1. Sharing its understanding of key competitors and intelligence about likely reactions, conducting sanity checks, validating financial model assumptions, and providing feedback on financial analysis results
2. Describing the scenarios created to develop long-term forecasts, simulating models of company operations, and providing sales and revenue projections based on alternative scenarios
3. Explaining the rationale of assumptions used and demonstrating the key revenue drivers, the order of their importance, and revenue sensitivity to changes in drivers
4. Performing the project economic evaluation required to pursue project financing options and validating or evaluating imputed costs in business plans relative to market norms

F. *Monitoring and reporting plan.* The support of the SDF team in this function consists of sharing knowledge of recent findings and analyses of the external and competitive environments and best practices. It begins with helping the project team to define appropriate variables as performance indicators to monitor and judge progress made toward achieving objectives. Then, its support helps to create a management dashboard and a scorecard to have instantaneous views of performance in key areas. In projects where there is a multitude of risks involved, the SDF team develops the entity's EWS to detect, evaluate, and communicate threats as they appear on the horizon. In some instances, it also lends its experience and provides valuable inputs to the variance analysis and reporting functions.

G. *Risk management plan.* The key SDF team activity in this part of the entity's business plan is to determine the residual uncertainty in the business plan, identify forecast risks involved, and investigate their sources. Once the nature of the risks is understood, it assesses the likelihood of, and quantifies the potential impact of risk occurrence on the project's expected value creation. The SDF team also assists in the development of actions and plans to mitigate, shift, share, insure, or accept business plan risks. And, just as important are the brainstorming sessions it leads to identify black swans and create contingency plans.

19.5 TARGET DEVELOPMENT SUPPORT

Targets are goals, metrics, and performance indicators established to be influenced or by management actions or changed by external events. They are part of a company's performance management system and usually take the form of performance measures intended to stretch management efforts to achieve certain operational objectives. Target setting is viewed as the means to drive increased performance, but it often has unintended consequences if they are not founded on well-considered principles. Target setting principles include the following:

1. Reasonableness and achievability within a balanced scorecard framework
2. Having minimal influence or triggering unfavorable consequences on related areas they can impact through interactions
3. Being validated and reported on a regular basis to ensure they are relevant

In addition to benchmarks, scorecards and dashboards are tools commonly used to measure and monitor performance using metrics to manage a project toward defined goals. The former tools are associated with alignment of effort and management of strategic issues while the latter are connected to measuring and understanding tactical performance issues. Managing performance at either level requires the integration of strategy, goals and objectives, initiatives, and metrics. The SDF team's process of creating targets for an entity created by a strategic project is shown in Figure 19.4, where its major activities are outlined. Notice that several analyses are required as well as research to establish the benchmarks needs and identify sources for them.

A benchmark is a point of reference by which company performance can be measured, and it is also used to measure a competitor's areas of operations according to specified standards in order to compare it with and improve one's own operations. Benchmarks and scorecards are useful tools for reporting on the quality of a process, the status of a project, performance on the outcome of a decision, efficiency of operations, and progress on future expectations. Target-setting benchmarks are associated with well-managed companies or those known for being best in one area or another, and the questions in which SDF team members provide inputs are the following:

1. What function, area, process, performance, and financial measure to benchmark
2. What types of criteria target decisions are expected to meet
3. What sources can be tapped to obtain reliable data and information

Three additional questions to be answered are to what extent external benchmarks can be adopted, what modification may be required to apply them to the specific company situation, and how to interpret the results of comparing company measures against benchmarks.

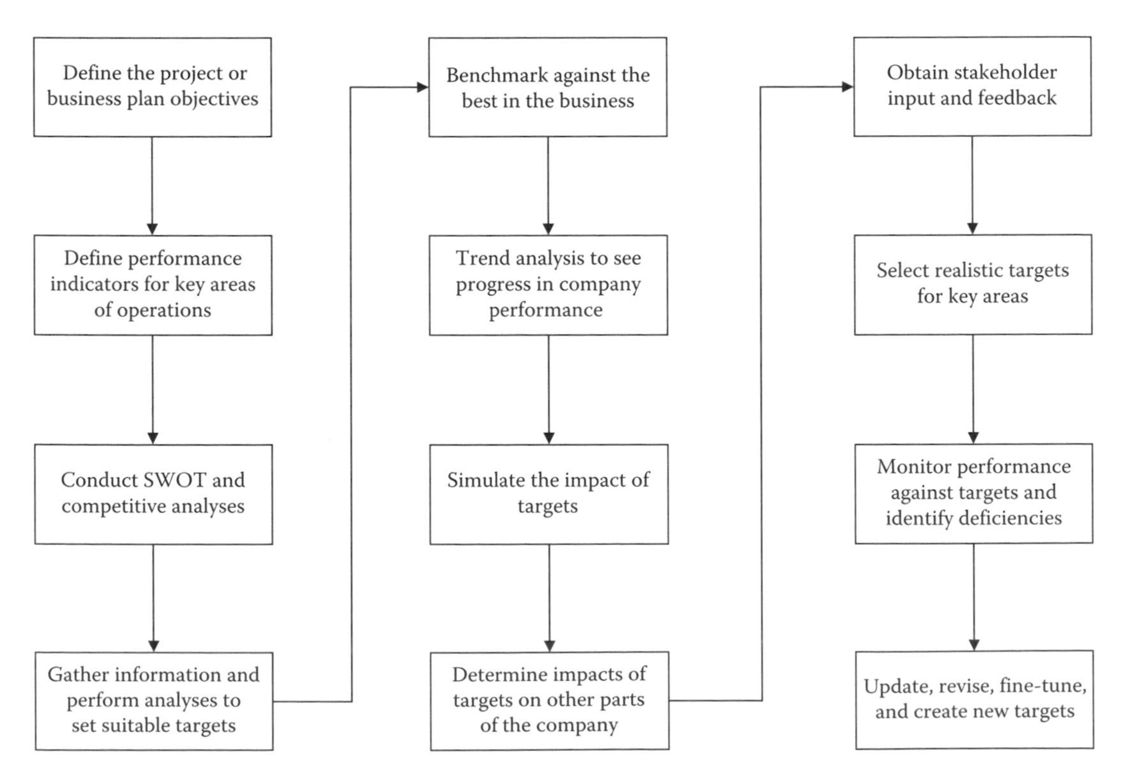

FIGURE 19.4 SDF approach to developing performance targets.

19.6 SOUND PRACTICES AND SDF TEAM CONTRIBUTIONS

Some kind of SDF support is present in all business case, business plan, and target development, and for that reason a large number of clients shared their views in our discussions on success factors and how SDF teams contributed to project success. They are listed separately in a summary format, but some of those success factors are also direct SDF team contributions to the strategic project evaluation and implementation.

19.6.1 Sound Practices and Success Factors

The sound practices and success factors identified by participants in business case, business plan, and operational target development include factors in the following categories:

A. *Decisions and conditions.* In this category, the practices considered essential to success are the following:

1. Everyone understands the project decisions to be made, why other decisions are not made, and possible options are laid out without attribution to team members.
2. There is sufficient project team interest to go forward with the project and strong project support, not a lukewarm consensus, and there are no strong objections from key stakeholders.
3. Only key aspects of the project economics receive attention in business cases and business plans, otherwise too many distractions are introduced that complicate decision making.
4. There needs to be stakeholder buy-in of business case recommendations, which provide a common basis to move forward and unwavering senior management and client support, involvement, and commitment.

B. *Analytics and evaluations.* Two aspects of the analyses and evaluations in business cases, plans, and targets considered important are that

1. The merits of the methodology and business operations model implementation must outweigh the problems likely to be encountered.
2. Sound strategic project rationale and clarity of purpose and all aspects of project assessment are fully addressed by SDF team analyses and evaluations.

Completeness of analyses, a balanced perspective of evaluations, and brief and to the point, not voluminous documents are important considerations. Also, the analyses performed must be factual, reliable, validated, and well considered and recommendations are fully supported with facts and evidence. In addition, balance of costs and benefits in the analyses and evaluations and reporting systems are important in developing business cases, plans, and operational targets.

C. *Validation and confirmation.* The validity of the assumption set in forecasting and financial models and high quality of data, methods, models, and scenarios is central to sound business cases, plans, and targets as is the reasonableness of the planning approach and use of industry benchmarks and reference points. Reasonableness of business plan objectives and targets and adequate management and resource support to meet targets is crucial, and the use of appropriate benchmarks against which to compare the findings of the business case process is vital. Just as important, however, is to confirm clearly articulated project goals and objectives, adherence to business case and business plan development processes, and use of best-in-class benchmarks to set operational targets for a new entity and to control its management team's expectations.

D. *Implementation competencies.* Here, sound implementation plans and realistic work plans to ensure timely and correct handoffs and smooth transition of responsibilities are most important, done in the spirit of teamwork, high levels of cooperation and coordination of activities, and management of internal politics. Superior technical qualifications and interpersonal skills are fundamental requirements to go along with brevity of business case and business plan documentation and the capacity to convince decision makers of their soundness. On the other hand, adequate qualified human resources availability and funding of project implementation and new entity operations support are essential. But unimpeded 4Cs with the parent company are crucial as are strong linkage of compensation to performance targets and appropriate management team incentives.

19.6.2 SDF Team Contributions

The SDF team processes, analyses, evaluations, and recommendations in the development of business cases, business plans, and performance targets judged as significant contributions to project success are in the following areas:

A. *Mindset and approach.* In this group of sound practices and success factors, a holistic approach from the feasibility study down to defining indicators, setting performance targets, and clarity of thought are SDF team characteristics. Providing focus to core business issues and continuity of purpose throughout project evaluation and implementation are highly valued contributions. Orientation toward producing and delivering convincing arguments for recommendations based on this team's strong relationships and links with counterparts in other organizations is also important. The knowledge transfer; efficient data interfaces; and information storing, processing, and sharing capabilities are decision success-enhancing contributions. Lastly, the performance threshold improvement recommendations and frequent assessments and judgments of the situation as projects move along the execution process are important because they allow for efficient course correction.

B. *Structure and process.* In this cluster of SDF team contributions, their help in defining a good project structure and creating process that fit that structure include the following:

1. Thorough treatment, probing of issues, and depersonalized discussions where viewpoints, perspectives, and the complexity of issues are considered and all opinions are examined, but more weight is placed on those supported by facts
2. Their research, evaluations, and analyses create a sound basis for quick and effective senior management decisions by holding regular stakeholder and senior management briefings to eliminate surprises
3. The SDF team's ability to distill lessons learned from past and current projects and use them in other business cases, business plans, and operational targets makes the job of project teams easier
4. Creating processes linking entity objectives with operational targets and inclusion of performance incentives to increase chances of targets being met

C. *Analytics and synthesis.* The methodical, structured, and disciplined approach of the SDF team's analyses and synthesis of findings is valuable and their notable contributions here include the following:

1. Independent, critical, and objective assessments of business operations, the external environment, and the company's capabilities vis-à-vis key competitors
2. Evaluation of scenarios, forecasts, forecast implications and risks, and support requirements in all phases along with seasoned synthesis without compromises, give and take, or watering-down of evidence
3. SDF team member willingness to share knowledge and experiences for the project team to structure profitable projects and FSR plans

4. Accuracy of assessments and clarity of descriptions with minimal distractions introduced by the volumes of data analyzed
5. Identification of information needed to make a business case decision, the quantification of qualitative information, and testing to make it usable
6. Persistence in continuous checking of facts, figures, analyses, and evaluations in the development of the business case, business plans, and performance targets
7. Development of FSR plans and excellence in forecasting project management, which are the most valued contributions

D. *General support.* In this class of contributions, the SDF team's readiness to provide general support and no objections to client requests for exploring alternative methods is noteworthy and includes the following:

1. Helping to make the decision that felt right when emergence of consensus forecast has taken place and presented to clients and senior management
2. Soliciting fairness opinions from external experts to add credibility to forecasts and the business case quality, which provide added comfort to decision makers
3. Checking the basis of assumptions and conducting sanity checks on the results of evaluations that produce positive results along with multiple layers of sanity checks and the forecasting due diligence
4. Validating and verifying processes, models, data, and forecasts to enhance the factual base, increase confidence in the recommendations, and speed up the decision process
5. Setting strategic return on investment targets in the business plans for the new entity, which include financial performance, market position, technical leadership, and progress toward development of competitive advantage
6. Creating a management dashboard and an EWS capability to manage risks more effectively

Part IV

Implementing Strategic Decision Forecasting

20 Best-in-Class Strategic Forecasting Teams

This book is about helping strategic forecasters provide excellent support to decision makers so they can make sound strategic decisions effectively on a consistent basis. In strategic decision making, forecasting involves describing alternative future states, what it takes to get there, and how those descriptions are created through long-term forecasts. And because all decisions are based on forecasts, whether explicit or implicit, good strategic decisions are based on sound long-term forecasts. Thus, if a company can consistently generate sound long-term forecasts, the right strategic decisions are made consistently. And, making strategic decisions consistently leads to getting a competitive advantage, which is partly traced to the actions of a competent strategic decision forecasting (SDF) team.

All strategic business decisions involve forecasting based on some type of environmental assessment, market research, competitive analysis, analytical modeling, and scenario development and planning. The results of these activities form the foundation for tactical and strategic decisions and planning activities around them. Marketing and sales cannot plan programs and activities without demand forecasts, which are the starting point of financial analysis and planning and the front end of supply chain management. Tactical forecasts may have some influence on corporate strategy updates, but strategic forecasts are key inputs to the development of corporate strategy. Strategic forecasts determine the outcome of R&D projects, product portfolio management (PPM) adjustments, and large capital expenditure (capex) or project financing approvals. Also, they are of crucial importance in successful restructure and turnaround projects, and long-term demand forecasts form the basis of strategic partnership valuations, licensing and co-marketing projects, and M&A valuations. These relationships are displayed in Figure 20.1, which depicts forecasting as the foundation of all planning. And, all evidence supports the conclusion that the better the forecasts, the more effective the planning activity, which, in turn, necessitates well-structured, staffed, and managed SDF teams generating reliable forecasts.

SDF teams are collaborative forecasting organizations and their processes are based on teamwork, free information exchange and knowledge sharing, and close coordination with project team activities. In all projects, SDF team members create forums for project teams to combine product, industry, environmental, competitive intelligence, corporate strategy, market knowledge, analytical forecasting models, and knowledge of future events to create plausible scenarios and develop long-term forecasts. In short, these forums are created to bring together all corporate resources and collective knowledge. Therefore, from the SDF team perspective, the reasons that good forecasts are really needed are to build the team's own confidence in the forecast, to develop all-around credibility, and to manage key project stakeholder expectations. Over time, reliable strategic forecasts help clients and the SDF team to command resources needed, to manage project issues and politics more efficiently, and to plan and navigate through future uncertainty more effectively. Thus, reliable forecasts enable SDF team to build best-in-class organizational competencies and position SDF team members to become trusted advisors in strategic projects.

In addition to setting the stage for all future planning and creating a forum to bring together diverse business functions, participation of best-in-class SDF teams in strategic projects yields benefits such as the following:

1. Helping the project team to manage future uncertainty and risks effectively
2. Determining objectively the value and consistency of corporate strategy

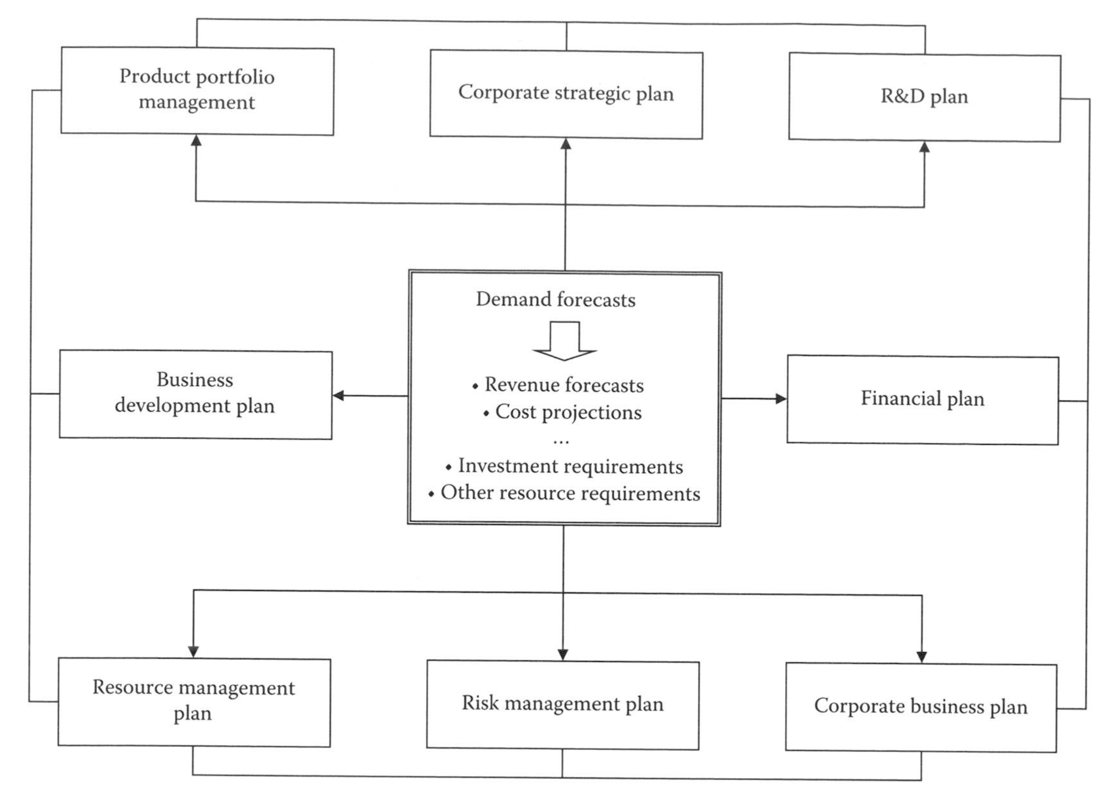

FIGURE 20.1 The view that forecasting is the foundation of all planning.

3. Minimizing subjective inputs and distortions to the decision-making process
4. Assessing alternative strategies, projects, and initiatives objectively and selecting the best options
5. Enhancing the speed and accuracy of evaluations and reducing costly unprofitable investments
6. Improving credibility with industry analysts on the company's future plans

20.1 GOVERNANCE OF SDF TEAMS

The usefulness of strategic forecasting rests with the utility clients get and communicating it to the organization, facilitating the creation of more accurate company financial predictions, and conserving scarce capital resources. Its usefulness also extends to assessing the direction of the business, providing a basis to value strategies and programs, and reducing duplication of effort. Chapter 4 addressed the charter of SDF teams, which provides sound and reliable support to strategic decisions, the needs of which, in turn, define the scope and functions it performs, and the extent of its activities in support of strategic decisions, projects, and growth opportunities. Best-in-class SDF teams are structured to provide high-quality client support and meet senior management expectations concerning actionable decision recommendations. A summary of client requirements and the SDF team support functions performed to meet those requirements are shown in Figure 20.2.

The value proposition of best-in-class SDF team includes a few key elements, such as the following:

1. Identifying alternative growth opportunities and minimizing uncertainty and mitigating project risks
2. Creating the evidence base for client decisions and reducing time to decision making

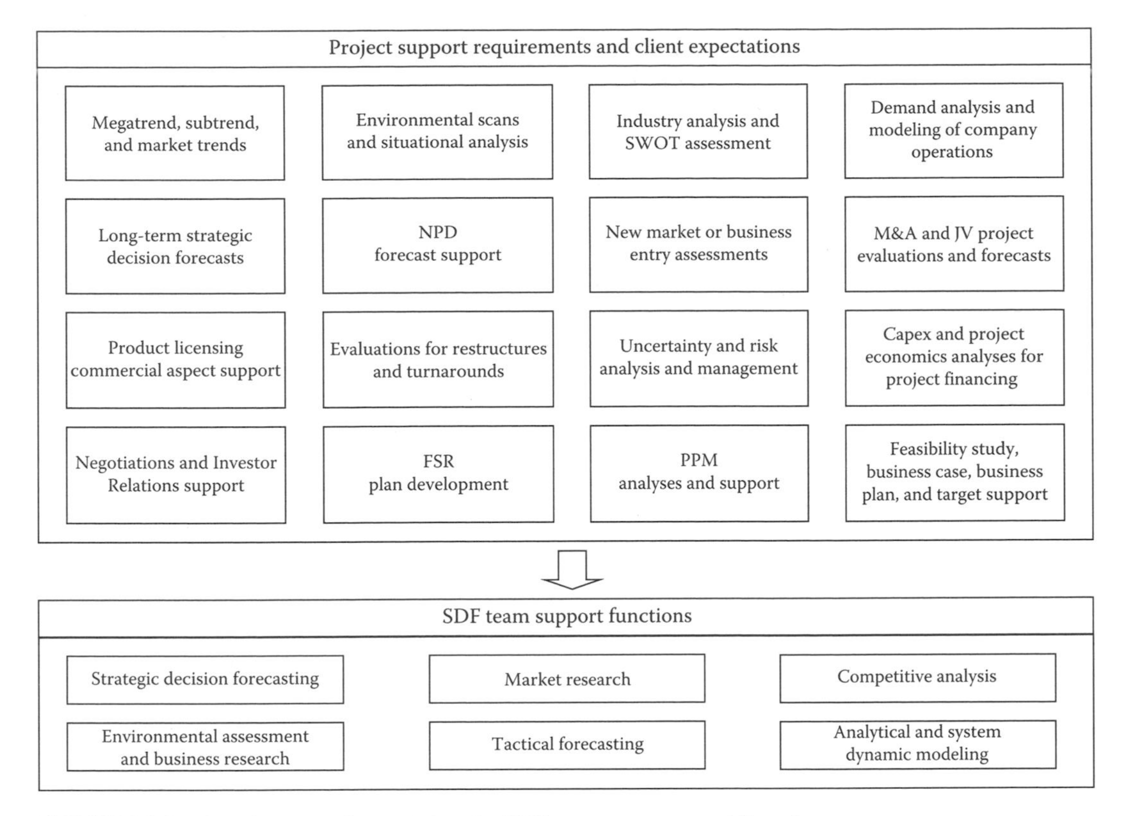

FIGURE 20.2 Requirements determining the SDF team structure and functions.

3. Developing performance measures and creating project dashboards and early warning systems (EWSs)
4. Enhancing the SDF teams' function from technocratic to serving as client trusted advisors

Under this SDF team value proposition, the scope of strategic forecasting entails five major functions:

1. Developing intelligence, insights, and knowledge databases
2. Creating the common assumptions for the project and being the central agent of assessments, analyses, and evaluations
3. Modeling company operations, assessing project uncertainty and risks, and creating future-state scenarios
4. Synthesizing the results of findings and developing long-term forecasts and recommendations
5. Creating performance indicators, monitoring project results, and reporting

The SDF team's business definition was discussed in Chapter 4, but little was said about reporting structures and what makes a team best in class. An effective structure is derived from the client support requirements shown in Figure 20.3, where the SDF team is part of a Business Research and Analytics group. Under this structure, the SDF team reports directly to the Strategic Planning or to the PPM organizations and in a few cases to the Corporate Planning group. And, depending on the support needed, it reports on a dotted-line basis to corporate senior management or the R&D, Marketing and Sales, Business Unit heads, Finance, Business Development, Corporate Planning, and Investor Relations groups.

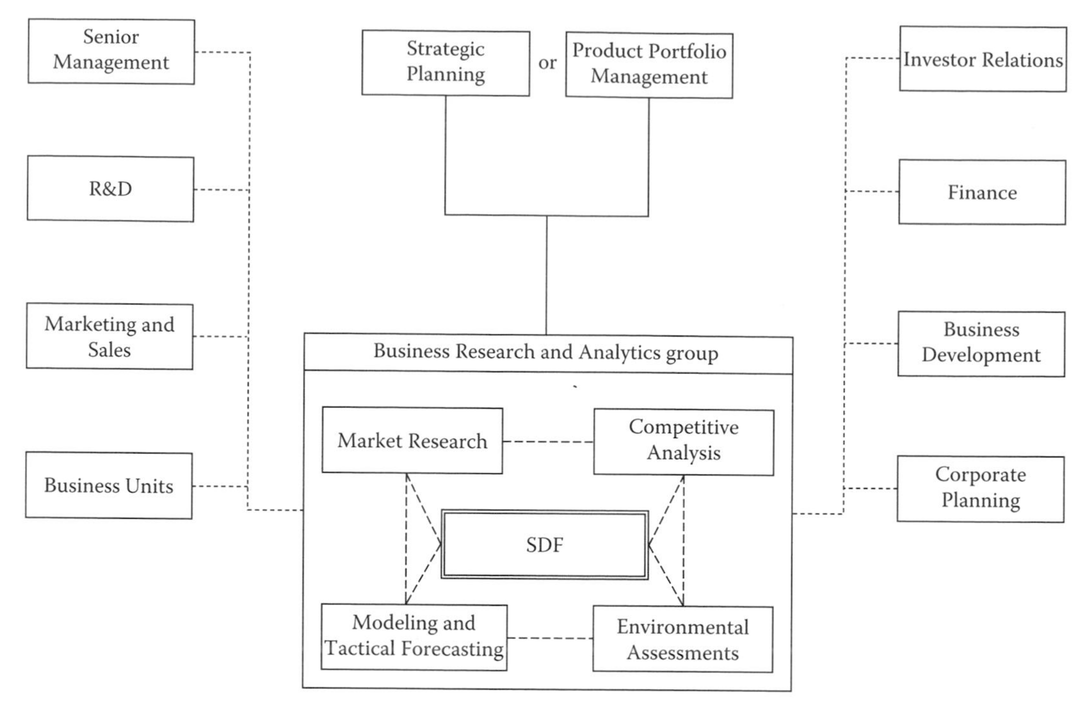

FIGURE 20.3 Sample SDF team governance structure.

Professional SDF teams are a relatively new paradigm, and credible long-term forecasts come out of teams that have structures similar to the example structure shown in Figure 20.3, which usually includes

1. A competitive analysis subteam with thorough knowledge of the industry, the main competitors, and the own and competitor products and services. This team has a good understanding of the company's operating environment and its markets and handles all issues pertaining to competitor activities.
2. A market research subteam that engages in the collection and evaluation of data about customer preferences for existing and new products or services and information about the effectiveness of marketing programs. The function of this subteam is to profile trends in consumer markets and changing consumer behavior, work with market research vendors, and evaluate their findings.
3. A modeling, demand analysis and tactical forecasting subteam composed of highly qualified quantitative forecasting experts that produce short- and medium-term forecasts. This team handles supply chain and logistics-related issues and, at times, lends modeling support to strategic decision forecasters.
4. An environmental assessment subteam with expertise in monitoring and assessing the effects of events, megatrends and subtrends, and PESTLED (political, economical, social, technological, legal, educational, and demographic), industry, and market developments on the company's future performance.
5. A strategic forecasting team made up of seasoned and broad-thinking managers with a good understanding of the industry, long-term forecast development and applications, and decision support experience. This team works closely with and leverages the expertise of the other subteams to develop strategic decision forecasts and conduct project evaluations.

From the discussion of SDF applications in Chapters 8 through 19, it is clear that the SDF team's responsibility spans across all areas in its charter. The scope of its operations makes it accountable for project managing the strategic forecast function and the knowledge and insight creation management process from start to finish. However, the nature of the support provided to various clients and functional areas is determined by the relationships with them, senior management, counterparts in other functions, and the nature of the decision to be made. Also, the business problem to be solved, the project type, the growth opportunities and investment proposals presented, and the resource constraints present influence that support. For these reasons, the various functions, activities, and tasks required to meet the SDF team's charter mandates deal with methodology, processes, analyses and evaluations, and tools.

20.2 FORECASTER RESPONSIBILITIES AND QUALIFICATIONS

To eliminate the gaps and shortcomings of the current state of forecasting support for strategic projects, SDF team members first seek to understand the company's business, the decision maker's needs and the decision rationale, the corporate strategy, the risk tolerance of the firm, and the perspectives of decision stakeholders before attempting to model the workings of the company business. Since SDF teams project manage the forecasting function, they are de facto keepers of assumptions, internal and external data, competitive intelligence and business insights, and the company's Intranet database and knowledge center. And, in that capacity, they play a major project coordination and communication role.

The strategic decision forecaster–client interaction is focused on collaborating, jointly developing inputs to and using the forecast solution, assessing forecast implications for company resources, evaluating competitor reaction to a project, and managing forecast risks. Also, a major responsibility of SDF teams is to provide assistance in describing future-state scenarios and develop forecasts and actionable recommendations to achieve them. In that context, the main skills and experiences needed to be effective in the roles they play and the functions they perform include the following qualifications:

1. Forward thinking, outward looking, ability to work with uncertainty, and willingness to deal with changes
2. Able, humble, and effective in interacting successfully with all levels of management
3. Efficient in developing clear vision and reasoning coupled with proactive, participative, and attentive listening
4. Asking questions, proficient in pattern and trend recognition, and capable of dealing with uncertainty
5. Deciphering interactions among many variables, choosing forecasting methods, and developing models suited to particular decision problems
6. Possessing knowledge and experience in statistical, numerical modeling, financial analysis, and related principles and processes
7. Understanding the limitations of strategic forecasting and appreciating the power of relationships in decision making

Strategic, future-focused thinking about what the company needs to accomplish and how to do it determines the core competencies needed: that is, the expertise, organizational skills, and systems. Also, maintaining focus on the long run, sorting out the important elements from piles of data of different market segments, and establishing their individual significance to company performance are expected of SDF teams. Such qualifications also require the following:

1. Ability to multitask, flexibility, situational awareness, decisiveness, and being well organized to deal well with pressures of tight deadlines and high expectations
2. Passion for strategic forecasting, knowledge creation and sharing, and capacity to visualize, conceptualize, and ask "what-if" questions for continuous learning and helping clients use forecasts properly
3. Understanding the client business, their personal needs and expectations as well as those of senior managers and key stakeholders and managing expectations and organizational issues affecting forecasts
4. Ability to apply critical analysis capabilities to financial planning and decision making in the context of constraints, uncertain opportunities, and rapid change
5. Interpersonal skills and personal relationships to be a business partner, a sounding board, a trusted advisor, and a credible resource of guidance to clients and other stakeholders
6. Expertise to be a coach and instructor on issues relating to principles of analytics, modeling, forecasting, the methods and tools used, and how they can be applied successfully

Observing with a critical eye and assessing objectively the external environment, the industry, and megatrends and subtrends and their impacts on company operations is crucial to understanding the problems surrounding the decision to be made and helping clients identify and clarify alternative approaches to meeting their needs. That expertise entails industry knowledge, modeling and forecasting skills, understanding the stakeholder organization's culture, managing existing client contacts and relationships, and ability to create knowledge using functional relationships. Creating analytical frameworks and models to develop an SDF solution and provide answers to questions raised, using industry best practices and benchmarks are other parts of the needed SDF team member expertise. Performing root cause analysis to generate fact-finding questions and get clarifications is a skill necessary in breaking complex problems into their component parts, analyzing them, and synthesizing the findings to create a range of forecast scenarios. Also, the SDF team members' ability to apply knowledge of business markets, trends, and judgment to determine project value creation and select the most likely option to achieve that value is another important capability.

Understanding of system dynamics elements, identifying and doing visual mapping of issues, and determining system relationships are additional qualifications for modeling and performing scenario simulations. Also, experience in determining the right methodology and identifying the right analogs to use along with skill in identifying, obtaining, and validating internal, external, and special purpose data, and creating insights and long-term forecasts is paramount. Competence in creating plausible scenarios and performing meaningful "what-if" simulations for identified and random events and incorporating forecast adjustments due to their occurrence is important to describe future states and understand what is involved in attaining them. Additional SDF team expertise is required in

1. Managing the collaboration with strategic forecasting vendors and external experts.
2. Evaluating and critiquing vendor work and the analysis and findings of the Market Research, Competitive Analysis, and Environmental Assessment groups.
3. Managing the forecasting function and shepherding forecasts internally to obtain consensus effectively.
4. Briefing senior management on the particulars of the forecasts and obtaining their concurrence.

Again, developing distinct plausible scenarios describing business operations and alternative business futures based on sound methodology, expert analysis and evaluations, effective processes and practices, and experienced project management is vital to creating sound long-term forecasts.

All SDF team members must possess the required interpersonal skills in order to provide a high degree of comfort to decision makers about the range of project outcomes based on their recommendations. They must have the ability to identify and manage project uncertainty using controllable levers and have readily available environmental impact, market change, and competitor response plans. Establishing close working and personal relationships with counterparts in internal planning organizations and external industry analysts and experts and leveraging these relationships to create valuable intelligence and insights is a prerequisite. But, a highly valued quality is skill in sharing knowledge with project team members on methods, data, analyses, models, results, forecasts, evaluations, scenarios, and the synthesis of the results with clients in a simple and easy-to-understand manner. Along the interpersonal skills line, effective communication, coordination, cooperation, and collaboration (4Cs) with forecast stakeholders to resolve conflicting interests and perspectives and arrive quickly at consensus solutions are required talents.

At the technical level, SDF team leaders expect team members to be able to deal with project uncertainty, strive for forecast certainty, and then question it and mistrust it. This process of getting good forecast results in SDF solutions comes with considerable amount of confidence with which forecasters communicate to clients. Over long forecast horizons, black swans do occur; and while there are no ways to predict when they appear, forecasters need to support project stakeholders prepare for such eventuality. Here, the views of SDF teams are shaped by two principles:

1. Interpreting results of analyses and findings of evaluations should take place in light of reality checks in order to develop reliable recommendations and communicate them convincingly to decision makers.
2. Models work well in evolutionary environments, but break down in revolutionary environments. Hence, they need to be properly adjusted and enhanced to be relevant.

A different set of skills and qualifications required of SDF team members includes forecasting function management skills. That is, the ability to build compatibility in cross-functional teams, curve out a participation niche, and plan and execute a forecasting plan. Also, adeptness in anticipating major problems, defining deliverables, and assigning responsibilities to project stakeholders is expected. However, ability to communicate effectively, deliver consensus forecasts in an efficient manner, and being part of change management is paramount to success. Experience in cost–benefit analysis, knowledge management and transfer, project evaluation, financial reporting, business case and business plan development and evaluation results in SDF team members being consulted in every project. Additionally, proficiency in developing management dashboards, balanced scorecards, and EWSs; applications of benchmarking best practices; and the ability to link forecasting to activity-based management are highly valued skills. Furthermore, Six Sigma, business process mapping, process improvement, total quality management (TQM) skills and reorganizations, and turnaround experiences are prized qualifications. Last and equally important is the SDF team members' ability to deal with and shun extreme reactions to unavoidable failures in analysis and forecasting for strategic decisions and criticisms that go with such failures.

Strategic forecasting is a highly collaborative approach, which involves forecasters and diverse stakeholders doing meaningful scenario development, planning, and assessment jointly. This is an intensive issue management effort, which requires creating forums to bring together and incorporate all knowledge residing in the company in scenarios and long-term forecasts. Strategic forecasting also requires an intensive strengths, weaknesses, opportunities, and threats (SWOT) and demand analyses from which SDF team members draw impact conclusions and incorporate

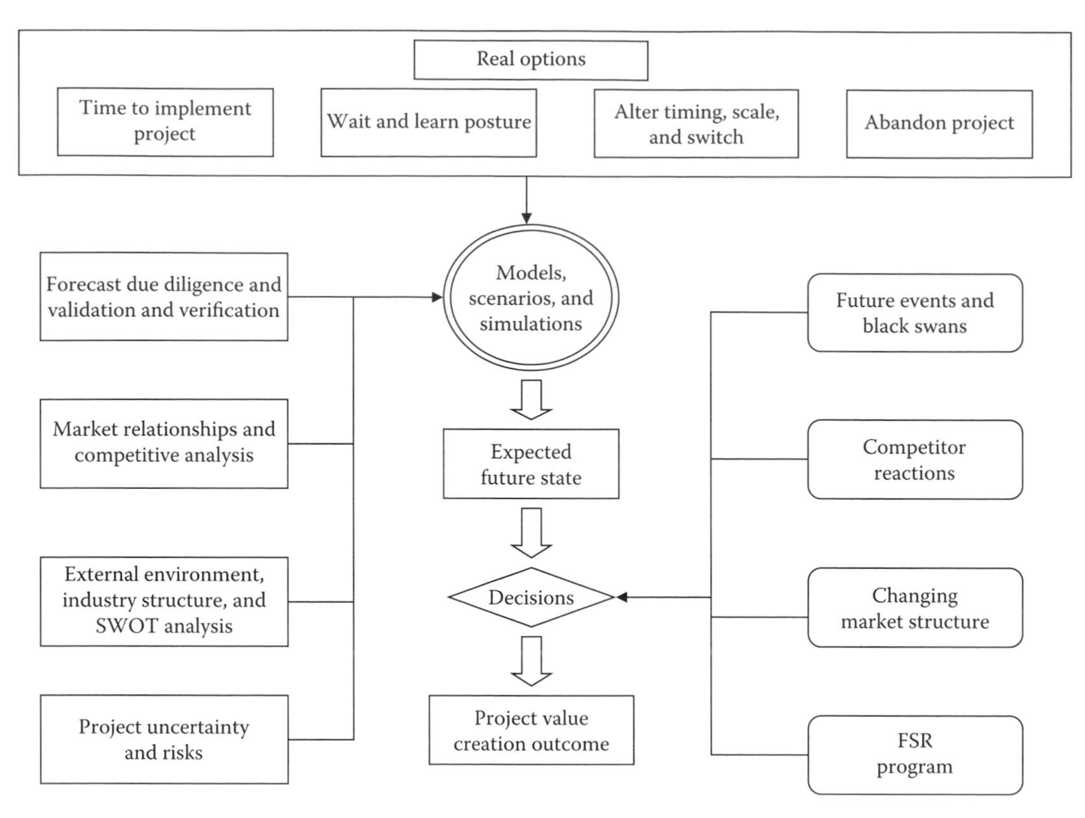

FIGURE 20.4 Real options, causal, and system dynamics framework.

them in forecasting models. In their capacity as trusted advisors, SDF team members must be skilled in evaluating multidimensional implications of forecasts and in the identification, analysis, and management of project uncertainty and risks. To do that, professional competence, confidence, and composure are crucial at all times.

One SDF team member technical qualification that is always assumed, but one that needs to be ascertained, is experience in combining causal and system dynamics modeling with real options thinking to enhance models, forecasts, and the chances of future-state value creation by projects are indeed being realized. The approach used by SDF team members to implement that integration and its key elements are outlined in Figure 20.4. The requirement of highly developed SDF team member communication and interpersonal skills have been stated repeatedly, but the important skills often overlooked have to do with articulating the SDF team's best-in-class vision and communicating the benefits of this vision. This entails SDF team member ability to

1. Present effectively the basic SDF guiding principles to the client, key stakeholders, and senior management.
2. Define the scope of the strategic forecasting function and its goals and objectives and assign roles and responsibilities to project team members.
3. Identify not only processes, inputs, outputs, linkages, and dependencies but also client needs, biases, and expectations.
4. Obtain stakeholder agreement on roles and responsibilities and reach agreement on forecast assumptions effectively.

Summary of SDF team backgrounds: *It is worth repeating that effective SDF teams require associate backgrounds to include training, skills, and qualifications in the following areas:*

- *Ability to create personal relationships and manage expectations, understand what needs to be done, conceptualize, and model company operations, and produce sound forecasts*
- *Identification of trends and market structures, quantification of market interactions, knowledge creation and dissemination to clients, and professional delivery of forecasts and presentations*
- *Effective communications and good understanding of what is involved in the strategic planning, financial planning, business development, new product development, portfolio management, and business planning functions*
- *Ability to obtain project team agreement on objectives, numbers, assumptions, models, and implications*

These are crucial in getting to consensus forecasts efficiently.

To communicate the best-in-class benefits to the wider organization, SDF team members need to convince clients and key stakeholders that their interests are aligned with those of the SDF team. Subsequently, they need to demonstrate that SDF forecast solutions are the best long-term forecasts in the sense that they

1. Incorporate the perspectives of the company, the client, and the forecaster and are, therefore, well balanced.
2. Represent the most-informed decision-making foundation created under the most effective processes.
3. Come from engaging top talent and best practices throughout the forecasting process.
4. Have clients coached on the proper use of long-term forecasts to help to ensure that future states are realized and project risks are minimized.
5. Lead to the prospect of creating some type of competitive advantage over time and with appropriate management of the SDF function.

The primary support services provided by best-in-class SDF teams are summarized in Table 20.1 and shown by major project sponsor organization. Notice that most of the general client support is rendered to all project sponsor organizations to one degree or another.

TABLE 20.1
SDF Team Support Provided by Project Stakeholder

Client support: problem definition; decision analysis, reconciliation, execution; external environment evaluation, development and management of assumptions; competitive, industry, market, SWOT analyses; uncontrollable events and black swans identification, megatrend and subtrend evaluation; data and information collection, modeling and testing; scenario development and planning, communication of forecasts, uncertainty and risk assessment and management, project IVV and postmortem

Strategic Planning support: business environment and situational analysis, examination of alternative strategies, strategy development inputs, strategy evaluation, structural change and market trend identification, development and testing of assumptions, strategic objective sanity checks, assessment of reorganization and restructuring impacts, identification of strategic plan uncertainty and risks, program monitoring

(continued)

TABLE 20.1
Continued

Product Portfolio Management support: external environment assessment and projections, evaluation of marketing and sales projections and external forecasts, factor relationship identification and measurement, industry benchmarks, inputs to product uncertainty and risk assessment and management, capital allocation decision support

Finance and Corporate Planning support: development and updates of assumption set, assessment of financial analysis implications, financial planning assumptions, inputs to corporate business plan, project economics for financing, industry benchmarks and sanity checks of financial targets

Business Development support: Business Development strategy, sanity checks and management of expectations, M&A and joint venture analysis; technology, product, and service licensing commercial aspects support; new market or business entry analysis; feasibility study, business case, business plan, and target creation; performance measures, evaluation of alternatives

R&D Support: Market research and competitive analysis, inputs to R&D strategy, external environment assessments, industry benchmarks, analogs, technology evolution, trend assessment, scenario development and planning, new product development state forecasts, sanity checks, assessment of forecast implications, assessment of external forecasts, evaluation of new product targets, assessment of future opportunities and support levels required

Investor Relations and restructuring and turnaround support: forecast assumptions, methodology and models used, project uncertainty and risks, forecasts, implications, and sanity checks performed, preparation for industry analyst meetings, investor guidance support, competency and capability gap analysis, situational analysis, industry benchmarks, analogs, standards, competitive analysis, process mapping and evaluation, management of expectations, forecasts and validation

20.3 BEST-IN-CLASS ATTRIBUTES AND PRACTICES

The best-in-class SDF team attributes and quality elements entail distinction in all areas of operation, but most importantly in

1. Organizational structures, functional links, and integration, and personal relationships.
2. Organizational culture, core values, and mindset.
3. Approach, frameworks, methods, techniques and tools, and processes used.

A. *Quality and effectiveness.* Quality elements include the background, training, skills, and qualifications of SDF team members followed by forecast performance metrics, knowledge management systems, and associate rewards and recognitions for their contributions in the strategic decision-making process. For SDF teams to be effective, appropriate governance structures are required along with ongoing contacts with business intelligence agents, close functional linkages, people networks, strong personal relationships, and enduring senior management support. The right culture, values, and mindset of the SDF team are very important in producing quality support along with a balanced leadership style in intelligence integration and forward thinking. Additionally, the shared corporate values of excellence in client support, continuous innovation, and forecast performance improvement must be exhibited at all times by SDF teams.

B. *Approach and practices.* In the approach, methods, and practices area, best-in-class attributes include sound SDF processes and practices; appropriate methodologies; reasonable assumptions; the right data and models; and maximum use of organizational intelligence, collaboration, and coordination. Inclusion of sanity checks, benchmarks, and comparison with industry norms are required as are identification and management of project uncertainty and risks. Inputs to and evaluations of forecasts by other experts and forecast triangulation are always useful because they incorporate all knowledge available and ensure consistency with history and forecasting expert experiences. SDF team support to corporate, business development, R&D, and portfolio management strategy development is essential to build and maintain competitiveness through ongoing assessment of environmental and market conditions. Also, ready access to and understanding of market data, competitors, and industry dynamics is expected, and business process mapping, modeling, and client briefings on the state of the business are vital SDF team capabilities to produce sound forecasts.

C. *Performance measures.* To know that an SDF team is operating as best in class, meaningful performance measures for the forecast, the SDF team, and the project need to be defined and monitored. Some of the performance metrics often used include the following:

1. Efficient development of forecasts; that is, minimum turnaround cycle and within budget
2. Timely delivery of solutions and forecast acceptance by clients and client and senior management satisfaction
3. Forecast accuracy appraised by percent-error deviations and by broader, Six Sigma measures
4. Efficiency of project implementation and effectiveness of the forecast realization plan
5. Progress in achieving project milestones and objectives relative to plans
6. Value created by the project using net present value (NPV) and strategic return-on-investment measures

Other performance measures, depending on the magnitude of project impact, include quality of the forecasting due diligence, sanity checks, forecast risk assessment and minimum residual uncertainty, and future-state realization (FSR) planning. Naturally, ongoing TQM and continuous forecast reliability improvement are expected to be present performance measures along with indicators of SDF team productivity gains.

D. *Knowledge management.* When it comes to knowledge management and systems, SDF teams are expected to lead the creation of database systems to house market data and play a key role in the knowledge management function. This is done by defining, populating, and updating the company's Intranet repository of data. This database contains intelligence, insights, specific project assumptions, findings of analysis, and results of evaluations. In addition, it includes decisions made and decisions not made, alternatives considered, and lessons learned from postmortem analyses. Acquisition of data and demand analysis, modeling, system dynamics, and statistical forecasting software is a shared responsibility with the tactical forecasting group. And, an active SDF team role in project postmortems and the application of past project lessons learned in the development of long-term forecasts is also expected.

E. *Migration to trusted advisor.* The development of the spectrum of required SDF team competencies and expertise takes forecasters from being technically skilled to having research capabilities, to accumulating industry dynamics expertise, to understanding forecast impacts, to making impacts on strategic decisions. In parallel, the migration from technocrats to trusted advisors takes strategic forecasters on a journey through a sequence of transformations from the competence area where low politics are present to the advisory role where confidence, collaboration, composure, and high politics dominate the scene. And one of the main enablers of SDF team performance is adequate rewards and recognitions for those highly performing team members.

A number of performance awards and recognitions are in place in companies with established SDF teams, such as the following:

A. *Internal awards and recognitions.* In addition to cash prizes for SDF team support excellence in strategic projects, the common forms these awards and recognitions take include the following:

1. Annual monetary awards to the SDF team for high forecast accuracy over several periods
2. Client satisfaction recognition and appreciation awards to the SDF team for support excellence

3. Incentives for new solutions, approaches, and continuous innovation to enhance the efficiency of forecasting and decision-making processes
4. Encouraging professional development and enhancing the image and standing of the SDF team internally
5. Career path rotations to Strategic Planning, Business Development, Portfolio Management, and other client organizations
6. Assignment of SDF team members in forecasting education, client special tasks, and senior advisor positions
7. Invited speaker program to client and business unit meetings concerning their support in strategic projects

B. *External awards and recognitions.* In this area, we find that the following options are often used to reward strategic forecasting excellence:

1. SDF team member participation in industry analyst conferences and support to presenting company officials
2. SDF team member presentations in professional conferences, industry association workshops, and special meetings of experts
3. Publication of papers in professional journals of nonconfidential material to attract responses with new ideas
4. Participation in invited speaker programs and workshops in different industries to learn from their experiences
5. Membership in industry and academic advisor programs to identify potential talent

A synopsis view of looking at best-in-class SDF team attributes and quality elements is shown in Table 20.1, which cites some best-in-class strategic forecasting components and the facilitating elements present in the SDF teams benchmarked.

Best-in-class strategic forecasting is not simply forecasting; it is a decision-making tool because of the foundation which its support creates for the client decision-making processes. Its broad scope and areas of coverage add significant value, which can be measured at each step of the SDF processes. Figure 20.5 shows why strategic forecasting is a decision-making tool for the client organizations it serves, for the scope of coverage and the analyses and evaluations it performs, and the value added in every project measured by select performance indicators. The best-in-class SDF team attributes and quality elements are necessary, but they always need to be supplemented by a differentiation approach, which demonstrates the SDF teams' value-adding contributions in strategic projects.

Being best in class is a necessary but not sufficient condition for being recognized as such. For that reason, SDF team members must demonstrate the superiority of their approach versus the deficient practices in the current state of forecasting. And, an effective way of doing it is by differentiating their mindset, approach, processes, solutions, and support they provide to project teams for strategic projects. The differentiation approach depicted in Figure 20.6 is used by some best-in-class SDF teams to rate and showcase performance, which is characterized by factors such as the following:

1. Clarity of purpose, forecast ownership and accountability, and efficiency of developing forecast solutions and recommendations
2. High-level client satisfaction and integration with other functional areas to solve client decision problems
3. Ease of doing business and high level of 4Cs
4. Practical, cost–benefit balanced approaches in creating forecasting solutions

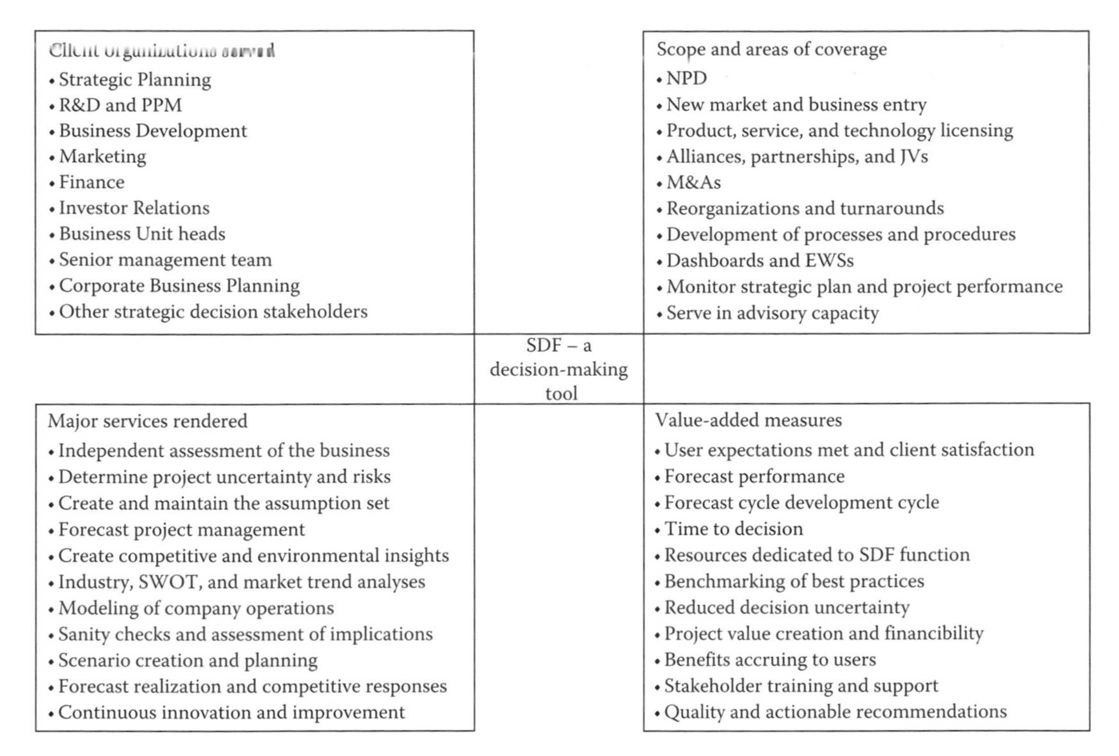

FIGURE 20.5 Elements of a strategic decision-making tool.

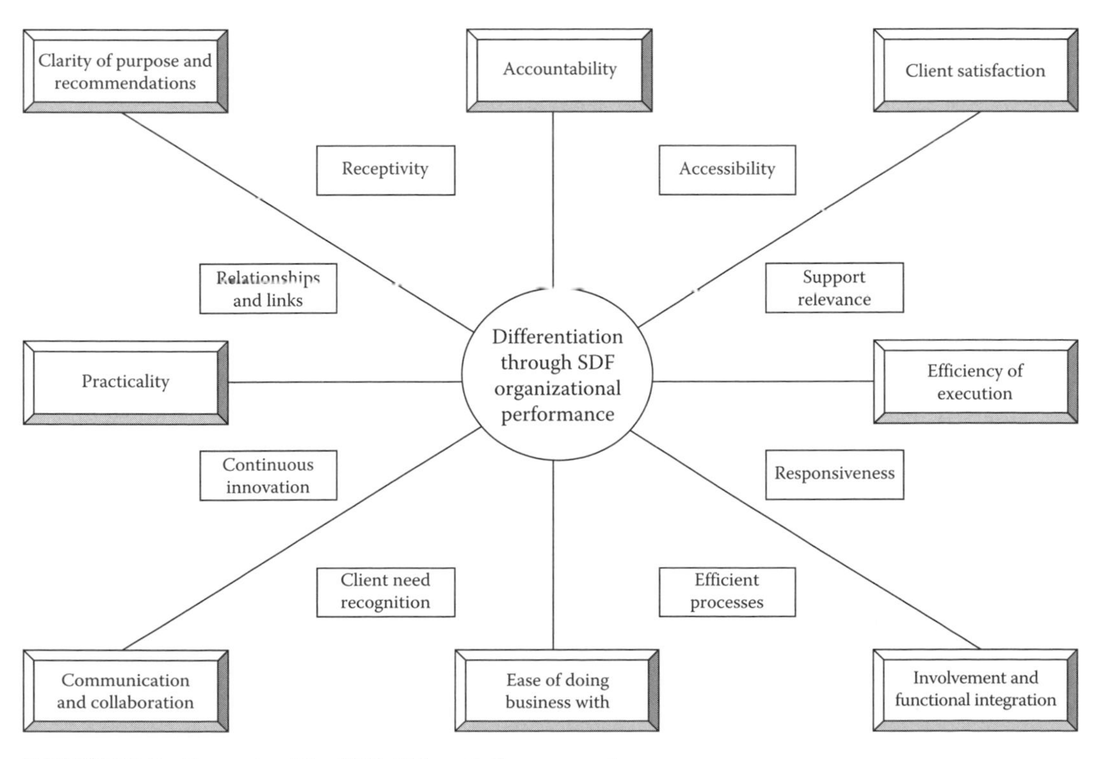

FIGURE 20.6 Synopsis of the SDF differentiation approach.

The SDF team member's personal and process quality attributes enabling this differentiation are based on several factors, which include the following:

1. Functional links and strong personal relationships that facilitate high levels of 4Cs
2. Accessibility of SDF team members to help clients and key project stakeholders solve their problems
3. Early recognition of client needs and responsiveness to client and project team needs
4. Relevance of the support they provide coupled with practicality and professional conduct
5. Complete and efficient processes created to fit the project particulars and address related issues
6. Receptivity of SDF team members to incorporating project stakeholder input, feedback, and new ideas
7. Continuous innovation and improvement to increase SDF team productivity and the efficiency of the decision-making process

20.4 SDF PROCESS AND PRINCIPLES

Each strategic decision and project are unique in their environmental context, company needs, strategic plan and project goals and objectives to be met, organizational capabilities, and client needs and expectations. Accordingly, SDF processes are tailored to fit specific project needs, some of which are described in Chapters 8 through 19. There is, however, a general and uniformly applied process in the preparation phase, which varies little in approach by project and includes the elements and activities shown by each strategic decision forecast development phase:

Important note on SDF processes and principles: *The best-in-class elements and the principles guiding the SDF team support are listed here and are intended to be a central reference for strategic forecasters. The sequence of activities is not fixed and the order in which activities are performed is determined by immediate project needs and time and resource constraints.*

A. *Preparation phase process*. The following is a list of process activities performed in preparing for the development of a strategic decision forecast. In this phase, SDF team members need to

1. Understand the company operations, evaluate the current competitive situation, and identify and eliminate knowledge gaps in those areas.
2. Identify company needs, the client perspective on needs and requirements, biases, and expectations, and determine differences and potential issues.
3. Obtain clarity on project definition, rationale, goals and objectives, and alternatives examined, and understand the business problem the proposed project intends to solve.
4. Identify customer preferences and market needs and determine how and to what extent the proposed project will impact them.
5. Determine the levels of corporate and client risk tolerance vis-à-vis project uncertainty and risks.
6. Develop a mental model of the business, determine how things get done in the company, and validate that model.
7. Determine what needs to be known to solve the problem and the forecast solutions required to do it effectively.

29. Develop actionable recommendations for the submission of the consensus future-state forecasts to the responsible decision makers.
30. Reiterate the appropriate uses of the forecasts and the need for monitoring performance and adjustments to the underlying assumptions and scenarios in outer forecast years.

C. *Leveraging other organizational capabilities.* Basic tactical forecast process components are reviewed in the forecast development phase by the SDF team to extract insights, which may be useful in strategic forecasts. This review is a joint effort with the tactical forecasting subteam of analyses and evaluations performed in developing short- and medium-term forecasts including the following:

1. Market research for customer and market need identification conducted for a lifecycle project of tactical nature
2. Competitive intelligence review and validation of findings vis-à-vis historical data
3. Short- and medium-term market opportunity assessment approach and forecasts
4. Own and competitor product evaluations and attribute ranking matrices
5. Validation of order of entry matrix market shares in the industry, analog proximity to the project, and resulting market performance
6. Market segment targeting, product pricing strategy, evaluation of distribution and sales channels, advertising and promotional spending, and expected impacts over the next two to three years
7. Market research and Delphi approaches to determine product uptake and diffusion models as well as cannibalization and stealing from other products
8. Quantitative demand analysis, elasticity estimates, sensitivity studies, and tactical forecasts
9. System dynamics modeling applications, sanity checks conducted on the tactical forecast assumptions and forecast implications, and key lessons learned
10. Intra and intercompany collaborative short-term forecasts, forecast triangulation, and tactical forecast performance

D. *Post-forecast development phase process.* By now, if the decision is made to proceed, project implementation has begun. The SDF team's activities following the development of the required forecast solutions involve a few more steps to complete the process. The activities in these steps are as follows:

1. Creating a project dashboard in conjunction with the project team and identifying the variables to monitor and appropriate forecast performance measures to track
2. Implementing an EWS to detect and communicate the emergence of threats
3. Tracking forecast performance by key elements and conducting variance and error contribution analyses to understand the causes of the deviations, confirm the signals from the EWS, and determine possible underlying currents of change
4. Reviewing assumptions and scenarios in light of actual performance, revising models and forecasts in light of actual data, and assessing their implications for expected project value creation
5. Determining when and how to implement FSR plans, monitoring their effectiveness, and escalating the need for strategic actions
6. Performing a project postmortem review, communicating important findings and lessons learned to project stakeholders and senior management, and applying lessons learned to future projects
7. Ensuring that all data and knowledge gained in the project are uploaded to the Intranet repository and made available to project stakeholders on a need to know basis

E. *SDF principles.* Best-in-class SDF teams apply Six Sigma to principles to strategic forecasting for good reasons. First, unfreeze the prevailing forecasting function mental model and then

8. Identify what is known and unknown, what is knowable and unknowable, and what will take to make them knowable.
9. Review corporate strategy and evaluate goals and objectives in light of market realities and the proposed project to ensure consistency.
10. Determine deficiencies and limitations in the current strategic forecasting function and figure out how SDF practices can help to eliminate them.
11. Help clients define the strategic project or decision forecasting requirements and obtain agreement on realistic goals and objectives, timelines, and deliverables.
12. Determine the resources needed to develop sound forecast solutions versus the currently allocated resources and expected future financial constraints.
13. Define measures of forecast, SDF team, and project performance, identify the level of support needed to achieve them, and obtain project team agreement.
14. Determine which events are controllable and those outside the company's control and how they may be leveraged by internal and external agents.
15. Ensure that project stakeholders have a basic understanding of forecasting processes and conduct as many learning sessions as needed.
16. Identify project stakeholder predispositions, expectations, biases, and competing interests and how they get manifested in internal politics so that they can be dealt with appropriately.
17. Engage the project team in the forecast project management preparations and obtain insights from their experiences in past strategic projects.
18. Create and communicate the SDF processes, define clear roles and responsibilities, assign them to specific team members, and obtain their agreement.
19. Develop an SDF team strategic intent statement, share with the project team, and get their pledge of support.
20. Brief and engage members of the Competitive Analysis, Market Research, Environmental Assessment, and Tactical Forecasting subteams to assist in the information and data collection, validation, and the analyses efforts.
21. Review earlier company project experiences with strategic forecasting and learn from those experiences to eliminate process defects and conflicts.
22. Unfreeze the existing forecasting function mental model by taking stakeholders on the strategic project forecasting process journey to enhance coordination of activities and their cooperation.
23. Obtain client concurrence and project team support for the SDF team's plans and requirements to develop long-term forecasts.

B. *Forecast development phase process.* There are several SDF team member activities in the phase where forecast solutions are actually developed, which include several steps. The first activity is to identify or confirm what needs to be known and what is knowable and unknowable, controllable and uncontrollable, and at what cost. Notice that the order of the steps that follow allows for some flexibility in the execution of those activities:

1. Create an Intranet repository to populate with data, information, and results of analyses, evaluations, business intelligence and insights from the project.
2. Conduct an evaluation of megatrend and subtrend impacts on the company's business and specific products, services, and technologies.
3. Perform a thorough PESTLED analysis to determine important elements, how they are likely to affect the company in future years, and how the company should address them.
4. Undertake an in-depth industry analysis and market trend assessment to detect structural changes, discontinuities, and their implications for the company.
5. Lead the SWOT analysis to determine the company's competencies and capabilities vis-à-vis those required to execute the project successfully.

6. Develop the common assumption set to be used in the creation of scenarios, long-term forecasts, subsequent evaluations, and planning activities.
7. Perform a market analysis to identify key business drivers, causality, and feedback effects.
8. Identify relevant industry experiences and appropriate benchmarks, practices, standards, and analogs and how to apply them correctly in the project.
9. Conduct a forecasting due diligence that includes review of earlier experiences, market assessment, competitive intelligence, data, and models and scenarios.
10. Reevaluate the uncertainty level and risks associated with the project and communicate findings to the project team, the client, and senior management.
11. Investigate, in conjunction with the project team, if there are other alternatives and more attractive projects consistent with market realities.
12. Obtain, validate, analyze, and adjust as needed the data required to support model and scenario development and project decisions.
13. Reduce project uncertainty to residual uncertainty through analyses and evaluations, development of strategic posture, and identification of project risks.
14. Model the company operations based on the analyses and evaluations to this point and project the company's future under the status quo scenario.
15. Identify likely future events, their timing, potential impact, and black swans that could impact the project.
16. Determine any own product line cannibalization impacts and likely impacts on competitor operations.
17. Create and describe three plausible scenarios of how the company will get from the current- to the indented future-state and conduct simulations, sensitivity analyses, hypotheses testing, and "what-if" evaluations.
18. Test the stability of the plausible scenarios with logical consistency and reality checks, unexpected external shock events, and black swan occurrences.
19. Conduct a critical and objective validation and verification of the entire process, assumptions, models, and data and information used.
20. Adopt an action–reaction–action methodology to develop long-term forecasts based on the three scenarios and assess their resource implications, the risks and uncertainty involved, and the likelihood of each scenario's achievability.
21. Run client-directed scenarios, assess their implications on resources, review results with clients, obtain their feedback, and make warranted scenario adjustments.
22. Triangulate the scenario forecasts with forecasts from other methods and externally developed long-term forecasts and evaluate the consequences of long term over and under forecasting on project success and the company's operations.
23. Identify the forecast risks associated with each scenario, their sources, timing and likelihood of occurrence, and potential impacts.
24. Estimate the support levels needed in the implementation and beyond phases to execute the project per each scenario selected.
25. Subject the three scenario forecasts to further stakeholder and project team review; obtain input, feedback, and comments; update if need be; and reach agreement on the baseline and some version of high and low views.
26. Present and explain the baseline, high, and low future-state views to senior management and obtain their views and inputs to arrive at a consensus forecast on which to base the final decision.
27. Validate, verify, confirm, and conduct sanity checks on the major factors impacting forecast realization.
28. Assess likely competitor reaction and impacts to the company's project and prepare an FSR plan, which includes risk management, forecast realization, and competitor reaction modules.

to probe into the forecasting group's business practices. Second, ask hard questions and use data and benchmarks to develop tailor-made SDF solutions for a project. More importantly, Six Sigma principles are used because of the need to project manage the strategic forecasting function more effectively. How? By focusing on eliminating major gaps and faults through tested best practices based on understanding, measuring, and improving organizations, processes, modeling, scenario development and planning, and FSR plans.

Application of Six Sigma principles to strategic forecasting drives SDF teams to improve forecast performance, to enhance productivity, to increase client satisfaction, and to raise the forecasting bar. Over the years, a collection of SDF principles has been assembled from the experiences of several experts in the subject. They include the following advice and guidelines:

1. Know what you want strategic forecasting to be and never stop learning.
2. Clients do not know how to help you and you need to show them how to help you so you, in turn, can help them and provide superior support to project stakeholders.
3. Remember that people sources are key for hidden information: about 70% comes from them and only 30% from printed documents.
4. You do not know all that needs to be known in a strategic project, but you will always learn if you search; and the more you search, the more you will uncover.
5. Know yourself and manage from a plan, be aware of forecasting pitfalls because they are everywhere, and strive for certainty but then distrust it.
6. Deal with forecasting problems not with symptoms, prioritize and do the most important things first, delegate to other project team members when appropriate, and share knowledge.
7. Know your clients, their needs and concerns, and the goals they hope to achieve. Always stay in tune with your clients and be of service to them continuously.
8. Be respectful and good to people you rely for support, talk the client's language, and in dialogues with management put aside egos.
9. Demonstrate commitment to the business and be good at what you do. It is OK to amaze clients with your extensive knowledge and experiences, but remain humble.
10. Defend the validity of your work and do not let SDF solutions to be misused or misrepresented; they are too valuable to be subjected to such treatment.
11. Educate clients on the shortcomings of forecasts and be prepared to explain how they were created and why they are the best possible forecasts.
12. Be communicative, professional, and civil and avoid burning bridges and making enemies.
13. The first SDF team duty is loyalty to corporate value creation; the second is loyalty to the interests of the client; the third is loyalty to the project team; and the fourth loyalty is to the forecasting profession.
14. Maintain a broad context awareness, a sense of history, and respect for what key competitors are capable of doing.
15. Have respect for the evidence and a feel for the jugular with objectivity and a balanced perspective.
16. Keep your feet on the ground and set limits to the scope of SDF team's involvement in strategic projects.

20.5 SDF PERFORMANCE FACTORS AND MEASURES

The SDF teams' consistent excellence in the support they provide to each project is accomplished by applying the five elements of the Six Sigma approach in their methodology and processes: defining the projects goals and deliverables to clients, measuring the current performance of their processes, determining the root causes of shortcomings and limitations, eliminating gaps and limitations and improving the forecasting process, and once improved, controlling the performance of the SDF process to desired levels. A meaningful self-assessment of performance begins on the one hand with the evolution of project definition and SDF goals and objectives, and on the other hand with alignment of knowledge to achieve those goals and objectives. This enables the SDF team to set performance goals using forecasting best practices and define performance metrics that align with the forecast performance goals. Thus, measurement criteria and metrics are defined to be used in judging SDF performance.

The tool used to identify gaps, deficiencies, and limitations that cause poor strategic decision forecasts is the fishbone diagram, which focuses on the major causes of such forecasts. The three major sources of poor strategic forecasts are the following:

1. The organizational structure under which the SDF team operates and the quality of SDF team member qualifications, skills, and training
2. The functional linkages of the SDF team and its internal and external contacts and personal relationships
3. The processes used to develop long-term forecasts, the methods, models, scenarios, the types of data used, and the forecast realization plan created

To implement improvements of SDF team performance requires going from criteria to actual performance metrics using a sequence of steps starting with determining what is important and critical to the strategic forecasting function, identifying what aspects of forecasting are critical to clients, and meeting their expectations. Then, performing the data gathering and analysis, due diligence, scenario development and planning takes place. Having completed these activities, the SDF team conducts an important process validation and verification, and proceeds to the development of long-term forecasts. Lastly, the FSR planning effort takes center stage.

Determining what facets of forecasting project clients value is very important in judging SDF team performance. These aspects include client satisfaction, accuracy, quick turnaround, development of consensus, verifying and clarifying inputs and outputs of the SDF process, and meeting major client support expectations. Also, figuring ways to measure how well forecasting is meeting project needs and project team requirements and expectations is crucial to improving SDF team performance. This is accomplished by

1. Working backward through the process to establish activities and metrics crucial to improving forecasting performance and meeting major expectations.
2. Making sure all measurements are linked to client-valued functions, sound scenarios and forecasts, and actionable recommendations.

How well the SDF team is performing its support function involves the evaluation of its practices along several dimensions: assessment of the current state of internal affairs and individual process component and an evaluation of each step, process component rating, and overall performance scoring. It also involves assessment of process component deviations from best practices, forecasting function management and forecast realization planning, and development of recommendations to address process deficiencies and limitations. When management dashboards are used to assess project and SDF team performance, the idea is to think about parsimony and keep the number of measurements small, but use metrics that cause project stakeholders to act in the best interest of the business. It is important to select indicators that give needed information on factors that affect client satisfaction and the bottom line, and avoid metrics for which data are not readily available, are incomplete, or controversial.

Other factors contributing to successful measurement of project performance through dashboards are staying away from complex metrics that are hard to explain to others and not using metrics that create excessive overhead and costs. Some of the guidelines SDF teams use to develop useful and practical dashboard metrics include getting clients and senior management involved in the process, representing the performance metrics visually, and selecting indicators that are impactful and respond quickly to changes. Three related guidelines to ensure appropriate assessment of performance via dashboard metrics are using metrics that gauge and drive only important activities, limiting the number of metrics to five, and taking corrective action based on readings of metrics relayed by the EWS and displayed on the dashboard.

On some occasions, the SDF team's performance is measured by the extent to which the EWS created has been successful in managing project uncertainty and risks effectively. EWSs developed by SDF teams are created to detect the emergence of risks prior to their occurrence and to minimize their impact through sharing of prior knowledge. But, this also requires continuous monitoring of forecast driver variable performance, scenario evolution, external events, competitor reactions, environmental factor shifts, and project implementation efficiency. Effectiveness of EWSs is judged along the following elements:

1. Adopting a leading signal of threat identification, timely notification of responsible stakeholders, and immediate action prior to threats and risk events materialize
2. Building awareness among project stakeholders of how the EWS operates, determining individuals responsible for taking actions, and assigning specific responsibilities to them
3. Picking a few EWS indicators to monitor, such as those on the management dashboard, and identifying trigger levels of risk event occurrence
4. Monitoring forecast performance and progress of those key indicators displayed on the business management dashboard
5. Ensuring instantaneous data feeds, reporting, and verification of accuracy of information in order to be able to detect signals from volumes of data and validate the impending threat occurrence
6. Disseminating clear and understandable warnings to responsible stakeholders on the nature, the cause, and the potential impact of threats about to materialize
7. Acting quickly and providing a joint recommendation with the client based on the guidelines of FSR plans

A useful way to evaluate SDF team performance in a given project is to evaluate how well the inputs to the development of the forecast solution correlate with client requirements. This is done using a variant of the Six Sigma XY matrix, shown in Figure 20.7, which rates strategic forecasting process elements along project and client requirements. Notice that the XY matrix evaluation allows for project specific elements to be rated along the forecasting and quality client requirements axes. An alternative and valuable SDF team performance monitoring and enhancement tool is its balanced scorecard whose approach and key components are shown in Figure 20.8.

The value of the best-in-class forecasting team balanced scorecard is that it focuses on four key elements: the use of best practices, ongoing value creation, cost–benefit balance, and client assessment of the support provided. Notice the central importance of the SDF team's business definition, its mission, vision, and strategy for each project, and the importance of the client's views, the internal functional links and processes, and knowledge and learning performance metrics. Also, observe that the inputs for client assessment and evaluation of ongoing value creation come directly from clients, whereas the evaluation of use of best practices and balance in the cost–benefit approach originate with the manager of the SDF team.

Another similar approach to measure SDF team performance is the senior management and client evaluation along the eight major categories shown in the spider chart example of

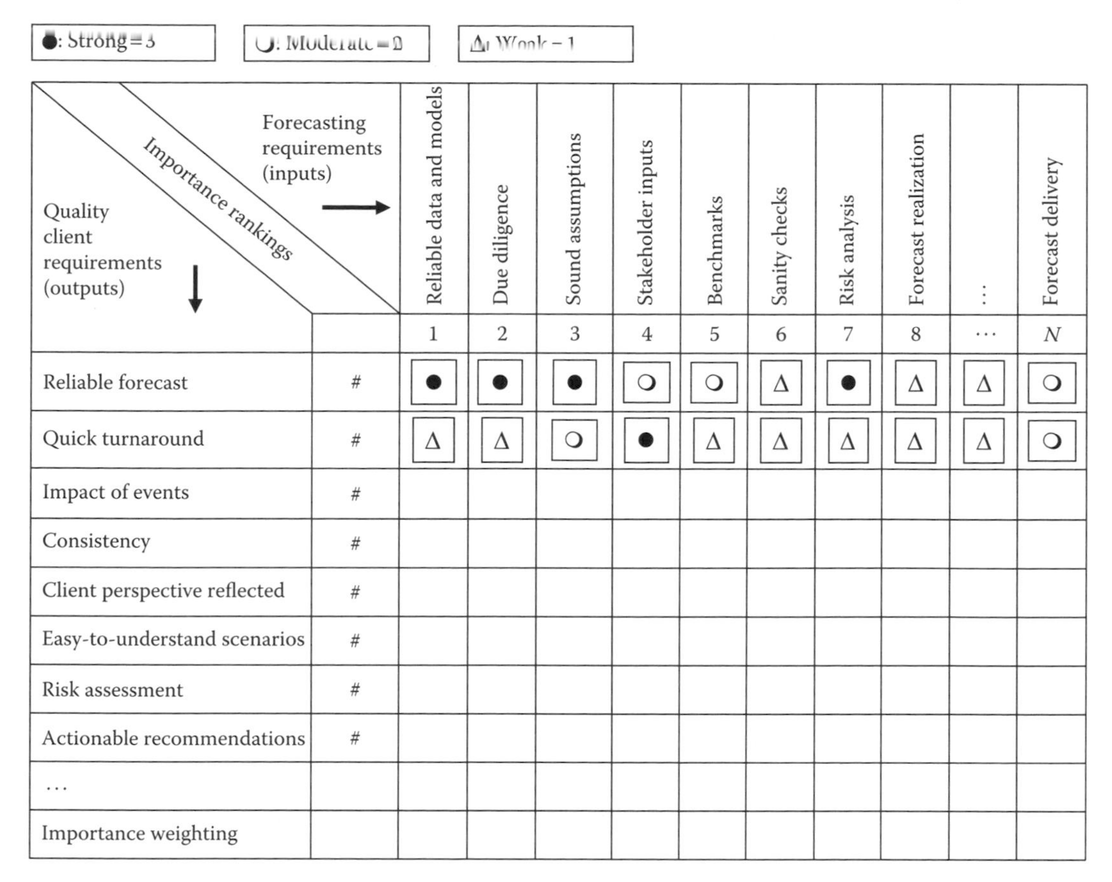

●: Strong = 3 | ❍: Moderate = 2 | Δ: Weak = 1

Quality client requirements (outputs) ↓ / Importance rankings / Forecasting requirements (inputs) →		Reliable data and models	Due diligence	Sound assumptions	Stakeholder inputs	Benchmarks	Sanity checks	Risk analysis	Forecast realization	...	Forecast delivery
		1	2	3	4	5	6	7	8	...	*N*
Reliable forecast	#	●	●	●	❍	❍	Δ	●	Δ	Δ	❍
Quick turnaround	#	Δ	Δ	❍	●	Δ	Δ	Δ	Δ	Δ	❍
Impact of events	#										
Consistency	#										
Client perspective reflected	#										
Easy-to-understand scenarios	#										
Risk assessment	#										
Actionable recommendations	#										
...											
Importance weighting											

FIGURE 20.7 Sample of SDF performance evaluation matrix. (Adapted from Brue, G., *Six Sigma for Managers*, McGraw-Hill, New York, 2002.)

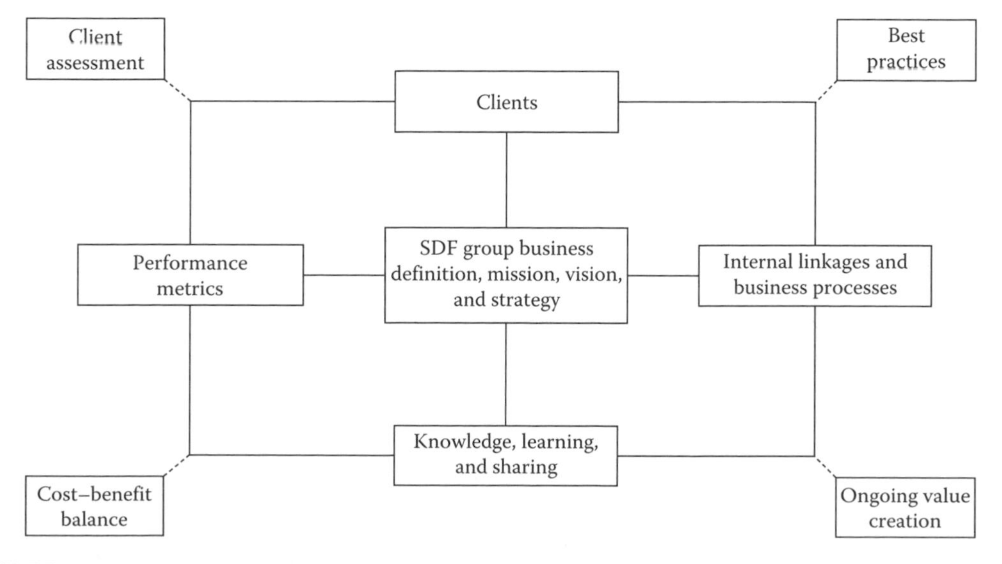

FIGURE 20.8 Best-in-class SDF team balanced scorecard framework.

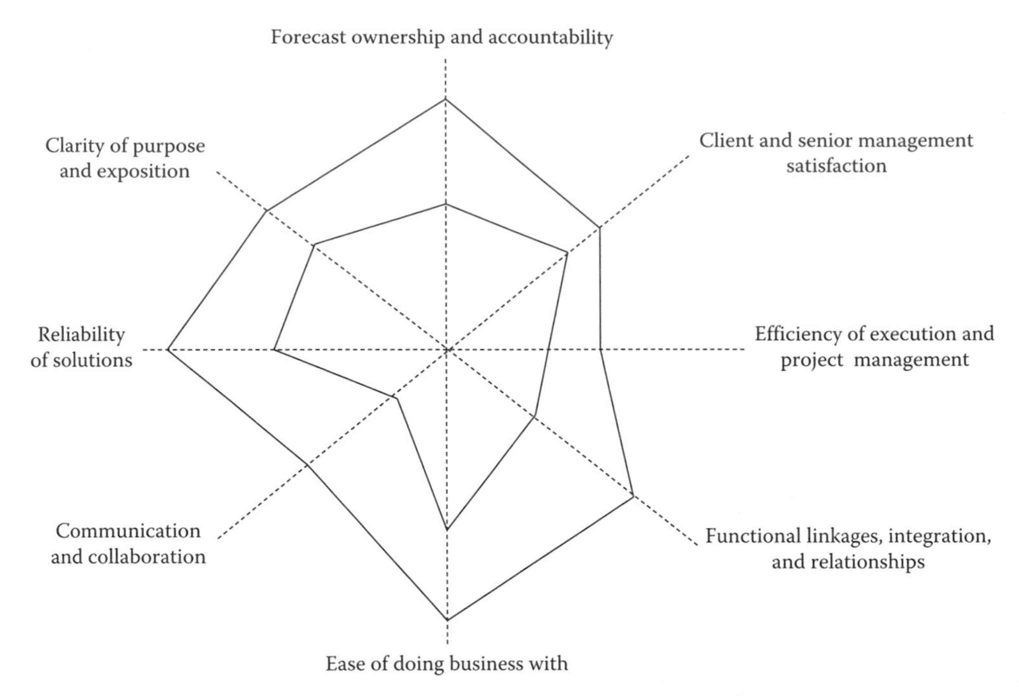

FIGURE 20.9 SDF team performance evaluation by senior managers and clients.

Figure 20.9. A key feature of this approach is that senior management and client assessments are made at different times and are recorded on each dimension on a 1–10 scale. Plotting performance along each dimension and connecting those points gives a good indication of overall project performance and client satisfaction at a given point in time. This provides clear indications on the progress the SDF team is making in achieving its continuous improvement objectives and the areas where more improvement is needed.

20.6 SUCCESS FACTORS AND SDF TEAM CONTRIBUTIONS

The unique characteristics of best-in-class SDF teams are summarized in Table 20.2. Success factors in establishing and managing best-in-class SDF teams articulated by subject matter experts, clients, and SDF team leaders include organizational, qualification, functional link and relationship, and process and analysis dimensions. They are classified under the following categories, but notice that SDF team contributions are subsumed in best-in-class practices because they are one and the same.

A. *Team member mindset*. The SDF team members' outlook and mindset lay the foundations for successful projects and include the following success elements:

1. Practicing a culture of integrative thinking, quality improvement, and innovation and providing dedicated personal attention to one client at a time and exclusive focus on their needs
2. Maintaining highest standards of integrity and honesty and providing innovative, top-notch, reliable, and practical solutions
3. Supplementing client organizations in a flexible, customized, coordinated, collaborative, and interactive manner
4. Application of a forward thinking mindset to solving business decision problems, knowledge and intelligence validated with sanity checks

TABLE 20.2
Uniqueness of Best-in-Class SDF Teams

Communication, cooperation, collaboration, coordination	Innovation mindset and culture	Modeling methods and application of tools
SDF organization business	Clearly articulated project goals and objectives	SDF governance
Associate skills and qualifications	Value- and principle-driven behavior	Complete and interconnected processes
Benchmarking best practices and guidelines	Balanced scorecard and Six Sigma approach	Knowledge sharing through Intranet repository
Project grounding and forecasting due diligence	Scenario creation and planning black swan identification	Monitoring and reporting, dashboard and early warning system
End to end forecasting project management	Future-state realization planning	Megatrend, PESTLED, industry, market trend, and SWOT analyses
Verification and validation of processes, analyses, and evaluations	Confluence of unique SDF organization experiences	Uncertainly and risk identification and management

Note: PESTLED, political, economic, social, technology, legal, educational, and demographic; SDF, strategic decision forecasting; SWOT, Strengths, Weaknesses, Opportunities, and Threats.

5. Learning from the best, introducing best processes and practices, and helping clients grow and become more effective over time
6. Making strategic decision-making based on SDF solutions a disciplined, easy to implement activity for senior management
7. Passion for establishing the SDF function as a source of sustainable competitive advantage, tested forecasting expert judgment, and balanced approach rewarded with team member incentives, recognitions, and rewards tied to forecast performance

B. *Team member qualifications.* SDF team member qualifications essential to high support performance including competencies and skills have been discussed earlier. But, because of their critical importance to project success, they are captured here as well and include the following:

1. Ability to understand corporate needs, create common assumptions, and demonstrate the value of SDF
2. Knowledge of what, when, and how to use the different forecasting methods and models
3. Ability to create relationships, inclusion, judgment, and, above all, credibility sanity checks
4. Experience in modeling and long-term forecasting techniques and capacity to translate collective knowledge to model inputs
5. Facility to create reasonable assumptions and obtain agreement on the assumptions and the rationale behind them
6. Skillfulness in the management of forecasting politics and client expectations and professional presentation of forecasts with convincing backup evidence and data
7. Forecast ownership and managing the entire forecasting process and expertise in intensive business research and evaluation of the business environment and company capabilities
8. Ability to bring structure and discipline to the decision-making processes, and experience in qualitative and quantitative demand analysis and forecasting
9. Experience in independent, critical, and objective assessments of the business; expertise in environmental, industry, and SWOT analyses; in-depth forecasting due diligence; and competitive response planning, which are important success factors

10. Aptitude and skill in developing scenarios, testing hypotheses and assumptions, and identifying the key factors to achieving a future state
11. Business research, forecasting, and problem-solving experience shared with project team members and lead roles in forecasting for strategic projects
12. Proficiency in the application of innovative business decision tools, Six Sigma principles, and TQM
13. Expertise in investment opportunity assessments, technology transfers, and licensing project valuations along with experience in project negotiations, project implementation, and performance monitoring support
14. Ability to be a trusted advisor to clients and senior management in the areas of risk management, business transformations, and strategic project issue resolution
15. Benchmarking studies and introduction of best processes and practices to achieve major productivity enhancements in strategic forecasting and decision making

C. *Functional links and relationships.* The functional links and personal relationships critical to successful strategic forecasts are crucial in the new SDF paradigm. Their value is recognized in companies that have established strategic forecasting teams, which are characterized by the following practices:

1. Relationship management with economic analysis, strategic planning, business development, financial planning, R&D, portfolio management, marketing, and business planning organizations
2. Close SDF team working relationships with client organizations and senior managers to obtain their commitment and support and ensure SDF team participation in all client strategic decisions
3. Contacts with external experts, advisors, and opinion influencers, such as consultants, industry association representatives, Wall Street analysts, and academic researchers
4. Unimpeded communication, cooperation, coordination of activities, and free knowledge and information exchange
5. Ease of doing business with the SDF team, which results in efficient coordination of effort in all phases of a project and getting to decisions quickly
6. Emotional support and increased level of comfort for clients to navigate through project uncertainty and management of risks

D. *SDF team effectiveness.* Success factors required for high SDF team performance are created by input from senior managers, clients, and SDF team managers and include the following elements:

1. Appropriate SDF team governance structure and business definition, working in an environment that fosters self-motivation, self-assessment, and self-confidence
2. Clearly articulated SDF team objectives in the project, deliverables, and timelines and client and project team clear understanding of forecasting process and effective management of client support expectations
3. Strong senior management support, adequate resources allocated to the function, and cooperative stakeholder participation in the forecasting process
4. Correctly recognized, perceived, and unmet market, corporate, and client needs and highly coordinated and collaborative project assessments
5. Clear, complete, balanced, and effective decision-making processes and procedures, which incorporate sanity checks, triangulation of forecasts, client and senior management inputs, and appropriate extraneous forecast adjustments
6. Delineation, understanding of dependencies, and ownership of project stakeholder responsibilities, and effective handoff of deliverables and tasks

7. Adopting practical, sound, and best-in-class forecasting process, practices, and procedures and professional delivery of forecast presentations with caveats and necessary supporting data and evidence
8. Sound management dashboards and EWSs linked with FSR plans and customized reporting of project and SDF team performance

E. *Analytics and tools.* The analyses, evaluations, techniques, and tools considered sound practices and success factors in best-in-class SDF teams include the following elements:

1. Identification of business opportunities and project evaluations and identification, evaluation, and management of project uncertainty and risk
2. Assessments focused on customer needs, product attributes, pricing and cost–benefit analyses, project economics and valuations, project competitive impact, and forecast realization plans
3. Evaluation of megatrends, the external environment, industry developments, and company SWOT
4. Sound project feasibility studies, business cases, business plans, target development, adequate preparation, and positioning for successful product launches
5. Evaluation systems measuring project success, management dashboards, EWSs, and balanced scorecards
6. Detection of market signals from the noise of voluminous data by focusing on essential and critical evaluations and seasoned and balanced analyses
7. Obtaining independent, objective, and efficient third-party business assessments to give more evidence and added assurance to clients and senior management
8. Strategy and strategic, business, and financial plan evaluations; portfolio management and product development strategy; and development of positions, negotiations strategy, planning, and support
9. Reducing project uncertainty to residual uncertainty and development of strategic posture, risk management, and project structuring and development based on real options
10. Benchmarking studies and introduction of best practices in forecasting function management, process and productivity assessments, gap analysis, development of targets, and operational performance measures

F. *Reliable forecast solutions.* SDF solutions that clients, experts, and SDF team managers believe to be critical in creating reliable strategic decision forecasts include the following properties:

1. Correct assessment of corporate needs and risk appetite and the business environment
2. Independent, critical, and objective evaluations of the company business and assessments of internal capabilities and competencies relative to key competitors
3. Good market, industry, and competitor data and intelligence together with quality of data collection, verification, analysis, and adjustment
4. Close monitoring, evaluation, and selection of relevant analogs and historical data for long-term forecasts
5. In-depth forecasting due diligence that touches on all areas affecting product attributes and quality, unit sales, pricing, marketing, and advertising decisions
6. Selection of appropriate methodology, models, driver variables, and scenarios taking the company from the current to a desired future state
7. Simple, clearly defined, and easy-to-explain forecast performance measures
8. Long-term forecast scenarios based on sound assumptions and consistent with actual financial results and client and investor expectations
9. Holistic, knowledge, intelligence, knowledge-intensive, and balanced approach throughout the project

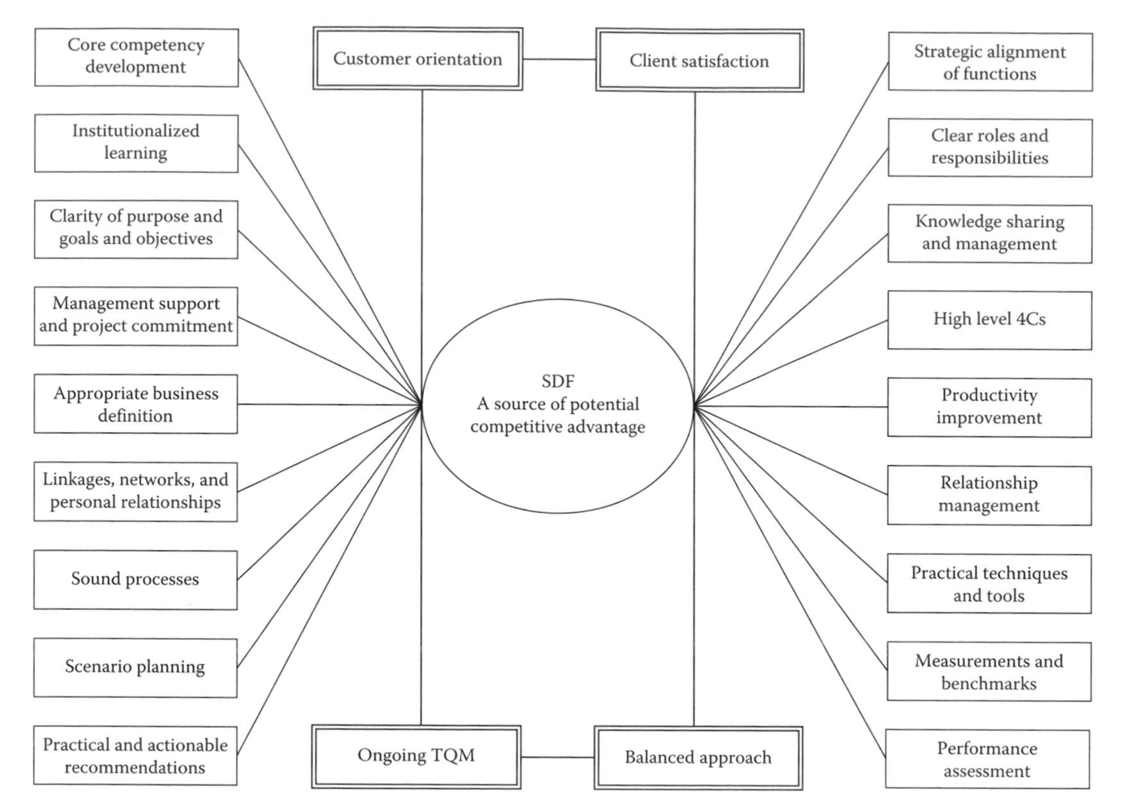

FIGURE 20.10 Best-in-class SDF teams: A source of competitive advantage.

10. Use of Six Sigma principles and guidelines that bring additional discipline to the forecasting and decision-making processes
11. Decision support systems, processes, and database management capabilities to facilitate sharing of data, intelligence, evidence, and knowledge

When all the useful practices and success factors mentioned earlier are present, the company is in a position to build a basis to create a competitive advantage. The success factors that make an effective SDF team a source of potential competitive advantage are illustrated in Figure 20.10. The presence of these factors in best-in-class organizations and the pillars of the SDF team performance are believed to lead to SDF team differentiation. The four pillars of SDF team excellent performance are focus on customer satisfaction, client satisfaction, a balanced approach, and ongoing TQM.

21 SDF Organization Implementation Challenges

In competitive environments, senior managers expect to have standardized approaches and models to create long-term forecasts to support various types of strategic decisions. However, it is naïve for people to be looking for cookbook type of instructions on how to do strategic forecasting. The reality is that all projects are different and each decision needs a unique forecasting solution, which, in turn, requires the right mindset, processes, and principles applied and flexibility of strategic decision forecasting (SDF) team members to fit the situation. Strategic decisions require a long-term mindset and thinking; that is, focus not on getting immediate results, but on creating the conditions for company growth. The role of the SDF team, therefore, is to help decision makers identify future-state alternatives, what is needed to get there, create options, pick the best solution, determine project uncertainty and risks, and identify ways to manage them.

Typical barriers encountered in establishing SDF teams and processes include a corporate culture of expedience, skepticism about the effectiveness of strategic forecasting, and fear of failing if adopting the new paradigm. Other obstacles to overcome are views such as "we know our business best; we don't need help from outsiders," "we have our own business processes; the current forecasting process works fine," and, of course, the "not-invented-here" syndrome. Additional SDF team and process implementation challenges are inertia and low priority assigned to forecasting. On the other extreme, one finds unreasonable client expectations of what SDF teams can deliver upon their creation and elimination of external forecasting support that has been employed for years.

Companies that undertake strategic projects frequently understand the value of professional SDF teams, but they also realize that they face significant implementation challenges in the presence of wrong corporate culture, attitudes, mindset, and festering organizational politics and conflicts. The broad nature of the SDF team scope and responsibilities and the definition of forecast quality and ownership of the forecast add to implementation challenges as do project stakeholder knowledge gaps and the scarcity of qualified strategic forecasting experts. A different set of challenges are due to routine and frequent outsourcing of SDF and client and key project stakeholder job reassignments. Also, difficulties in getting buy-in to create or restructure strategic forecasting processes, implement performance measures, and assess cost–benefit considerations and resource constraints are common problems.

Other, lesser challenges to SDF team creation and function implementation, which affect the governance structure, the business definition, and the scope of its support activities to strategic projects, include the following:

1. Strategic decision analysis and forecasting paradigms vary by industry, the internal and external environments, different forecasting needs, forecast drivers, and project particulars, which make it difficult to determine beforehand the type of SDF team best suited for the company needs.
2. Project clients, forecast users, and, sometimes, forecasters have the Keynesian view that "in the long run we are all dead" and long-term forecasts do not really matter because forecast performance is not monitored that far out. So, costly SDF teams are not really necessary.
3. The creation of an SDF team requires priorities to be reset away from trivial activities and reports and, instead, focus on critical success factors. This requires some unwelcome reorganization decisions.

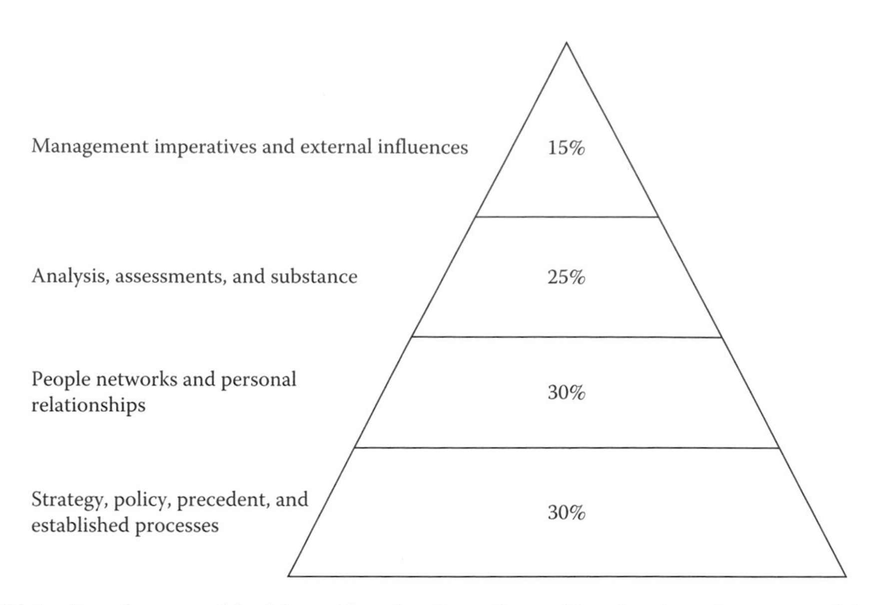

FIGURE 21.1 Key elements of decisions. (Based on Long Range Planning Associates research.)

4. To a large degree, strategic decisions are based on people relationships and factors, to a lesser degree on structured processes, and to an even lower degree on the substance of analysis and evaluations. This is demonstrated in Figure 21.1; and in such environments, SDF teams are not considered crucial to project success.
5. Strategic projects take years to plan, implement, and obtain the expected benefits. In the meantime, the environment changes, the company changes, technology changes, consumer tastes change, and products change. Therefore, in the minds of cost-conscious managers, good initial forecasts do not really matter. But, this is false.

In the sections that follow, we identify the challenges by main categories; determine the reasons for their occurrence; and discuss concepts, ideas, and solutions, which when properly executed are effective in overcoming resistance to SDF team and function implementation and the creation and monitoring of performance measures.

21.1 ATTITUDES AND ORGANIZATIONAL MINDSET

Why long-term forecasts fail so miserably is a common and justifiable question, which deserves a good answer. While there are several reasons for this phenomenon, the major reason that strategic forecasts fail is because of corporate cultures being not conducive to developing methodical, well-balanced, and well-thought-out long-term forecasts. Culture is the personality of the company and it is made up of beliefs, values, norms, assumptions, and employee behavior. Wrong corporate cultures for SDF team creation and process implementation are those where

1. There is little alignment with proven effective SDF team values and forecasting function control is exercised through bureaucratic rules.
2. Groupthink and the motive for strong unanimity dominate the need to consider other alternatives.
3. Uncooperative and noncollaborative behaviors are dominant due to conflicts of interest.
4. Short-term orientation values adhere to current practices more than new ideas and approaches to decision problem solving.

The organizational attitudes and mindsets that are impediments to implementing successful SDF teams and functions originate with a culture of expedience that produces wrong governance, business definition, and a work environment that does not encourage creativity, innovation, testing new ideas and approaches. With such conditions goes a value system that is not consistent with best-in-class SDF teams and team member performance. Then, again, there are unclear policies and organizational priorities, and undue client pressure to conform to preconceived and biased notions of forecast quality. Also, unreasonable client expectations are surfacing time after time, such as no client involvement in the creation of assumptions, but having influence on the level of forecasts or immediate turnaround of forecasts and revisions, but no involvement in supporting and monitoring forecast performance.

Sometimes, a forecaster and client narrow focus and mindset on producing single-number long-term forecasts for decision-making purposes is a major issue to overcome as is their inability, fear, and unwillingness to venture outside the comfort zone of existing approaches, models, and established relationships. On other occasions, organizational risk appetite inconsistent with the project uncertainty and risks and the decisions to be taken makes management feel uncomfortable with such prospects. In some companies, selective adopting SDF team recommendations, picking and choosing SDF solutions, and not taking advice on the use of forecasts are a big problem. Also, assigning forecast responsibility to the SDF team with minimal senior management support and not allocating sufficient resources to do it right are difficult challenges to overcome. An unfortunate occurrence in environments of high concerns over proprietary information disclosure is retaining tight control of long-term forecasting and future-state realization (FSR) planning until things go wrong and then blaming SDF teams for failures. And, on rare cases, conceding too much responsibility to the SDF team, well beyond management of the forecasting function that encroaches into client functional areas, is an issue to be managed carefully.

Successful SDF teams have worked out and employed a number of solutions to compensate for wrong corporate cultures, and they all start with realizing that failures in the current state of strategic forecasting paradigm have their roots in such cultures. Critical project postmortem reviews are helpful and reveal that SDF function implementation problems are eliminated by changing organizational attitudes and mindsets. Also, objective assessment of factors creating challenges to establishing successful SDF teams identifies wrong cultures as a major cause and root cause analysis exposes wrong corporate values and policies as another source, both of which need to be reevaluated.

Benchmarking best-in-class strategic forecasting organizations effectively identifies gaps between them and the company's strategic forecasting group that can be eliminated through changes in culture elements responsible for long-term forecast failures. Also, reviews of organizational mission, vision, and values determine problem areas that, once addressed, they allow for creation of successful SDF teams and functions. Positive results are also obtained by managing the negative aspects of internal politics produced by the excesses of highly competitive organizations. In all cases, however, demonstrating the value of SDF team contributions to the success of strategic projects is necessary to overcome resistance to establishing the SDF function.

21.2 ORGANIZATIONAL POLITICS

As SDF team members move from being technocrats dealing with low levels of politics to being trusted advisors, they encounter the presence and effects of high politics. The intensity of political posturing increases going from forecasters to clients, to business unit heads, to senior management. Notice though that organizational politics is situational; that is, it varies by level of participants, department, and project. As a general rule, it is behavior caused by differing goals and agendas and the larger the organization, the higher the politics. While organizational politics is often neutral, it is not necessarily bad but bad politics leads to irrational conclusions and wrong decisions.

Culture defines to a large degree the political environment, corporate values, and risk tolerance and organizational politics are a natural byproduct of the corporate culture. Their occurrence is motivated by competition to obtain resources and influence over decisions, implementing personal agendas, and leading the company in certain directions according to different interpretations of management and corporate needs. When the intensity of organizational politics rises to cut-throat competition levels, they affect SDF team implementation and its functions. This is reflected in impediments to creating an SDF team manifested in the following:

1. Lack of senior management commitment, involvement, and support to the project unless their own requirements are fulfilled
2. Pronounced differences of client, forecaster, vendor, and senior management interests and perspectives
3. Impaired communication channels and poorly coordinated efforts across different functional areas and project phases
4. Internal politics limiting or delaying SDF team participation in strategic decisions and inability to reach consensus forecasts in a timely and efficient manner

In the midst of environmental changes, the results of these impediments are drawn out decision processes and poorly managed strategic projects characterized by limited coordination, collaboration, cooperation, and communication (4Cs) among project stakeholders.

In most cases, organizational politics determines the home of SDF teams, their business definition and governance structure, the scope of roles and responsibilities, and representation in strategic projects. The way these issues are settled, in turn, decides the client organizations to be supported, the areas of SDF team coverage needed, and the services to be performed. Politics often puts emphasis on expected benefits to certain project participants and not on corporate benefits and value-added measurements. In the presence of intense organizational politics, there is disagreement about who owns the forecast, getting agreement on SDF team member participation, and conflicts with other organizations that claim ownership of strategic forecasting. Because of all these problems, it is hard to find the right home for the SDF team, and there is ongoing review and redefinition of the function, all of which make forecast consensus difficult to achieve.

Not only reaching consensus forecast development becomes difficult to accomplish in the presence of heavy politics, but sharing forecasts with stakeholders is fraught with challenges in adjusting and negotiating baseline forecasts. Limited senior management involvement, support, and reviews and feedback result in less effective forecast solutions. Also, the development of project negotiation positions and trade-offs becomes cumbersome and requires many more iterations to quantify their impact. Triangulation and convergence to a well-considered forecast solution becomes difficult as is getting agreement on the consequences of under versus overforecasting. In some instances, politics and not evaluations and judgment determine the selection of scenarios used to forecast. Then, the consensus forecast does not serve its primary purpose to drive the strategic decision; instead, political positions determine what forecasts are used in corporate plans and the support to be contributed by each stakeholder organization to execute the project.

The approach used by SDF teams to identify bad politics is to listen closely to the forecast users' response when the best possible forecast is presented to them and to understand the underlying meaning of the response. Responses loaded with political considerations include statements such as the following: "the forecast seems too low," "the forecast is wrong," "this is not what we were expecting," "it needs to be fixed," "your sources must be wrong," or "my organization will not buy it." Getting to the meaning of those responses is the job of SDF team members who must determine whether the forecast solution answers client questions, whether it shows that there is misalignment of interests, or if it requires reallocation of resources. They also need to determine if responses to forecasts are based on power and influence considerations, if they are indicative of organizational dynamics, or if ego and pride elements are involved. Additionally, it is important to establish if forecast user expectations are not met, or if the responses are simply intended to make the SDF team look bad. This is a formidable task; but if handled properly, it yields positive results.

The SDF team must always have close relationships with senior management, client groups, and other planning organizations company-wide; and for that reason, it is crucial to avoid pitfalls and navigate carefully through the minefields of organizational politics. The first step in doing that is to map the organization, determine how to communicate effectively and converse based on the decision makers' behavioral styles. While this is ascertained in every project, the SDF team must understand the vehicles of influence, the inner circles, and who the key influencers are. However, the first priority of the SDF team is to do its homework to understand the political situation, recognize the various individuals' behavioral type, determine their interests and objectives, and earn the right to participate in the strategic project and sit at the decision table.

> ***A note on behavioral types:*** *The major behavioral types are the promotional, supportive, analytical, and controller/driver type. Natural personality preferences according to the Myers-Briggs indicators are classified as extraverts or introverts, sensing or intuitive, thinking or feeling, and judging or perceiving. Personality types also classified as dominant director, interactive socializer, steady relater, and cautious thinker. Knowing and understanding personality types enables the SDF team to position its collective natural personality preference as an indicator of strength and success* (Meyers and Meyers 1995).

Management of organizational politics dictates that SDF team members do not develop ideological, philosophical, and personal ties to certain positions and that they do not take the manifestations of politics personally. They are well advised to understand the limits of their influence, to adopt a lose attitude, and not forget their careers. They are also counseled to first identify the key stakeholders in the strategic decision and who stands to gain or lose and get the proper people involved in defining their role in the project, SDF implementation processes, and performance measures to be monitored. It is advisable for the SDF team to get all project stakeholders involved early; and if they do not really belong in the progression of project execution events, they will drop out. Then, it becomes possible to align the project team behind clear goals and objectives.

The guidance of seasoned SDF team managers to team members is to show to project stakeholders that they are not happy delivering solutions that do not conform to their expectations and to demonstrate that they want project stakeholders to succeed, but without sacrificing corporate objectives. Understanding that forecasting is not just about numbers, SDF team members need to choose selectively who is involved in their areas of responsibility, to make it about the right things to do in the process, to show why new forecast processes and performance measures are needed, and how they are developed. By focusing on relationships, enabling project stakeholders to do their jobs better, and the value creation bottom line, SDF team members increase senior management and client acceptance. This translates into support to introduce new process and performance measures and secure needed resource allocations to achieve their objectives.

21.3 BROAD SCOPE AND RESPONSIBILITIES

By their nature, forecasts create conflict because they support decisions across organizations led by managers who have different needs and agendas and diverse priorities and interests. The broader the scope of the SDF team participation in a project, the greater the chances of conflicts arising. One view of the nature of the forecasting challenge is shown in Figure 21.2 where some ever-present influences impact every aspect of the strategic forecasting areas of responsibility. These challenges and influences are organizational politics, culture, and mindset; hidden stakeholder agendas; client and user lack of understanding of forecasting basics; project resource constraints; time pressures; and client and senior management expectations. And, they all take place in the presence of continuous change in the operating environment.

In order to provide continuity in the decision–implementation–value harvesting phases of a project, SDF team responsibilities are broad and span across several functions. These responsibilities are stated in terms of retaining the historical context of the project, continuity of purpose and project rationale,

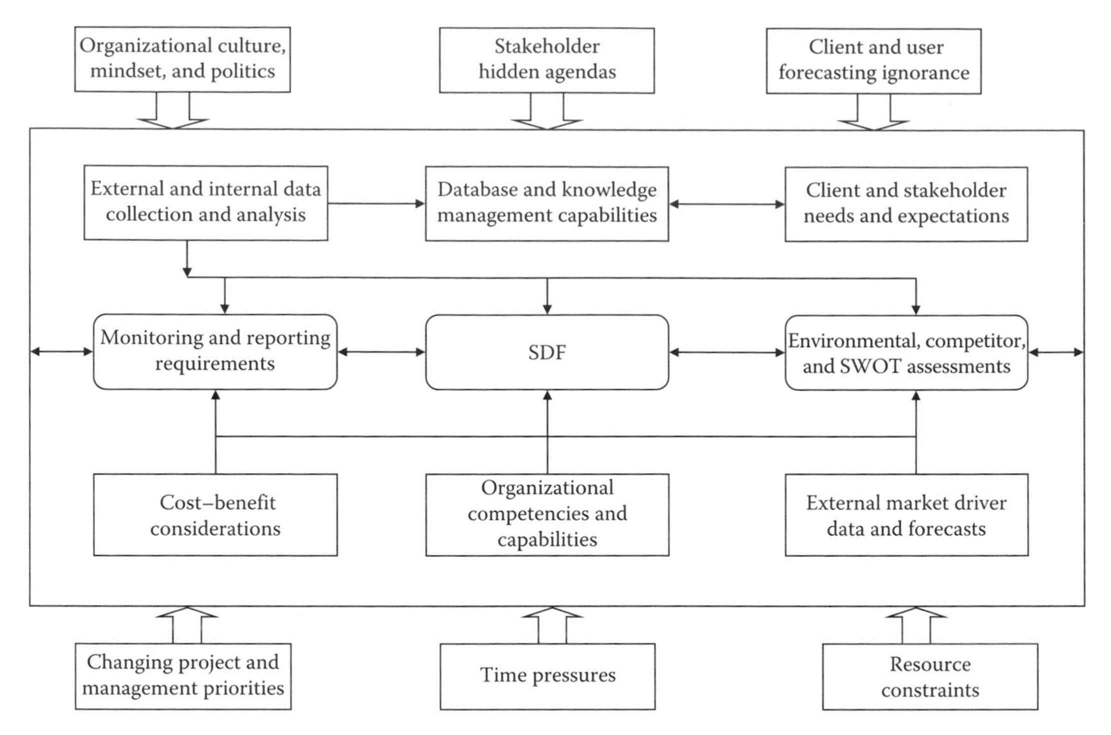

FIGURE 21.2 SDF challenges in the midst of constant change.

and focus on project goals and objectives. To fulfill these responsibilities, SDF team members create, update, and maintain the common assumption set; manage the strategic forecasting function; control changes in the future-state outlook; and manage project stakeholder expectations and reactions. They are also responsible for providing the organizational memory and institutionalized learning from past projects. They do that by creating, managing, and keeping records of data, assumptions, analyses, evaluations, scenarios, models, forecasts, options created, and recommendations. Once they define forecast performance indicators, they monitor support commitments and the performance of the decisions made. They also engage in monitoring, measuring, and enhancing SDF team performance and produce useful but human resource-intensive management reports.

Notice that the challenges of the SDF function are further complicated by changes in the political, economical, social, technological, legal, educational, and demographic (PESTLED) environment, industry structure, company and competitor SWOTs (strengths, weaknesses, opportunities, and threats), and market and consumer preferences. The interaction effects of these changes are very difficult to decouple and predict.

Operating under the challenges of high pressures from external influences on the strategic forecasting function tends to produce inadequate long-term forecasts. This is particularly true when continuous environmental changes dominate the scene. To cope with this situation, SDF teams employ approaches that help overcome challenges due to the broad scope of their function, such as the following:

1. Assessing the current state of strategic forecasting and identifying major gaps, costly problems, and issues, which render the current process ineffective
2. Investing time to educate clients and senior management on approach, process, challenges of external changes on the forecasting function, and its effectiveness

3. Demonstrating tangible benefits to them from well-structured, supported, and managed teams engaged in continuous assessments and monitoring
4. Creating a senior management-sponsored SDF team governance structure and vision, mission, values, and strategy, which allow for developing a business definition that both senior management and potential clients fully support

It is crucial that sponsors of the SDF team creation obtain senior management and client general guidance at the start of each strategic project and their commitment for resources and support to perform the SDF function appropriately. What helps a great deal is obtaining agreement on clearly defined roles and responsibilities of project participants, the deliverables, handoffs and presentations, rework loops, and other process particulars to avoid confusion and finger pointing. Other factors SDF team members need to be mindful of in order to manage the scope of its involvement are the following:

1. Reaffirming senior management and potential client supported SDF team involvement in a project
2. Assessing patterns and trends of SDF team, forecast performance, and client acceptance in prior projects
3. Modifying the scope of the group's responsibilities to fit the prevailing corporate environment and current needs

21.4 FORECAST QUALITY AND OWNERSHIP OF FORECAST

What is a good forecast? This question is raised all the time and the definition of what constitutes a good strategic decision forecast reflects differences in project stakeholder perspectives, needs, and interests, which, in turn, determine the SDF approach and solutions developed. The senior management perspective of a good forecast to form the basis for a strategic decision is that it meets their own and corporate needs and interests with respect to consistency with corporate strategy and policies, risk tolerance, and efficiency considerations. A common client view of a good long-term forecast is that the underlying scenarios are reliable enough for a project to create value and meet their needs consistent with current resource constraints. The SDF team's idea of a good long-term forecast includes quality elements such as collaborative in nature, reflecting all corporate knowledge, reasonable and well-balanced, passing several screens and sanity checks, being reliably accurate, and having been efficiently developed. And, because consensus has to be reached, the resulting "good forecast" has to have many common elements of these participants. Hence, the mix of good forecast elements is determined by politics and how well the SDF team is positioned in the organization.

Who owns the strategic forecast is another source of challenges when establishing the SDF team and its charter and what defines a "good forecast" also determines forecast ownership. In most projects, forecast ownership is divided between the client, the senior management, and the forecaster in the following manner:

1. The client is responsible for providing information and the required support, helping with creating some business assumptions, and suggesting scenarios to be analyzed and sanity checks to subject the forecasts to.
2. The role of senior management is to share insights, make available required resources, provide feedback and adjustments to forecasts, and determine consistency with their risk tolerance.
3. The forecaster's role is limited to selecting the appropriate approach or method, collecting and analyzing data, developing models, generating forecasts, and updating forecasts as clients and senior management deem appropriate.

When the two basic questions of what constitutes a good forecast and who owns the forecast are not addressed properly, conflicts and new challenges arise. In best-in-class SDF teams, these challenges are resolved in a satisfactory manner when the definition of a good forecast integrates the perspectives of the three agents. This creates the conditions for high levels of 4Cs. But, the roles of senior management, clients, and SDF team members are redefined as follows:

1. Senior management shares insights, makes available required resources, provides feedback and adjustments to forecasts, and determines consistency of project uncertainty with their risk appetite.
2. Clients share ownership responsibility with the SDF team in developing assumptions, creating scenarios, performing sanity checks, determining project uncertainty and risks, and creating FSR plans.
3. Forecasters are active decision participants. They bring together and project manage the forecasting process, data and information, methodology, models, scenarios, analyses and evaluations. They lead the forecast development process from project definition to synthesizing findings, monitoring forecast performance, and providing actionable recommendations.

21.5 ORGANIZATION KNOWLEDGE GAPS

SDF is a fairly new function; hence, acceptance of its processes and performance measures meets resistance from tactical forecasting groups and clients alike. Being an untested entity, earning acceptance is difficult due to challenges starting from the very definition of the SDF function. In the current state of limited knowledge, any request for a forecast from upper management or any forecast more than two years out are considered strategic forecasts. And, the key conclusion of forecasting experts confirms James Bryce's statement that "three-fourths of the mistakes a man makes are made because he does not really know what he thinks he knows." This is particularly true in the case of strategic forecasting, why it is needed, and how it can contribute to project success. Knowledge gaps around the SDF function are determined by comparing the company's current state of strategic forecasting processes and practices against those of benchmarked best-in-class organizations and identifying major limitations, faults, and problems. Then, the challenge becomes to determine the desired level of support the SDF team provides, finding different ways to fill identified gaps, determining associated costs, and prioritizing the gaps to fill within a specified time period.

Lack of strategic forecasting knowledge among senior managers and clients translates into inability to align the SDF processes with those of project stakeholders and planning organizations. This, in turn, makes it difficult to perform client expectations analysis, to identify and understand client needs, determine client satisfaction components, and manage client expectations effectively. Project stakeholder knowledge gaps result in erroneous goals, objectives, and target setting, ineffective planning, and incorrect prioritizing of forecasting initiatives. On the other hand, lack of SDF team member knowledge of financial analysis, strategic planning, business development, new product development (NPD), portfolio management, corporate planning, and business planning makes it difficult to communicate effectively with clients and other project team members.

How to create, manage, build applications from projects, and share knowledge effectively is a skill generally lacking. This presents a challenge for the SDF team in creating the needed processes for different types of strategic decision projects and assigning specific responsibilities to other project team members. Lack of knowledge is reflected in the clients' inability to measure SDF team and forecast performance, manage SDF team involvement in projects, and limiting the resources needed to do it effectively. When all is said and done, lack of organization knowledge is about ignorance of the nature of strategic forecasting; it is fear of the unknown strategic forecasting function. This is a source of SDF team uneasiness and one of the causes of failure to attract, retain, and motivate experienced strategic forecasters.

There are several lessons learned by SDF teams from the field of Six Sigma to address lack of knowledge. To start with, in year one focus and attention is directed to addressing the SDF team infrastructure requirements followed by establishing and publicizing its internal and external communication plan. It is vital to design the communications plan to demonstrate SDF team value added by addressing organizational needs, fears, misconceptions, and knowledge gaps. Developing an ongoing list that registers projected and actual savings of time to decision, improvements in client satisfaction, forecast reliability, and help provided is important to show how well different options meet these objectives. Also, developing a common metric and a reporting system that evaluates and updates the status of all projects semiannually contributes significantly to increasing the knowledge base of how to enhance SDF team efficiency.

Resolving challenges due to knowledge gaps is affected by moving from performance criteria to metrics and building a knowledge creation and management function through

1. Developing the company Intranet database to house the data and intelligence created by the SDF, Competitive Analysis, Market Research, Tactical Forecasting, and Environmental Assessment subteams along with assumptions used and findings of past project evaluations.
2. Sharing reviews of past company and competitor projects, researched industry analogs, averages, norm, practices, and standards, and validations of their applicability to the current project.
3. Confirming what aspects of forecasting are critical to clients and what aspects of strategic forecasting clients value, such as satisfaction with support provided, forecast accuracy, and quick turnaround.
4. Learning from external experts' experiences knowledge and applying findings from academic research and consulting evaluations in the current project.
5. Performing project postmortem analyses and sharing findings and success stories, and conducting regular project team briefings.
6. Demonstrating to clients and project team members the basics of the SDF function and the value of key analyses, assessments, evaluations, tools, and the practical approaches introduced.
7. Showing team members what is important to the forecasting function, identifying value drivers to monitor, defining performance measures, and creating management dashboards and early warning systems (EWSs).
8. Summarizing the essential client expectations and corresponding SDF team value-adding functions performed and then verifying and clarifying.
9. Determining how well the SDF team is meeting client expectations and making sure all measurements are linked to client valued functions and bottom-line results.
10. Putting knowledge management to work by updating information on the Intranet repository revising it as needed and providing guidance on its applications.

SDF teams are organized to create value in each project through effectiveness, quality and quantity of contributions, client and customer satisfaction, and ongoing efficiency improvements. However, value creation also requires integration of sound processes and adequate knowledge on the part of project team members of the steps to move forward. Thus, to create an effective SDF team requires its presence, knowledge of its functions, and understanding of its capabilities by all project stakeholders. This is accomplished by

1. Developing the SDF team identity, defining its business so as to add value by reducing the organization's knowledge gaps, and articulating its capabilities.
2. Sharing demand analysis, modeling, and scenario planning experiences.
3. Keeping knowledge sharing simple so that clients know where to find it and understand its relevance to the SDF solutions in project needs, recommendations, and their implications.

21.6 AVAILABILITY OF QUALIFIED SDF TEAM MEMBERS

In most cases, the scarcity of experienced strategic forecasters presents a difficult challenge to establish and staff a whole SDF team. The lack of qualified strategic decision forecasters is due to factors such as the following:

1. The strategic forecasting discipline that is not currently being taught in academia or by professional forecasting associations
2. The small number of forecasting groups currently engaged in and using strategic forecasting processes and practices
3. Rarity of forecasting function expertise and project management capabilities among forecasters, which are crucial in developing strategic forecasting solutions
4. Inability of many forecasters to identify, quantify, and incorporate extraneous information, such as environmental, industry, and SWOT analyses into assumptions and computational models
5. Exclusive focus on the business of forecasting at the expense of understanding what drives business decisions and how forecasting impacts decision making
6. Unwillingness of companies to invest in long-term forecasting because it is costly to attract qualified forecasters and retain them

Two key elements in having best-in-class SDF teams are that they have top performers who are worth pursuing because they make all the difference, and remembering that in the long run, quality is always the cheaper alternative. There is a pressing need to recruit qualified candidates who possess all the critical skills to be effective, but identifying and attracting such top forecasting professionals is hard. Shortages of forecaster qualifications are most pronounced in the areas of strong interpersonal skills, relationship development and management, process mapping, forecast project management, real options, systems dynamics, scenario development and planning, and FSR planning. Candidates having the right mindset and experiences, broad industry knowledge, skilled in identifying feasible future states and creating forecast realization plans, and providing support needed to achieve them are rare. Such individuals have either been promoted to higher-level positions or hired by competitors. If the requirement of experience or significant exposure to other planning functions is added, that is, strategic planning, business development, NPD, portfolio management, and financial planning, then there are very few forecasters with all of those qualifications.

Some solutions to compensate for the lack of qualified strategic forecasters that have produced positive results include schemes used to attract and retain them, such as the following:

1. Creating the right SDF team culture, work environment, and value system in order to retain qualified members
2. Encouraging innovation, free knowledge exchange and communication, integrative thinking, and ongoing improvement
3. Drawing from the pool of the Competitive Intelligence, External Assessments, and Tactical Forecasting subteams
4. Cross-pollinating with other groups and training on the basics of their functions and secondments to and from other planning groups for the duration of a project
5. Recruiting SDF team members and team managers from best-in-class companies in the industry and other industries as well
6. Creating an SDF curriculum for tactical forecasters and members of the other Business Research and Analytics organization's subteams
7. Differentiating the SDF team by performing value-adding functions for small client groups and providing immediate assessments of major changes, reducing future uncertainty to residual uncertainty, and helping to manage project risk

8. Working with professional forecasting associations to develop SDF curricula and training programs
9. Demonstrating continuous SDF team value creation for the corporation and for clients in internal and external meetings and conferences
10. Creating SDF team member progression plans for two years with corresponding rewards and recognitions for excellence of support in a project

21.7 OUTSOURCING OF STRATEGIC DECISION FORECASTS

A major challenge to creating and implementing successful SDF teams and performing required analyses and evaluations for strategic decisions is the wide practice of outsourcing the SDF function. This occurs due to

1. Senior managers and clients not really understanding what it takes to generate reliable long-term forecasts.
2. Nonexistent, newly created, or poorly performing SDF teams or questionable forecasting team member experience.

A more important reason for strategic forecast outsourcing, however, is lack of senior management and client faith in the SDF team's ability to produce reliable strategic decision forecasts. The client's inability to influence subjectively the forecasting outcome to their liking when forecasts are developed internally is another reason for outsourcing. Also, because several strategic project decisions need to be addressed simultaneously, while there are severe and short-turnaround timelines placed on resource-constrained SDF teams, outsourcing is viewed as the best solution.

It is also true that illusions of more objective strategic decision forecasts being produced by external experts, coupled with client doubts of the SDF team's objectivity in conducting objective assessments of the business, tilt the decision toward outsourcing. In other instances, it is a strong client desire to control assumptions and dictate the forecast supporting strategic decisions that cause outsourcing of this important function. However, the perceived benefits of strategic forecast outsourcing are offset by the issues created by external advisors not understanding company operations well and using off-the-shelf approaches and models instead of customized solutions. Also, uncompensated knowledge transfer to external experts, conflicts of interest, and more significantly, proprietary information being compromised and transferred to competitors should tip the choice toward internal strategic forecasting.

In the presence of intense politics and conflicting interests, outsourcing of the strategic forecasting function may be unavoidable; but in most cases, actions can be taken to limit outsourcing to rare instances. Some of the approaches used by SDF teams to overcome the tendency to outsource strategic decision forecasts have been initiatives and programs such as the following:

1. Creating functional links and strong personal relationships with client groups, senior management, and project stakeholders
2. Demonstrating competence with the creation of sound processes, databases and intelligence, descriptive models, forecast realization plans, and enhanced forecast performance
3. Educating clients and senior management on the value of best forecasts being highly collaborative, integrative, balanced, and not contaminated by internal politics
4. Collaborating with external strategic forecasting experts to create long-term forecasts, assess the extent of their value creation, and learn from their experiences
5. Showing examples of the superiority of the SDF teams performing functions normally not performed well by external experts, such as conducting validation, verification, forecasting due diligence, sanity checks of processes, models, and scenarios used to develop forecast solutions and recommendations

6. Creating individual forecast and SDF team performance metrics, management dashboards, and EWSs for each strategic project
7. Adopting practical, well-considered, cost–benefit-driven approaches to creating internal forecast solutions and comparing them to outsourced solutions
8. Evaluating and adjusting externally produced forecasts when it is considered necessary to outsource and integrating them with SDF forecasts
9. Helping the project team with functions not outsourced, such as FSR planning, which involves creating the first-, second-, and third-order competitor responses
10. Striving for and assuming a trusted advisor role to senior management, clients, the other planning groups, and the Investor Relations group

21.8 MANAGEMENT AND CLIENT REASSIGNMENTS

Decision makers and managers responsible for implementing strategic decisions based on SDF solutions and key stakeholders move on to new assignments and better positions all the time. This means that there is no real project ownership end to end. New people inherit the original forecasts and the results of earlier decisions in which they had no input to. They try to manage the situation going forward with limited understanding of what led to the decision, how it was made, what factors influenced it, and what commitments were made at that time. The problems related to senior management, client, and other project stakeholder frequent churn are caused for a number of reasons, the most notable being

1. Long implementation periods with unclear signs of success are high-risk propositions that have limited project stakeholder support.
2. Changing corporate strategy and priorities, drastic industry changes, and revolutionary technologies that present different prospects.
3. Better positions or promotional opportunities or project team member rotational assignments two to three years.
4. Temporary or permanent reassignments due to corporate restructures and reorganizations or separations from the company.
5. Frequent or not well-planned SDF team member rotations, which create difficult challenges in the implementation of FSR plans.

The major issues resulting from high levels of senior management, client, SDF team member, and other project stakeholder churn that need to be overcome in order to achieve the stated project goals and objectives are the following:

1. Providing continuity of purpose and ensuring high levels of 4Cs in all phases of the project
2. Ensuring that there are no breaks in senior management commitment and resource support pledged at the start of the project to achieve the original project objectives
3. Changing assumptions and success criteria at major handoffs of responsibilities without SDF team concurrence while holding only the SDF team responsible for the original long-term projections
4. Ensuring fact-based, well-considered, and reasonable changes of assumptions, scenarios, long-term forecast revisions, and FSR plans

Best-in-class SDF teams have found ways to address these challenges and turn them to their benefit starting with ownership of forecast assumptions going to the SDF team and assigning joint forecast monitoring responsibility with the client organization. Making the information, analyses, evaluations, findings, and the learning from all project phases and postmortem analyses available on the company Intranet helps in transitioning responsibility to new project stakeholders. And, it is especially useful to welcome new stakeholders to the project, understand their knowledge levels and needs, and brief them thoroughly on the key elements of the project and the particulars of the forecast that supported decisions up to that point.

Meeting with new project stakeholders to share how the original forecast solution was developed and getting their buy-in and support is essential to overcome other project stakeholders' reassignment challenges. At the same time, the SDF team needs to engage in close 4Cs with the new managers responsible for the project, relate progress to date relative to the projected state, and obtain their feedback and support. Another helpful practice to lessen the impact of management and client reassignments is for SDF team members to familiarize new project stakeholders with the management dashboard and EWSs and help them understand the project uncertainty and risks involved and how to manage them. New project participants need to be familiar with management dashboard indicators and EWS signals because they may be involved in deciding actions to implement to offset those risks. That is, they need to know how the FSR plan was created and how and when to put it into effect. They also need to understand the usefulness of the SDF team scorecard and the forecast performance measures that go with it. In all instances, SDF team members need to share project knowledge and insights with new participants and ensure continuity of personal relationships.

21.9 RESOURCE CONSTRAINTS AND COST–BENEFIT CONSIDERATIONS

Best-in-class SDF team governance is a superior structure of processes, rules, and policies that govern decision making around forecasting issues, operation, and control of the team. It provides the structure used to set goals and objectives, determine the means required to achieve them, and monitor the SDF team's performance. However, a best-in-class governance structure is costly to implement. The major costs are personnel costs, development of infrastructure needs, time investments, team member training and development, performance rewards, and senior management commitment.

The SDF infrastructure is determined by what clients want, when they want it, how they want it, and the price they are willing to pay, but during recessionary periods SDF team cost containment dominates other considerations. In addition to ongoing payroll, personnel-related costs include identifying and hiring costs, secondment expenses, and organizational cross-pollination-related expenditures. The costs for SDF team infrastructure building include acquisition of information technology assets, modeling and forecasting software, development of the Intranet database, and linkages with databases of other planning groups for data and information exchanges. Additional costs for infrastructure building include planning regular client reviews, monitoring of the strategic forecasting effort, and making the SDF team's suppliers part of the strategic forecasting process to minimize the impact of erroneous inputs on forecast errors and duplication of effort. Lastly, senior management time investments in strategic forecasting to ensure high levels of 4Cs are costly and difficult to maintain. However, the need to demonstrate why sound decisions need good forecasts, why SDF team participation benefits the company, and why the right infrastructure is needed ranks above cost considerations and is crucial all around.

Applying a Six Sigma-based methodology and tools to strategic forecasting and building them into SDF plans and the SDF plans into company-wide strategic and business plans are high upfront cost propositions. Also, scheduling and hosting periodic and annual reviews with clients, project team stakeholders, and senior management to assess forecast performance are expensive as are planning and sponsoring events that recognize and reward SDF team contributions and achievements. Other cost items include developing incentive plans for client and the SDF team to ensure continued collaboration and support, finding available SDF team leaders to work on several engagement projects a year, publicizing the corporate benefits realized by the participation of the SDF team, and obtaining stakeholder inputs to determine and fund the following year's continuous improvement objectives.

There are considerable costs associated with time investments in assessing the current state, benchmarking best-in-class organizations, SDF team building and business definition, process creation, education of clients and senior managers, and identification of client expectations. But, a good part of costs is driven by the creation of sound forecasting solutions and project performance monitoring and reporting. Training and development expenses include costs related to creating workshops in SDF processes and methods for members from other Business Research and Analytics subteams and developing learning sessions for potential SDF team members, clients, and project stakeholders. Expenditures for SDF team member attendance of industry conferences and participation in industry analyst meetings are also included in this category. Also, SDF team member performance rewards, recognitions, and career path planning involve some costs as well.

Overcoming the resource constraints to creating and implementing effective SDF teams is done on one-by-one basis and ensuring that cost–benefit analyses encompass all relevant factors. The essence of SDF team involvement in strategic decisions is the improvements it makes in areas such as changing the way key stakeholders look at the decision process holistically, understanding the vital few factors causing forecasting errors, and acting to eliminate those problems and rework. Improving the effectiveness of the SDF processes is another major benefit brought about by a change in the culture of the strategic forecasting team. This results in fixing forecast problems and resolving issues permanently, delivering higher quality forecasts as judged by clients, leveraging organizational knowledge to enhance performance, and reducing forecasting function costs and turnaround cycles.

To conquer the challenges due to high cost concerns, SDF teams must demonstrate to senior management and clients alike that the quality strategic decision forecasts are not only a source of better decisions, but that they could lead to competitive advantage. They are cheaper in the long run because they reflect the collective organizational knowledge, they are faster to reach consensus on, and they require far fewer revisions. Also, the benefits from implementing an SDF team and function exceed costs because they include support for FSR plans, management dashboards, and EWSs that result in lessening project uncertainty and risks and reduce time to decisions. Ordinarily, such costs are charged to the project sponsor group; hence, the need for strong relationships. Furthermore, skeptics of SDF team contributions being valued higher than their costs need to be reminded of cost savings by not wasting resources on forecast rework and revisions and savings from freeing other project team members of time-consuming responsibilities and activities in which they are less proficient than SDF team members.

21.10 IMPLEMENTING PROCESSES AND PERFORMANCE MEASURES

Establishing SDF teams goes along with implementing new processes and performance measures that necessitate changes, which are usually met with resistance. This is because clients prefer to obtain forecasts when needed, in the most expeditious manner, and without interference from other groups and outside spectators. More importantly, it is because there is no agreement among clients and other planning groups on how to establish new forecasting processes, what roles and responsibilities to assign to the SDF team, and how to influence and measure its performance. This resistance creates challenges, which are reflected in project stakeholder statements made in our discussions, such as the following:

1. We have good people, the current practices work, and new processes are not necessary. We are getting along fine.
2. Strategic project stakeholders have established processes to develop forecasts when needed and they are adequate for their needs.

3. Client organizations know best what is a good forecast and how to judge the performance of the project and the forecasting group.
4. SDF teams, processes, and performance metrics may be needed in other companies; but if we ever need them, we will create them.
5. We do not need help from individuals outside the current arrangement. We find the existing forecasting processes and performance metrics adequate.
6. We would support creation of new strategic forecasting processes and performance measures if the SDF team reports directly to our organization.
7. The promise of SDF teams seems too good to be true. We do not believe it can deliver all the benefits it promises in a way that meets our needs, despite of what performance measures say.
8. We would sponsor the SDF team and support development of new processes and performance appraisals if the focus of that group was on solving our problems and not be spread across all business units.
9. We support creation of new SDF processes and performance indicators if they do not involve additional involvement of our people, costs, responsibilities, resource commitments, and the need for coordination.
10. Trying out SDF tools, processes, and performance metrics may be a good idea, but not right now; perhaps in the next strategic project.

A. *Barriers to overcome.* The barriers underneath the above statements that the SDF team needs to overcome in order to implement needed processes and performance measures require addressing the issues and challenges caused by

1. Multiple business units and corporate organizations with divergent interests and objectives among project stakeholders participating in the project.
2. Limitations of tactical forecasting performance measures being routinely transferred to SDF projects.
3. Inconsistent planning assumptions and inputs across different stakeholder groups when the role of forecasting is treated as a sales and financial planning function.
4. Management of forecasting politics and disagreement on what performance measures to focus on and who owns which functions.
5. Lack of 4Cs in different forecasting activities and project execution make it more difficult in silo organizations.

B. *Links with other functions.* Questions and mistrust about the need to integrate forecasting with other planning functions, which functions to link together, how the integration process will work, and how to use the SDF tools and techniques are substantial issues to surmount. Other concerns are the inability of clients and senior management to assess when the SDF team has achieved integration with planning organizations and how to maximize benefits and minimize limitations of those links. But, these are nothing compared to the problem of organizational politics and partisan warfare-related issues that contaminate the development of sound SDF processes, solutions, and performance measures. Further, challenges associated with migrating from tactical to SDF modeling and forecasting such as additional time, cost, and cooperation requirements take time and considerable effort to resolve. Their resolution is needed for appropriate definition and use of performance measurements and effective monitoring processes, EWSs, and variance analysis.

Addressing the currently limited involvement and support of senior management in making forecast realization possible is really beyond the SDF team's ability to change in a short period. Also, root cause analyses of long-term forecasts that can reveal limitations traced to organizational links, processes, confusions and barriers, and forecaster skills and qualifications may be unacceptable and politically incorrect if failure factors are revealed. Lastly, wrong SDF governance structures and weak links with other planning functions lead to forecasts succumbing to broader organizational pressures, biases, and interests. Thus, improper forecaster, key stakeholder, and user alignment and weak relationships lead to poor strategic forecasts.

C. *Forecast ownership.* Strategic forecasting and decision-making processes must be sound and complete in order for long-term forecasts to be consistently reliable and lead to achieving expected results. And, value creation by strategic projects is only possible when internal SDF team confusion about forecast ownership and responsibilities and barriers and project stakeholder impediments are removed. Thus, the starting point to overcome some challenges is to get the entire organization ready for a disciplined approach to forecasting for strategic projects. What helps here are activities such as the following:

1. Assessing the organization's readiness for professional strategic decision forecasting and what comes with it
2. Surveying the forecasting knowledge base, processes, and tools currently used
3. Ensuring open communication channels in all directions and close coordination of effort
4. Doing some planning around communication and education about SDF
5. Identifying projects to apply the new processes and performance measures to demonstrate the effectiveness of the SDF team
6. Creating the infrastructure needed, identifying partners in the implementation effort, and showcasing benefits

D. *Benchmarking.* Some other ways SDF teams address the challenges to implementing new processes and performance measures begin with benchmarking against best-in-class practices and communicating effectively the expected benefits to clients. SDF teams articulate their vision and business definition and obtain senior management approval, backing, and support. After that, assessment of resources, training, and performance reward options as well as creation of the organizational structure to perform necessary functions takes place. Next is identifying and hiring top-notch professionals and creating a step-wise implementation plan of the new processes and performance measures. Last, and most important, is internalizing relevant best practices by SDF team members and key forecast stakeholders.

The approach used in developing SDF teams and processes begins with rating current forecasting practices and assessing the impact of poor long-term forecasting performance by benchmarking and demonstrating the benefits of best practices. An assessment of resource requirements is performed to determine feasibility of appropriate implementation; and if it is confirmed, the enhancement of internal links begins in earnest. Attracting and retaining the best associates in the business is an ongoing requirement to create a successful SDF team and ensure peak performance and credibility. In the meantime, SDF teams need to obtain strong endorsements from senior management, clients, and other forecast stakeholders. Recognitions by other forecasting entities, company planning groups, industry analysts, external consultants, and professional forecasting associations are also helpful.

E. *Productivity improvements.* Once the SDF team is in place, it embarks on a step-wise implementation of productivity improvements. To make strategic forecasting more comprehensive and inclusive and compare current forecasting practices to benchmarked best practices is accomplished through creation of effective processes and the right performance measures. And identifying the right performance measures satisfies the need to create effective tools to rate the long-term forecast and SDF team member performance and using them to enhance the SDF team's decision support capabilities. The productivity improvement process creation always entails a review of the internal current state of affairs, of individual process factors, and of performance components.

TABLE 21.1
Sample of SDF Team Member Personal Assessment Factors

Ability to Develop Internal and External Relationships	Score
Reasoning and problem-solving skills and analytical ability	...
Planning, budgeting, project management, and customer service skills	...
Conflict management, leadership, and people management skills	...
Ability to work at all organizational levels	...
Creativity and ability to adapt to new ideas and change	...
Supportive, committed, team-oriented, objective, and positive	...
Oral and written communication skills and presentations	...
Dedication to thoroughness and high-quality standards	...
Knowledge of industry and company business	...
Business savvy, strategic thinking, and sound judgment	...
Ethical, honest, and cost–benefit consciousness	...
High level of motivation, professionalism, and innovation	...
Initiative, flexibility, dependability, and adaptability	...
High productivity and accuracy of assessments	...
Total scores	...

Source: Long Range Planning Associates, Vero Beach, Florida.

This is a multidimensional evaluation, factor rating and overall scoring, and detection of gaps and factor deviations tool used for developing recommendations, implementation planning, and project execution. Notice that the current-state review and assessment extends over processes and procedures, analyses, evaluations, and sanity checks performed. It also includes an evaluation of functional links and personal relationships; knowledge creation, management, sharing, and client satisfaction.

So what is the significance of the productivity improvement effort? It results in obtaining senior management commitment and ongoing support, managing client expectations early on, and securing the help and backing of clients and key stakeholders. It helps project teams to identify options available, create more reliable forecasts, learn with time, and revise forecasts of future states according to changes in the scenarios underlying forecasts far more effectively. This is essential in using real options thinking to enhance forecast accuracy and helping to reduce project uncertainty, in understanding the implications of a forecast on the value being created, and continuously demonstrating the SDF team's value added. To that extent, productivity gains help to exhibit value created by new processes and performance measures internally and externally, to institute SDF team member recognition programs, and to evolve the strategic forecasting function to one of being a client-trusted advisor.

F. *Performance measures.* Forecast performance measures ordinarily include not only gauging the actual forecast number relative performance, that is, accuracy, but also factors such as forecast turnaround time requirements, the quality of assumptions and inputs, the validity of methods and models used, and the degree of forecast scenario realization. SDF team member performance measures are along rating dimensions such as forecast project management, resources used to develop forecasts to support a strategic decision, and client satisfaction levels. Performance measures for the SDF team as a whole include items such as levels of 4Cs effectiveness, knowledge management, and institutionalizing learning. However, the SDF team's

TABLE 21.2
Sample of General SDF Team Assessment Factors

Factor	Score	Factor	Score
Organizational residence and structure	...	Customer relationship management	...
Management appreciation and support	...	Functional links and personal relationships	...
Linkages with other organizations	...	Forecast adjustment and revision process	...
Number of forecasters in the group	...	Senior management and client support	...
Number of clients per forecaster	...	Extent of conflicts and resolution	...
Years since inception of SDF team (maturity)	...	Forecast ownership	...
Uses and applications of forecasts	...	Level of accountability for forecasts	...
Number of existing product forecasts	...	Variance analysis	...
Number of new product forecasts	...	Forecasting due diligence	...
SDF associate background	...	Industry analysis	...
SDF associate forecast training	...	Market and segment assessment	...
SDF associate industry experience	...	Data and information analysis	...
Appreciation of forecast implications	...	Demand analysis	...
SDF team manager background	...	Factors within company or SBU control	...
Client forecast knowledge and appreciation	...	Market research and competitive intelligence	...
Client expectations relative to forecasting	...	Inclusion of external information in forecasts	...
Forecaster–client interaction and relationships	...	Factors outside of company control	...
Forecasting processes and procedures	...	Forecast risks and management	...
Forecasting policies and practices	...	Consensus forecast development	...
Extent of forecast collaboration	...	Degree of forecast group independence	...
Forecasting software and systems used	...	Accuracy of forecast error reduction programs	...
Operational and forecast modeling capability	...	Management participation in forecasting	...
Scenario development and planning proficiency	...	Rewards and incentives for forecasters	...
Integration with all planning functions	...	Professional recognition (internal, external)	...
Client and user forecasting education	...	Job descriptions completeness	...
Subtotal scores	...	Total scores	...

Note: SBU, strategic business unit; SDF, strategic decision forecasting.
Source: Long Range Planning Associates, Vero Beach, Florida.

balanced scorecard is often supplemented with an evaluation of the personal factors and the general qualifications of SDF team members as shown in the sample assessments of Tables 21.1 and 21.2. These assessments are done periodically by SDF team managers and external consultants and are used not only to grade performance but also to identify areas of needed SDF associate improvements.

22 Summary and Conclusions

The aim of the book has been to share with readers beneficial findings and experiences in the new strategic forecasting paradigm, to illustrate its application to different types of strategic projects, and to discuss the organizations created and processes used to generate sound long-term forecasts that support strategic decisions. This is a book based on the author's practical business experiences augmented by those of his colleagues and it is about the following:

1. Reducing the widespread failures of the current state of long-term forecasting that support strategic decisions and making forecasters aware of better practices
2. Guiding forecasters and strategic project team members to create processes to address issues, perform required analyses and evaluations, and create sound long-term forecasts
3. Showing the organizations, approaches, methods, processes, and techniques to help forecasters, strategic project team members, and project sponsors in assessing, structuring, and implementing different strategic projects successfully
4. Demonstrating applications of the strategic forecasting paradigm to different types of large-impact projects to minimize project uncertainty and risk, reduce time to decision, and increase senior manager confidence in their decisions
5. Turning uniquely qualified forecasting technocrats into trusted advisors to decision makers

A large number of strategic projects fail with amazing regularity, and by now it is apparent that the absence of experienced strategic forecasting—which is a complex network of processes and activities—is a major root cause and source of failures. Success rates of strategic projects, however, are increased significantly when strategic decision forecasting (SDF) teams are engaged in strategic project evaluations early on, corporate strategy drives project objectives, and preestablished forecasting processes are followed. Also, strategic project teams need to have sufficient understanding of forecasting principles and experience in implementing strategic projects. Their activities need to be characterized by extensive cooperation and teamwork, world-class practices, and uninhibited 360-degree communications. Project stakeholders need to understand what is involved in developing long-term forecasts to support strategic projects and know their roles and responsibilities and what is expected of them. Continuity of purpose and effective transitioning of responsibilities between project phases are essential in maintaining focus on the goals and objectives and implementing a project flawlessly.

In this chapter, we summarize the challenges faced by SDF teams, which determine the important areas that strategic forecasters need to focus on to ensure project success shown in Table 22.1. First, we review the conditions for successful applications of the SDF paradigm and the main determinants of the SDF team's value creation in such projects. Then, the discussion evolves around the recurring themes in successful forecasting for strategic projects, which can also be viewed as critical success factors. After critical success factors and clarity of purpose and objectives, emphasis is given to preparation and following appropriate methods and preestablished processes designed to develop sound SDF solutions.

Because discipline in strategic decision making is a major requirement for a project to create value, we emphasize the key considerations and stress repeatedly the need for decisions to be based on forecasts supported by facts, evidence, independent validation and verification, and extensive sanity checks. Characteristics of excellent quality assessments and project teams are discussed and the need for sound, project-specific approaches and processes is raised again. Elements of effective

TABLE 22.1
Major Challenges Facing the SDF Team

• Lack of confidence in forecasting; testing the maintained hypotheses	• Flawless handoff of responsibilities	• Impact of wrong forecasts on decisions and decision makers
• Impact of wrong forecasts on the business	• Changing business objectives and internal priorities	• Conflicting stakeholder objectives and internal politics
• Dealing in uncharted waters	• Multiple options, scenarios, and futures must be evaluated	• Withstand internal and external forecast scrutiny
• Intensive sanity and reality checks; assessment of forecast implications	• Intensive communication, cooperation, coordination, and collaboration	• Effective project management capabilities
• Guard against undue forecast influence on strategy and decisions	• Mastering different business disciplines and functions	• Management of project stakeholder expectations
• Coupling of forecasting with risk management (they are inseparable)	• Absence of accepted SDF performance measures	• Long forecast horizon allows for occurrence of black swans
• Lack of forecaster training in the art of SDF	• Unrecognized need; SDF has had no home; done in decision-maker group	• No standard guidelines on how to forecast under high residual uncertainty
• Heavy time investment in maintaining links, networks, and relationships	• Lack of formal processes; wrong uses of long-term forecasts	• Leads to advantage in decision making only when SDF is properly deployed

Note: SDF, strategic decision forecasting.

strategic forecasting are distilled, and we conclude that functional link arrangements and close personal relationships are the way to strive for excellence in supporting project teams. The performance assessment of long-term forecasts, of the SDF team, and of the project involves many steps and considerations necessary to ensure project success.

Best-in-class SDF teams and the factors of effective implementation are reviewed briefly, and the need for client and senior management support for appropriate involvement of the SDF team is brought up because it is crucial in ensuring effective transitioning from forecasts to project implementation and to future-state realization (FSR). Without effective communication, cooperation, coordination, and collaboration (4Cs) in project managing the forecasting function, strategic projects are likely to fail; and for that reason, we review once again the key considerations when basing decisions on SDF solutions. Finally, we emphasize the need to benchmark and adapt best practices in creating effective FSR plans and end the chapter with a brief conclusion.

22.1 CONDITIONS FOR SUCCESSFUL SDF APPLICATIONS

The presence of SDF teams in strategic projects is not sufficient to guarantee project success. A number of other conditions that need to be satisfied in order for projects to create value and meet objectives according to expectations are as follows:

A. *Senior management support.* Senior management support is a key precondition to ensure allocation of adequate resources and acceptance of SDF team participation and solutions as is the case with every other new corporate initiative. For clients to obtain the kind of support needed to handle complex strategic project evaluations and to get forecast consensus effectively, early senior management education of forecasting processes and requirements is necessary as are regular forecast briefings and updates to obtain their feedback and input. However, senior management support must come with measured investment of their time, but with firm commitment for ongoing support of the SDF team's business definition and participation in strategic projects. Other key areas in need of senior management support are the ranking of project priorities, input and guidance in dealing with major issues, and help in removing impediments and resolving problems affecting efficient creation of sound SDF solutions.

B. *Functional links and personal relationships.* Since a good part of business decisions are based on people relationships, the two preconditions of SDF team effectiveness are strong functional linkages and close personal relationships with counterparts in all planning, client, and external expertise organizations. Another form of SDF team connection with other project stakeholders is through electronic systems, such as access to the corporate Intranet database and knowledge sharing. Growth of functional ties and personal relationships is affected via organizational cross-pollination and education through creation of learning forums and team member knowledge exchanges. Furthermore, personal relationships are strengthened with career path planning through client groups and SDF team member secondments.

C. *Appropriate SDF governance structure.* The first and one of the most important requirements to ensure success in SDF is finding the right home for the SDF team, and after that comes close cultural fit between the parent organization and this team. Depending on the nature of the company's operations, Strategic Planning and Product Portfolio Management are the most suited homes for the SDF team to reside within a larger Business Research and Analytics organization. This larger organization also contains the Market Research, Competitive Analysis, Environmental Assessment, and Tactical Forecasting subteams. A key condition in successful applications of the SDF paradigm to strategic projects is clearly defined and well-understood rules of engagement in different types of projects. The evaluations of the strategic forecast, SDF team member, and project performance based on sound and reasonable measures are important as are rewards according to objectives met.

D. *Satisfactory SDF team business definition.* A suitable SDF team business definition is essential and begins with a senior management sanction of the group's vision and mission. Once the right SDF strategy, processes, and plans are created, they lead to the right support being provided to the client and the project team. However, the scope of SDF team participation needs to be agreed upon at the start of project involvement to avoid scope creep and diffusion of focus. Once these conditions are satisfied, a clear and reasonable definition of SDF and project team member roles and responsibilities can be articulated. It is also important to plan the evolution of the SDF team's business definition to allow for higher levels of involvement as corporate needs change, its productivity increases, and its contributions are recognized.

E. *Right SDF team culture.* The right SDF team culture is the third essential part needed to complete defining the organization besides appropriate governance structure and business definition. Successful SDF teams are characterized by strong core values that guide the attitudes and behavior of team members who are known for having a mindset of effective knowledge creation, management, and sharing. However, all this is possible only in the presence of high levels of 4Cs. Again, a mindset of excellence, clarity of purpose, and focus on key factors are preconditions in building a culture of continuous innovation and improvement and providing excellent project support.

F. *SDF team member functional expertise.* High levels of interpersonal skills and excellence of SDF team member functional expertise is another layer of building blocks that enable the creation of long-term forecasts and which form a sound basis for strategic decisions. While critical analytics and integrative thinking skills are important, effective communications and relationship management are crucial. Excellence of functional expertise begins with substantial experience

effective communication and transition to implementation, which are essential to meet stated project objectives.

It is critical that only experienced SDF team members participate in strategic projects and team members without prior practical experience should become familiar with the basics of SDF analyses and evaluations needed in large impact decisions and strategic forecasting. Only trusted and seasoned experts, including SDF team members, should be assigned to the implementation team and the execution of the FSR plan. Adequate project preparation and adherence to process are important and project participants should know exactly what should be done, how to do it, and at what point in order for the SDF team to be effective. Over time and under competent management, the new strategic forecasting paradigm could create a basis for competitive advantage.

Bibliography

Akao, Y., ed., *Hoshin Kanri: Policy Development for Successful TQM*, Productivity Press, New York, 1991.

Allen, D. H., *Credibility and the Assessment of R&D Projects*, Elsevier Science, New York, 1972.

Anker, D., *Developing Business Strategies,* John Wiley & Sons, Hoboken, NJ, 1998.

Ansoff, L., Strategies for diversification, *Harvard Business Review*, vol. 35, September–October 1957, pp. 113–124.

Arioli, D., Bianchi, M., Creazza, A. and Cattano, F., *Results from an Audit in the Sales Forecasting & Demand Planning Process*, Italian Association of Logistics and Demand Planning, www.liuc.it/ricerca/clog/Audit_SF&DP.pdf.

Armstrong, J. S., Strategic planning and forecasting fundamentals, in Albert, K., ed., *The Strategic Management Handbook*, McGraw-Hill, New York, 1983, pp. 2-1–2-32.

Armstrong, J. S., Forecasting for environmental decision making, in Dale, V. H. and English, M. R., eds., *Tools to Aid Environmental Decision Making*, Springer-Verlag, New York, 1999, pp. 192–225.

Armstrong, J. S., ed., Standards and practices for forecasting, in *Principles of Forecasting: A Handbook for Researchers and Practitioners*, Kluwer Academic Publishers, Norwell, MA, 2001, pp. 679–732.

Armstrong, J. S. and Green, K. C., Demand forecasting: Evidence based methods, Working Paper No. 24/05, Monash University, VIC, Australia, September 2005.

Armstrong, J. S. and Grohman, M. C., A comparative study of methods for long range forecasting, *Management Science*, vol. 19, no. 2, 1972, pp. 211–221.

Asia Development Bank, *Early Warning Systems for Financial Crisis: Applications to East Asia*, Palgrave MacMillan, Hampshire, 2005.

Barney, J. B. and Griffin, R. W., *The Management of Organizations: Strategy, Structure, Behavior*, Houghton Mifflin Company, Boston, MA, 1992.

Barton, D., Global forces shaping the future of business and society, *McKinsey Quarterly*, November 2010.

Bing, G., *Due Diligence Techniques and Analysis: Critical Questions for Business Decisions*, Quorum Books, Westport, CT, 1996.

Bogue, M. and Buffa, L., *Corporate Strategic Analysis*, Free Press, New York, 1986.

Booz Allen Hamilton, *Value Measuring Methodology: How-to-Guide*, CIO Council, Best Practices Committee, Washington, DC, 2002.

Brennan, K., *A Guide to the Business Analysis Body of Knowledge*, International Institute of Business Analysis, Toronto, ON, 2009, p. 29.

Brittain, J., *Lessons from the Field: Applying Appreciative Inquiry*, Practical Press Inc., Plano, TX, 1998.

Brown, M. G., *Keeping Score: Using the Right Metrics to Drive World-Class Performance*, AMACOM, New York, 1996.

Brue, G., *Six Sigma for Managers*, McGraw-Hill, New York, 2002.

Caragata, P. J., *Business Early Warning Systems*, Butterworths, Wellington, 1999.

Chan, L. K. and Wu, M. L., Quality function deployment: A literature review, *European Journal of Operational Research*, vol. 143, 2002, pp. 463–497.

Chase, C., *Demand Driven Forecasting: A Structured Approach to Forecasting*, John Wiley & Sons, Hoboken, NJ, 2009.

Christensen, C. M. and Donovan, T., *The Process of Strategy Development and Implementation*, Division of Research, Harvard Business School, Allston, MA, 2000.

Clarke, A., *Situational Analysis: Grounded Theory after the Postmodern Turn*, Sage Publications, Thousand Oaks, CA, 2005.

Coates, J. F., The role of formal models in technology assessment, *Technological Forecasting and Social Change*, vol. 9, 1976, pp. 139–80.

Coates, J. F., Why study the future? *Research Technology Management*, vol. 46, no. 3, 2003, pp. 5–8.

Coates, V., Farooque, M., Klavans, R., Lapid, K., Linstone, H. A., Pistorius, C. and Porter, A. L., On the future of technological forecasting, *Technological Forecasting and Social Change*, vol. 67, 2001, pp. 1–17.

Cogan, R., *Critical Thinking: Step by Step*, University Press of America, Lanham, MD, 1998.

Cognos Plan-to-Perform Blueprints, *Strategic Planning and Forecasting*, Business Value Guide, vol. 2, Cognos Innovation Center, ftp://ftp.software.ibm.com/software/data/sw-library/cognos/demos/bp_od_blueprints/resources/bvg_strategic_financial_planning.pdf.

Cohen, A. R., *The Portable MBA in Management*, John Wiley & Sons, Hoboken, NJ, 2002.

Conklin, J., *Wicket Problems and Social Complexity*, CogNexus Institute, 2001–2003, http://cognexus.org.

Cooper, R. G., *Winning New Products*, Addison-Wesley Publishing, Indianapolis, IN, 1986.

Cooper, R.G., *Product Leadership: Creating and Launching Superior New Products*, Perseus Books, New York, 1999.

Cooper, R. G. and Edgett, S. J., Ten ways to make better portfolio and project selection decisions, *PDMA Visions Magazine*, June 2006, pp. 11–15.

Cooper, R. G. and Edgett, S. J., Your roadmap for new product development, New Product Development Institute, www.prod-dev.com/stage-gate.

Cooperrider, D. L., *Appreciative Inquiry: A Positive Revolution in Change*, Berrett-Koehler Publishers, San Francisco, CA, 2005.

Courtney, H., Kirkland, J. and Viguerie, P., Strategy under uncertainty, *Harvard Business Review*, vol. 75, no. 6, 1997, pp. 66–79.

Coyle, R. G., *System Dynamics Modeling: A Practical Approach*, Chapman & Hall, London, 1996.

Damelio, R., *Process Mapping*, The Bottom Line Group, Toronto, ON, 2007.

David, F. R., *Strategic Management*, 4th edition, MacMillan Publishing, New York, 1993.

De Bono, E., *Six Thinking Hats: An Essential Approach to Business Management*, Little, Brown and Company, New York, 1985.

Deschamps, E. A., Six steps to overcome bias in the forecasting process, *Foresight: The International Journal of Applied Forecasting*, no. 2, 2005, pp. 6–11.

Diamond, S., *Getting More: How to Negotiate to Achieve Your Goals in the Real World*, Crown Publishing Group, New York, 2010.

Doebelin, E. O., *System Dynamics: Modeling, Analysis, Simulation, Design*, Marcel Dekker, New York, 1998.

Drucker, P., *The Practice of Management*, Harper & Row, New York, 1982.

Drucker, P. F., Hammond, J., Keeney, R., Raiffa, H. and Hayashi, A. M., *Harvard Business Review on Decision Making*, Harvard Business School Press, Boston, MA, 2001.

Drury, D. H., Issues in forecasting management, *Management International Review*, vol. 30, no. 4, 1990, pp. 317–329.

Dweck, C. S., *Mindset: The New Psychology of Success*, Random House, New York, 2006.

Eckerson, W. W., *Performance Dashboards: Measuring, Monitoring, and Managing Your Business*, John Wiley & Sons, Hoboken, NJ, 2006.

Edgett, S. J. and Kleinschmidt, E. J., Optimizing the stage-gate process: What best companies are doing, Part 2, *Research Technology Management*, vol. 45, no. 5, 2002, pp. 43–49.

Egan, C., *Creating Organizational Advantage*, Butterworth-Heinemann, Woburn, MA, 1999.

Eversheim, W., ed., *Innovation Management for Technical Products: Systematic and Integrated Product Development and Production Planning*, Springer-Verlag, Berlin, 2009.

Fahey, L. and Randall, R. M., *Learning From the Future*, John Wiley & Sons, New York, 1998.

Figlewski, S. and Levich, R. M., eds., *Risk Management: The State of the Art*, Kluwer Academic Publishers, Norwell, MA, 2002.

Finkelstein, S., Whitehead, J. and Campbell, A., *Think Again: Why Good Leaders Make Bad Decisions and How to Keep it from Happening to You*, Harvard Business School Publishing, Boston, MA, 2008.

Fischhoff, B., What forecasts seem to mean, *International Journal of Forecasting*, vol. 10, 1994, pp. 387–403.

Fisher, A. and Scriven, M., *Critical Thinking: Its Definition and Assessment*, Edge Press, Point Reyes, CA, 1997.

Flaherty, J. E. and Drucker, P., *Shaping the Managerial Mind—How the World's Foremost Management Thinker Crafted the Essentials of Business Success*, Jossey-Bass, San Francisco, CA, 1999.

Forrester, J. W., *System Dynamics and the Lessons of 35 Years: A Chapter for the Systemic Basis of Policy Making in the 1990s*, Sloan School of Management, Massachusetts Institute of Technology, Cambridge, MA, 1991.

Foster, H. D., *Disaster Planning: The Preservation of Life and Property*, Springer-Verlag, New York, 1980.

Funk, S., *Crisis Management: Planning for the Inevitable*, AMACOM, New York, 1986.

The Futures Group International, Relevance tree and morphological analysis, AC/UNU millennium project, *Futures Research Methods*, v2.0, 2004.

Garcia, M. L. and Bray, O. H., *Fundamentals of Technology Roadmapping*, Strategic Business Development Sandia National Laboratories, Albuquerque, NM, 1997.

Garland, E., *Future Inc: How Business Can Anticipate and Profit from What's Next*, AMACOM, New York, 2007.

Gilad, B., *Early Warning: Using Competitive Intelligence to Anticipate Market Shifts, Control Risk, and Create Powerful Strategies*, AMACOM, New York, 2005.

Goldscheider, R., *Licensing Best Practices Set*, John Wiley & Sons, Hoboken, NJ, 2006.

Goldsmith, M., *Global Leadership: The Next Generation,* Financial Times-Prentice Hall, Upper Saddle River, NJ, 2003.

Goldstein, L. and Kearsey, B., *Technology Patent Licensing: An International Reference on 21st Century Patent Licensing, Patent Pools, and Patent Platforms*, Aspatore Inc., Boston, MA, 2004.

Gordon, M., Musso, C., Rebentisch, E. and Gupta, N., The path to successful new products, *McKinsey Quarterly Review*, January 10, 2010.

Gordon, T. J., *Cross-Impact Method*, Millennium Project, United Nations University, Tokyo, 1994.

Hax, A. C. and Wilde, D. L., The delta model: Adoptive management for a changing world, in Malone, T. W., Laubacher, R. J. and Morton, M. S., eds., *Inventing the Organizations of the 21st Century*, MIT, Boston, MA, 2003, pp. 173–203.

Hjorland, B., The foundation of the concept of relevance, *Journal of the American Society for Information Science and Technology*, vol. 61, no. 2, 2010, pp. 217–237.

Hoole, R. and Mandana, S., Three forecasting building blocks for supply chain excellence, *Chief Supply Chain Officer Magazine*, November 2005.

Hughes, W. H. and Lavery, J., *Critical Thinking: An Introduction to the Basic Skills*, Broadview Press Ltd., Orchard Park, NY, 2004.

Hunt, V. D., *Process Mapping: How to Reengineer Your Business Processes*, John Wiley & Sons, Hoboken, NJ, 1996.

Hunter, D., *A Practical Guide to Critical Thinking: Deciding What to Do and Believe*, John Wiley & Sons, Hoboken, NJ, 2009.

INFORTE, *Forecasting and the Enterprise—Best Practices for Operating an Effective Forecasting Function*, Inforte Corporation, Paris, 2002.

Ishikawa, K., *What Is Total Quality Control? The Japanese Way*, Prentice Hall, Englewood Cliffs, NJ, 1987.

Jain, C., *Practical Guide to Business Forecasting*, Graceway Publications, New York, 2001.

Jain, C., *Benchmarking Forecasting Practices*, Graceway Publications, New York, 2002.

Jain, C. and Malehorn, J., *Fundamentals of Demand Planning and Forecasting*, Graceway Publications, New York, 2012.

Jenkins, G. M., Some practical aspects of forecasting in organizations, *Journal of Forecasting*, vol. 1, 1982, pp. 3–21.

Kahn, K. B., An exploratory investigation of new product forecasting practices, *The Journal of Product Innovation Management*, vol. 19, 2002, pp. 133–143.

Kaplan, R. S. and Norton, D. P., *The Balanced Scorecard: Translating Strategy into Action*, Harvard University Press, Boston, MA, 1996.

Kendrick, T., *The Project Management Toolkit*, AMACOM, New York, 2004.

Kerzner, H., *Project Management: A Systems Approach to Planning, Scheduling, and Controlling*, John Wiley & Sons, Hoboken, NJ, 2009.

King, P. M. and Kitchener, K. S., *Developing Reflective Judgment: Understanding and Promoting Intellectual Growth and Critical Thinking in Adolescents and Adults*, Jossey-Bass, San Francisco, CA, 1994.

King, R., *Hoshin Planning: The Development Approach*, GOAL/QPC, Salem, NH, 1989.

Kotler, P. and Keller, K. L., *Marketing Management*, Prentice Hall, Upper Saddle River, NJ, 2006.

Kotov, R. P., The strategy wheel: A method for analysis and benchmarking for competitive strategy, *Competitive Intelligence Review*, vol. 12, no. 3, 2001, pp. 21–30.

Kotter, J. P. and Heskett, J. L., *Corporate Culture and Performance*, Simon & Schuster, New York, 1992.

Krishnan, V. and Ulrich, K. T., Product development decisions: A review of the literature, *Management Science*, vol. 47, no. 1, 2001, pp. 1–21.

Lewin, K., *Field Theory in Social Science*, Harper & Row, New York, 1951.

Lewis, J. P., *Fundamentals of Project Management*, AMACOM, New York, 2007.

Lipinski, H. and Tydeman, J., Cross impact analysis: Extended KSIM, *Futures*, vol. 11, no. 2, 1979, pp. 151–154.

Locke, E. A., *A Theory of Goal Setting and Task Performance*, Prentice Hall, Englewood Cliffs, NJ, 1990.

Locke, E. A. and Latham, G. P., Building a practically useful theory of goal setting and task motivation, *American Psychologist*, vol. 57, 2002, pp. 705–717.

Madison, M., *Process Mapping, Process Improvement, and Process Management*, Paton Press LLC, Chico, CA, 2005.

Mahajam, V., Muller, E. and Bass, F., New product diffusion models in marketing: A review and direction for research, *Journal of Marketing*, vol. 54, 1990, pp. 1–26.

Makridakis, S., Wheelwright, S. C. and Hyndman, R. J., *Forecasting Methods and Applications*, John Wiley & Sons, Hoboken, NJ, 1998.

Malcolm, M. R. and Kujundzic, N., *Critical Reflection: A Textbook for Critical Thinking*, McGill-Queens University Press, Montreal, QC, 2005.

Marks, R. E., Decision analysis: Games against nature, Lecture 5-1, 2004, www.agsm.edu.au/bobm/teaching/SGTM/id.pdf.

Martin, R. L., *The Opposable Mind: How Successful Leaders Win through Integrative Thinking*, Harvard Business School Publishing, Boston, MA, 2007.

Martino, J., *Technological Forecasting for Decision Making*, Elsevier, New York, 1983.

MasterCard, *Megatrends 20/20*, John Wiley & Sons, Hoboken, NJ, 2008.

Mauboussin, M. J. and Bartholdson, K., Measuring the moat, *Credit Suisse/First Boston Equity Research*, vol. 16, 2002, pp. 1–52.

McGuire, B., *Surviving Armageddon: Solutions for a Threatened Planet*, Oxford University Press, New York, 2005.

Mello, J. E., The impact of corporate culture on sales forecasting, *Foresight: The International Journal of Applied Forecasting*, no. 2, 2005, pp. 12–15.

Mentzer, J. T., The impact of forecasting improvement on return on shareholder value, *Journal of Business Forecasting*, vol. 18, 1999, pp. 8–12.

Mentzer, J. T. and Bienstock, C. C., Principles of sales forecasting systems, *Supply Chain Management Review*, Fall 1998, pp. 76–83.

Mentzer, J. T. and Moon, M. A., *Sales Forecasting Management: A Demand Management Approach*, Sage Publications, Thousand Oaks, CA, 2005.

Meyers, I. B. and Meyers, P., *Gifts Differing: Understanding Personality Type*, CPP Inc., Mountain View, CA, 1995.

Mileti, D., *Disasters by Design: A Reassessment of Natural Hazards in the United States*, Joseph Henry Press, Washington, DC, 1999.

Moon, M. A., What is world-class forecasting? A perspective on 20 years of research, *APICS International Conference Proceedings*, October 10–13, San Diego, CA, 2004.

Moon, M. A., Mentzer, J. T., Smith, C. T. and Garver, M. S., Seven keys to better forecasting, *Business Horizons*, September–October 1998, pp. 44–52.

Moore, K. D., *Effective Instructional Strategies*, SAGE Publications, Thousand Oaks, CA, 2009.

Morgan, M., *Creating Workforce Innovation: Turning Individual Creativity into Organizational Innovation*, Business & Professional Publishing, Sydney, NSW, 1993.

Naisbitt, J., *Megatrends: Ten New Directions Transforming Our Lives*, Warner Books, New York, 1982.

Narayanan, V. K., *Managing Technology and Innovation for Competitive Advantage*, Prentice Hall, Englewood Cliffs, NJ, 2001.

Narin, F., *Evaluative Bibliometrics: The Use of Publication Citation Analysis in the Evaluation of Scientific Activity*, National Science Foundation, Cherry Hill, NJ, 1976.

Nuefewld, W. P., Environmental scanning: Its use in forecasting emerging trends and issues in organizations, *Futures Research Quarterly*, vol. 1, no. 3, 1985, pp. 39–52.

O'Connell, J., Pyke, J. and Whitehead, R., *Mastering Your Organization's Processes*, Cambridge University Press, New York, 2006.

Ogilvy, J. and Schwartz, P., Plotting your scenarios, in Fahey, L. and Randall, R., eds., *Learning From the Future*, J. Wiley & Sons, Hoboken, NJ, 1998, pp. 57–80.

Oriesek, D. F. and Schwarz, J. O., *Business Wargaming: Securing Corporate Value*, Gower Publishing Ltd., Hampshire, 2008.

Ozer, M. A., A survey of new product evaluation models, *Journal of Product Innovation Management*, vol. 16, no. 1, 1999, pp. 77–94.

Parr, R. L. and Sullivan, P. H., eds., *Technology Licensing: Corporate Strategies for Maximizing Value*, John Wiley & Sons, Hoboken, NJ, 1996.

Pascale, R. and Athos, A., *The Art of Japanese Management*, Penguin Books, London, 1981.

Peters, T. and Waterman, R., *Search of Excellence: Lessons from Americas Best Run Companies*, Harper & Row, New York, 1982.

Porter, A. L., Roper, T. A., Mason, T. W., Rossini, F. A. and Banks, J., *Forecasting and Management of Technology*, John Wiley & Sons, Hoboken, NJ, 1991.

Porter, B. F., *The Voice of Reason: Fundamentals of Critical Thinking*, New York, 2001.

Porter, M. E., *Competitive Strategy: Techniques for Analyzing Industries and Competitors*, Free Press, New York, 1980.

Porter, M. E., *The Competitive Advantage of Nations*, Free Press, New York, 1990.

Porter, M. E., What is strategy, *Harvard Business Review*, November–December 1996, pp. 61–78.

Porter, M. E., The five forces that shape strategy, *Harvard Business Review*, January 2008, pp. 86–104.

Pries, K. H., *Six Sigma for the New Millennium*, Quality Press, Milwaukee, WI, 2009.
Project Management Institute, *A Guide to the Project Management Body of Knowledge*, 3rd edition, Project Management Institute, Newtown Square, PA, 2003.
Ramo, S. and Sugar, R., *Strategic Business Forecasting: A Structured Approach to Sharing the Future*, McGraw-Hill, New York, 2009.
Randers, J., *Elements of the System Dynamics Method*, Productivity Press, Portland, OR, 1980.
Rappaport, A., Linking competitive strategy and shareholder value analysis, *The Journal of Business Strategy*, vol. 7, 1987, pp. 58–67.
Rasiel, E. M. and Friga, P. N., *The McKinsey Mind: Understanding and Implementing the Problem Solving Tools and Management Techniques of the World's Top Consulting Firm*, McGraw-Hill, New York, 2002.
Rigatuso, C., *Aligning Strategic and Operational Planning with Balanced Scorecard Techniques*, Lenskhold Group, Manasquan, NJ, 2007.
Rittel, H. and Webber, M., Dilemmas in a general theory of planning, *Policy Sciences*, vol. 4, 1973, pp. 159–169.
Roberto, M. A., *Know What You Don't Know: How Great Leaders Prevent Problems before They Happen*, Wharton School Publishing, Upper Saddle River, NJ, 2009.
Rockart, J. F., Chief executives define their own data needs, *Harvard Business Review*, vol. 2, 1979, pp. 81–93.
Rogelberg, S. G., Barnes-Farrell, J. L. and Lowe, C. A., The stepladder technique: An alternative group structure facilitating effective group decision making, *Journal of Applied Psychology*, vol. 77, no. 5, 1992, pp. 730–737.
Romero, D., Galeano, N., Giraldo, J. and Molina, A., *Towards the Definition of Business Models and Governance Rules for Virtual Breeding Environments*, Springer-Verlag, Boston, MA, 2006.
Rose, K. H., *Project Quality Management: Why, What, and How*, J. Ross Publishing, Boca Raton, FL, 2005.
Rousso, M. and Lerner, N., Strategic technology planning as a catalyst for business process reengineering, *GIS/LIS*, 1994, pp. 666–676, http://libraries.maine.edu/Spatial/gisweb/spatdb/gis-lis/gi94083.html.
Sadgrove, K., *The Complete Guide to Business Risk Management*, 2nd edition, Gower Publishing Company, Burlington, VT, 2005.
Schimpff, S. C., *The Future of Medicine: Megatrends in Healthcare*, Thomas Nelson Inc., Nashville, TN, 2007.
Schwartz, P., *The Art of the Long View*, Doubleday, New York, 1991.
Sekhri, N., Chisholm, R., Longhi, A., Evans, P., Rilling, M., Wilson, E. and Madrid, Y., Principles for forecasting demand for global health products, Center for Global Development, Global Health Forecasting Working Group, Background Paper, 2006, www.cgdev.org/doc/Demand Forecasting/Principles.pdf.
Serrat, O., *The Five Whys Technique*, Knowledge Solutions, Asian Development Bank, Manila, 2009.
Sheldon, D. H., *Lean Materials Planning and Execution: A Guide to Internal and External Supply Management Excellence*, J. Ross Publishing Inc., Fort Lauderdale, FL, 2008.
Shim, J. K., *Strategic Business Forecasting: The Complete Guide to Forecasting Real World Company Performance*, CRC Press, Boca Raton, FL, 2000.
Shim, J. K., *Strategic Business Forecasting: Including Forecasting Tools and Applications*, Global Professional Publishing, Sterling, VA, 2009.
Slocum, M. S. and Lundberg, C. O., Technology forecasting: From emotional to empirical, *Creativity and Innovation Management*, vol. 10, no. 2, 2001, pp. 139–152.
Stacey, R. D., *Managing the Unknowable: Strategic Boundaries between Order and Chaos*, Jossey-Bass, San Francisco, CA, 1992.
Stacey, R. D., *Complexity and Creativity in Organizations*, Berrett-Koehler Publishers, San Francisco, CA, 1996.
Stacey, R. D., *Complex Responsive Processes in Organizations*, Routledge, New York, 2001.
Stacey, R. D., *Strategic Management and Organizational Dynamics: The Challenge of Complexity*, 5th edition, Prentice Hall, Upper Saddle River, NJ, 2007.
Stamatis, D. H., *Total Quality Service: Principles, Practices, and Implementation*, CRC Press, Boca Raton, FL, 1996.
Stamatis, D. H., *Advanced Quality Planning: A Commonsense Guide to AQP and APQP*, Productivity Inc., Portland, OR, 1998.
Stamatis, D. H., *Failure Mode and Effect Analysis: FMEA from Theory to Execution*, Quality Press, Milwaukee, WI, 2003a.
Stamatis, D. H., *Six Sigma Fundamentals: A Complete Introduction to the Systems, Methods, and Tools*, Productivity Press, New York, 2003b.
Stamatis, D. H., *10 Essentials for High Performance in the 21st Century*, CRC Press, Boca Raton, FL, 2011.
Sterman, J. D., *Business Dynamics: Systems Thinking and Modeling for a Complex World*, McGraw-Hill, New York, 2000.
Stewart, R. W. and Fortune, J., Application of systems thinking to the identification, avoidance and prevention of risk, *International Journal of Project Management*, vol. 13, no. 5, 1995, pp. 279–286.

Tague, N. R., *The Quality Toolbox*, ASQ Quality Press, Milwaukee, WI, 2005.
Taleb, N. N., The black swan: The impact of the highly improbable, *New York Times*, April 22, 2007a.
Taleb, N. N., *The Black Swan: The Impact of the Highly Improbable*, Random House, New York, 2007b.
Taylor, R. E., A six segment message strategy wheel, *Journal of Advertising Research*, vol. 39, no. 6, 1999, pp. 7–17.
Thomson, A., *Critical Reasoning: A Practical Introduction*, TJ International, Padstow, Cornwall, 2002.
Tran, T. A. and Daim, T., A taxonomic review of methods and tools applied in technology assessment, *Technological Forecasting and Social Change*, vol. 75, no. 9, 2008, pp. 1396–1405.
Triantis, J. E., *Project Financing Benchmarking Study*, AT&T Submarine Systems, 1994.
Triantis, J. E., *Creating Successful Acquisitions and Joint Ventures: A Process and Team Approach*, Quorum Books, Westport, CT, 1999.
Triantis, J. E., Collaborative forecasting: An intra-company perspective, *Journal of Business Forecasting*, vol. 20, 2001–2002, pp. 12–14.
Triantis, J. E., How to outsource successfully life cycle management forecasts, *Journal of Business Forecasting*, vol. 28, 2009, pp. 28–35.
Triantis, J. E. and Song, H., Building alliances with customers: Key to success in forecasting, *Journal of Business Forecasting*, vol. 4, 2002–2003, pp. 2–8.
Triantis, J. E. and Song, H., Pharmaceutical forecasting model simulation guidelines, *Journal of Business Forecasting*, vol. 26, 2007, pp. 41–49.
Truby, J., *The Anatomy of Story: 22 Steps to Becoming a Master Storyteller*, Faber and Faber Inc., New York, 2008.
Twiss, B., *Forecasting for Technologists and Engineers: A Practical Guide for Better Decisions*, IEE Management of Technology Series, Peter Peregrinus Ltd., London, 1992.
Van der Heijden, K., *Scenarios: The Art of Strategic Conversation*, John Wiley & Sons, Hoboken, NJ, 1996.
Vanson, J. H., Better forecasts, better plans, better results, *Research Technology, Management*, vol. 46, 2003, pp. 47–58.
Wack, P., Scenarios: Uncharted waters ahead, *Harvard Business Review*, vol. 5(a), 1985, pp. 73–89.
Wardak, A., Gorman, M. E., Swami, N. and Deshpande, S., Identification of risks in the life cycle of nanotechnology-based products, *Journal of Industrial Ecology*, vol. 12, no. 3, 2008, pp. 435–448.
Wodall, S. R., Strategic forecasting in long-range military force planning, Ph.D. Dissertation, Catholic University of America, Washington, DC, 1985.
Wooton, S. and Horne, T., *Strategic Thinking: A Step-by-step Approach to Strategy*, Kogan Page US, Sterling, VA, 2004.
Wright, J. F., *Monte Carlo Risk Analysis and Due Diligence of New Business Ventures*, AMACOM, New York, 2002.
Yu, O. S., *Technology Portfolio Planning and Management*, Springer-Verlag, New York, 2006.
Zwicky, F., *Discovery, Invention, Research through the Morphological Approach*, The MacMillan Company, Toronto, ON, 1969.

RELEVANT WEBSITES

7-S Framework of McKinsey, www.vectorstudy.com/management_theories/7S_framework.htm
Affinity Diagram, www.asq.org/learn-about-quality/idea-creation-tools/overview/affinity.html
Decision Tree Analysis, www.mindtools.com/dectree.html
Eight Disciplines of Problem Solving, www.quality-one.com/eight-disciplines/
FEMA, Risk Mapping, Assessment, and Planning, www.fema.gov/rm-main
Five Whys Method, www.iso9001consultant.com.au/5-whys.html
GE/McKinsey Matrix, www.quickmba.com/strategy/matrix/ge-mckinsey
Grid Analysis, www.mindtools.com/pages/article/newTED_03.htm
New Product Positioning Matrix, www.smartdraw.com/resources/glossary/positioning-matrix
Product Specific Need-Attribute Matrix, www.mindtools.com/pages/article/newCT_03.htm
Results Based Management, www.smarttoolkit.net/?q=node/387

Index

Note: Locators "*f*" and "*t*" denote figures and tables in the text

N

O

P

Q

R

S

T

U

V

W

Z